RECENT TRENDS IN BIOFERTILIZERS

RECENT TRENDS IN BIOFERTILIZERS

Editors

Bikas R. Pati

Professor
Department of Microbiology
Vidyasagar University
Midnapur, India

Santi M. Mandal

Assistant Professor
Department of Microbiology
Vidyasagar University
Midnapur, India

I.K. International Publishing House Pvt. Ltd.

NEW DELHI

Published by
I.K. International Publishing House Pvt. Ltd.
S-25, Green Park Extension
Uphaar Cinema Market
New Delhi–110 016 (India)
E-mail:info@ikinternational.com

ISBN 978-93-84588-65-6

Published by Krishan Makhijani for I.K. International Publishing House Pvt. Ltd., S-25, Green Park Extension, Uphaar Cinema Market, New Delhi–110 016 and Printed by Rekha Printers Pvt. Ltd., Okhla Industrial Area, Phase II, New Delhi–110 020.

Preface

We are grateful to all contributors for providing in-depth insights on biofertilizer technology and its recent developments. Soil fertility, an inherent capacity of soil, supports plant growth by ensuring adequate supply of nutrients and by creating suitable conditions for water uptake. A decline in soil fertility indicates a downturn in soil health by decreasing its level of organic matter, altering pH and ion exchange capacity, micro- and macronutrients, etc. Chemical fertilizers are now extensively used for increased production of crops. Continuous use of chemical fertilizers results in a gradual decrease in soil quality. Restoration of soil health and supply of nutrients from natural resources are possible through augmentation of the soil by the use of various microorganisms in the form of biofertilizers. Advanced biotechnology based biofertilizers are necessary to develop organic manure for upgrading the quality of agricultural soil. The bioorganic fertilizers not only restore the soil health but also increase the quality of crops. Biofertilizers contain a wide range of naturally chelated plant nutrients, carbohydrates, amino acids, trace elements, plant growth promoting vitamins and hormones. During the last decade, tremendous advancements have been made in the biofertilizer technology through the development of biotic and abiotic stress tolerant microbial strains. In this book detailed discussion on different aspects of the various types of biofertilizers already in use as well as those which are envasiaged to be used in future, are presented by leading academicians from India, Egypt and Brazil.

The book consists of sixteen chapters. In chapter 1, the author has placed emphasis on the production and supply of quality biofertilizers to be used by the farmers. Quality control of biofertilizers has an immense impact on the yield and quality of crops. It also reduces the investment thereby adding more profit to the farmer. Pros and cons on the status of biofertilization technology and biosafety of biofertilizers in Egypt has been discussed in chapter 2. Diversity of PGPR and characteristics of ideal strain for sustainable management of phosphorus level in soil have been dealt in details in chapter 3. The various aspects of liquid biofertilizer formulation, chemical additive use, and their methods of application and doses have been discussed in chapter 4. Further, the idea of molecular genetics of microbial phosphate solubilization has been presented in chapter 5. Chapter 6 deals with the assessment of the appropriate legume – *Rhizobium* symbiosis system for cultivation in diverse range of drought environments. Constraints facing the biofertilizer technology are illustrated in detail in chapter 7. Recent developments in bioinformatics tools and software which have significant contribution to the future of biofertilizer technology in enhancing the agricultural productivity have been well described in chapters 8 and 9. Chapters 10 to 15 deal with the role of arbuscular mycorrhizal importance in crops field; modified vermicomposting technology comprising rock soil nutrients and microbial enrichment; selection of biofertilizer strains as important factor for degradation of toxic contaminants; of the environment and sustainable bioresource management ensuring balance between optimum productivity and economy of the farmers has been explained in the last chapter.

Thus, this book provides an overview of the fundamental and technical aspects of biofertilizer technology which will help in reviving the soil health and make it further suitable for sustainable production.

Bikas R. Pati
Santi M. Mandal

Contents

1

Quality Control of Biofertilizers: Current Scenario and Trends

Surender Singh and Radha Prasanna*
Division of Microbiology, ICAR, Indian Agricultural Research Institute, New Delhi, India
E-mail: *radhapr@gmail.com**

ABSTRACT

Biofertilizers are an essential input in the integrated nutrient management strategies for various crops, especially with the advent of organic farming practices in agriculture. Judicious use of biofertilizers along with chemical and other organic sources of plant nutrients and proper management practices have shown promising results, not only in improving crop productivity and soil health, but also reducing chemical inputs in terms of fertilizers, pesticides, etc. These inoculants are available in market in various formulations for various crops, each targeting one or more nutrient deficiencies or their mobilization under different agro-climatic conditions. The recent inclusion of quality control measures for *Rhizobium, Azotobacter, Azospirillum*, Phosphate solubilizing bacteria (PSB), Potassium solubilizing bacteria (KSB), Zinc solubilizing bacteria (ZSB) and Arbuscular Mycorrhizal (AM) biofertilizers have helped to streamline production and supply of quality biofertilizers to farmers. This can reduce the costly investments made on chemical fertilizers and lead to ecologically sustainable and environment-friendly farming.

Keywords: Bacteria, Biofertilizers, Nutrients, Soil Health, Quality.

1.1 INTRODUCTION

The increasing use of chemical fertilizers and pesticides along with intensive agricultural production practices have led to slow but gradual depletion of organic matter of soil, disturbing the fine ecological balance, besides adversely affecting the surrounding environment and most importantly human health. Green revolution made the country self-reliant in food grain production (50.8 MT in 1950 to 264 MT in 2013), but unfortunately, very little emphasis has been directed towards conservation of natural resources, especially soil [2, 27]. The cumulative effect of over exploitation of natural resources has resulted in stagnation of crops yields in the last few decades. To overcome the adverse effects of chemicals in agriculture, and to keep pace with the ever growing population, there is a definite need to develop new strategies to ensure further growth in agricultural output. Adopting a strategy of integrated nutrient supply, i.e., using a combination of chemical fertilizers, organic manures and biofertilizers, can help in replenishing soil health to meet the nutrient demands. Biofertilizers and organic manures are essential components of this approach and are being promoted to harvest the naturally available biological mechanisms of nutrient mobilization/solubilization in soil, followed by stimulating their uptake by plants [28].

1.1.1 What are Biofertilizers?

"Biofertilizers can be defined as products containing carrier based (solid or liquid) living microorganisms which are agriculturally useful in terms of nitrogen fixation or nutrient mobilization, thereby helping in increasing the productivity of the soil and/or crops." Microbes included in these formulations not only mobilize or solubilize insoluble forms of N, P, K and Zn, but additionally promote growth by secreting various hormones or secrete polysaccharides and other carbon-containing compounds which improve soil health and quality [29]. The overall effect is reflected in the increased productivity of crop, enrichment of microbial diversity and soil fertility. Till recently, biofertilizers have been broadly defined as those which provide nitrogen or phosphate and in general promote plant growth (PGP). However, several new products like potassium mobilizing bacteria (KMB), zinc solubilizing bacteria (ZSB), *Acetobacter* and consortium mode multifunctional agents/biofertilizers have been developed in recent years (Table 1.1).

Table 1.1 Major biofertilizers and target crops

Biofertilizers	*Target crop*
Rhizobium	Leguminous crops like lentil, Moong, Pea, Arhar, Urad, Chickpea
Azotobacter	Wheat, maize, cotton, mustard and vegetables
Azospirillum	Cereal crops like wheat, maize, millets, sorghum, barley and sugar cane
Acetobacter	Sugar cane, Sorghum, Maize
Blue-green algae (BGA) and *Azolla*	Rice
Phosphate solubilizing microorganisms (PSB), Potassium mobilizing bacteria (KMB), Zinc solubilizing bacteria (ZSB) and consortium biofertilizers	All
Arbuscular mycorrhiza	Nursery raised crops and orchard trees, field crops

1.2 TYPES OF BIOFERTILIZERS

The steps involved in the production of biofertilizers are outlined in Fig. 1.1. The major characteristics of biofertilizers commonly deployed in agriculture are:

***Rhizobium* inoculants:** *Rhizobium* inoculants help in establishing efficient symbiotic association with pulses, leguminous oil-seed and fodder crops and fix 50 – 100 kg N/ha. A well nodulated legume crop also leaves sizeable amount of nitrogen in soil, which can meet a part of N-requirements of the succeeding crop in rotation [13]. An increase in yields of crops ranging from 10-70% due to inoculation with *Rhizobium* inoculants, over uninoculated control has been recorded depending on agro-climatic conditions, the variety planted and pest control measures used. Since *Rhizobium* is specific to each legume, only the crop-recommended inoculant should be used for each leguminous crop (gram, lentil, pea, soybean, groundnut, arhar, moong, urad, cowpea, berseem, lucerne, dhaincha or sunn hemp).

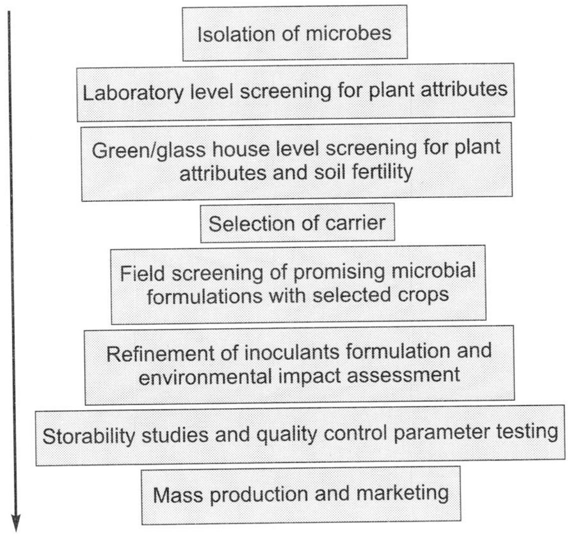

Fig. 1.1 Processes involved in the production of biofertilizer.

***Azotobacter* inoculants:** The use of *Azotobacter* inoculants is recommended in non-leguminous crops such as wheat, paddy, maize, barley, tomato, potato, cotton and mustard. *Azotobacter* fixes atmospheric nitrogen in soil and helps in saving chemical fertilizers by 15-20 kg N/ha [13]. Besides this, *Azotobacter* also secretes growth promoting substances which help in better seed germination and proliferation of roots, thus improving nutrient availability for plants. *Azotobacter* suppresses the growth of some saprophytic and plant pathogenic microorganisms in rhizosphere and therefore reduces the crop damage due to plant diseases. *Azotobacter* helps in maintaining better plant population, growth and improves yields of crops. In general, 10-20% increases in grain yields over uninoculated control have been recorded due to inoculation with *Azotobacter*.

***Azospirillum* inoculants:** These inoculants are recommended in non-leguminous crops like jowar, bajra, ragi and other millets like Italian millet, kodo millet, barn yard millet, small millet and oats [13]. Several fodder grasses also respond favourably to inoculation with *Azospirillum*. Increases in grain and fodder yields of millets due to its inoculation are almost equivalent to that attainable with 15-20 kg N/ha.

Phosphate solubilizing bacterial/fungal inoculants: Phosphorus is one of the important plant nutrients. Most of the phosphorus remains fixed in soil and as such is not available to plants. Inoculation with an efficient P solubilizing microorganism improves availability of phosphorus from insoluble form of phosphorus in soil and enhances use efficiency of phosphatic fertilizers such as superphosphate [13], by making them available to crops. A number of efficient phosphate solubilizing microorganisms, both bacteria and fungi, which can even mobilize P from rock phosphate and soil fixed phosphorus, have been reported. These microorganisms have been tested for their potential in many crops such as wheat, paddy, cowpea, soybean, lentil, gram and potato all over India under field conditions and increases in their grain yield were found to range from 10-50%. A combination of rockphosphate and phosphomicroorganisms has been found more effective than the rock phosphate alone. In general, 40% of superphosphate could be saved by combined application of rock phosphate and phosphomicroorganisms. Phosphomicroorganisms are recommended for use in all crops.

Arbuscular-Mycorrhizal inoculant: Arbuscular Mycorrhizae (AM) represent symbiotic associations between plant and specific group of fungi. The hyphae of AM form a bridge between plant and soil and serve to cater requirements of plant nutrients especially phosphorus and trace elements like zinc, iron, copper, cobalt, magnesium, molybdenum, etc [13]. AM colonization in plant roots increases plant growth, improves nutrient availability and reduces salt stress. They also check soil erosion, degradation and reduces losses caused by nematodes and plant pathogens. AM has been found very useful in crops like grape seedling hardening, coffee, tea, papaya, cocoa, oil palm, etc. Crops treated with AM inoculant show improvement in grain and fruit quality as well as in yield.

Blue-Green Algal (BGA) inoculant: Blue-green algal or cyanobacterial inoculants were developed so as to enrich the native flora of flooded rice paddies. It is well recognized that sustained fertility of rice soils is responsible for the consistent yields, which in turn has been attributed to the activities of cyanobacterial proliferation [31]. They not only provide nitrogen, but also organic carbon and growth promoting substances. Clay-based formulations of BGA inoculants are available for use in paddy cultivation. A 500 g pack of this inoculant is sufficient to inoculate one acre of paddy field. Inoculant application with a part of the recommended dose of phosphorus in 3-4 consecutive seasons has not only been found effective in rapid establishment of BGA but also in sustaining yield of paddy. If chemical fertilizers are not used, BGA inoculant gives a benefit of 20-30 kg N/ha, besides adding carbon to the soil. In the last decade, cyanobacterial inoculation has been found to be beneficial in various crops including wheat, tomato, cotton, leguminous crops and other vegetables as nutrient mobilizing and biocontrol agents [10, 18, 19, 21, 22].

***Azolla* inoculant:** *Azolla* is an ideal system under tropical conditions for biological nitrogen fixation in paddy fields. *Azolla* is a free-floating water fern which exists in symbiotic association with *Anabaena*. *Azolla* contributes 40-60 kg N/ha/crop depending upon the rate of *Azolla* application [17]. The important factor in using *Azolla* as a biofertilizer for paddy is its quick multiplication and subsequent rapid decomposition in soil and efficient availability of its nitrogen to the crop. *Azolla* has high rate of biological nitrogen fixation, which can improve use efficiency of applied fertilizers. Besides improvement in soil physico-chemical properties, the growth of *Azolla* is known to reduce the proliferation of weeds and pests.

Newly introduced biofertilizers: Different bacteria belonging to the genera *Pseudomonas, Agrobacterium, Bacillus, Rhizobium* and *Flavobacterium* are known to possess the ability to solubilize inorganic phosphate compounds, such as tricalcium phosphate, hydroxyapatite and rock phosphate [8]. Silicate bacteria were found to dissolve potassium, silicon and aluminium from insoluble minerals [1]. Similarly, microorganisms in the soil are able to solubilize 'unavailable' forms of K-bearing minerals, such as micas, illite and orthoclase, by excreting organic acids which either directly dissolve rock K or chelates silicon ions to bring the K into solution [5, 6]. These microorganisms are commonly known as potassium solubilizing bacteria (KSB) or potassium mobilizing bacteria (KMB) or silicate dissolving bacteria. Many formulations are available in market with ability to increase K availability in soil. Bennett *et al.* [6] reported that microorganisms produced various organic ligands during their metabolism. These included metabolic byproducts, extracellular enzymes, chelates and both simple and complex organic acids, which helped in dissolution of feldspar by decreasing the pH of the environment. The other possible hypotheses/ mechanisms to mobilize soil K reserve are due to biofilm formation on the rhizospheric mineral surfaces by certain bacterial strains [4].

Zinc (Zn) is one of the essential micronutrients required for optimum plant growth. Substantial quantity of applied inorganic zinc in soil is converted into unavailable form. Microbes are potential alternate sources that could cater to plant zinc requirement by solubilizing the complex zinc in soil. A term called zinc solubilizing bacteria (ZSB) was coined for those bacteria that are capable of solubilizing the insoluble zinc compounds/ minerals in agar plate as well as in soil. Several genera of rhizobacteria belonging to *Acinetobacter, Bacillus, Gluconacetobacter* and *Pseudomonas* are reported to solubilize zinc [12, 23, 33]. Microbes solubilize the metal forms by proton extrusion, chelated ligands, organic acid and oxido-reductive systems present on the cell surface and membranes [9, 11, 32]. Some of these bacteria also exhibit other traits beneficial to plants, such as production of phytohormones, antibiotics, siderophores, vitamins, antifungal substances, and hydrogen cyanide. In addition, production of inorganic acids such as sulphuric acid, nitric acid, and carbonic acid could also facilitate the solubilization [25]. A few fungal genera possess immense zinc solubilizing potential. Screening of these ZSB is usually done by incorporating insoluble forms like zinc carbonate, zinc oxide or sphelirite in medium. These bacteria have been found to increase yield of many crops including maize, mungbean and soybean [12, 26].

Use of multifunctional consortia: Several formulations comprising two or more synergistic microorganisms, mainly bacteria or cyanobacteria, are available which can provide major plant nutrients to crops. Such formulations may be cheap and also easy to apply compared to individual formulation for specific functions [15, 22-24].

Plant associated biofilms are often encountered as they possess a high ability to protect themselves from various types of environmental stress and microbial competition. Biofilms, which consist of polysaccharide sheath enclosed bacteria/fungi/algae, are known to play an antidesiccant role, especially in the rhizosphere soil [16]. The mucilaginous sheath increases the residence time of water, as compared to that on the bare rock surface, hence increasing the rate of potassium solubilization. The first report on developing biofilmed biofertilizers was using the fungal matrix-based biofilm (*Bradyrhizobium* - *Penicillium* spp.), which enhanced the nodulation and N accumulation in soybean. At IARI, a novel concept of using *Trichoderma* or cyanobacterium *Anabaena* sp. as matrices for generating biofilmed biofertilizers using agriculturally important microbes (N_2 fixer/P solubilizer/ PGPR) was developed [20, 30]. These biofilms exhibited enhanced metabolic activities and survival, compared to monocultures, and brought about significant improvement in disease reduction, plant growth, yields and soil fertility, when used as single formulation options for nutrient and pest management in several crops.

1.3 STATUS OF BIOFERTILIZER PRODUCTION IN INDIA

The first ever production of *Rhizobium* inoculants in India is reported to be in 1956 at the Indian Agricultural Research Institute (IARI), New Delhi. However, their commercialization started only in late 1960s when yellow seeded soybean varieties were introduced for the first time in India. The Division of Microbiology, IARI took a lead in the commercial production of soybean inoculants which encouraged the Government of India to stop import of soybean inoculants from USA. Realizing the importance and demands of inoculants, Government of India started a project "National Biofertiliser Development Centre (NBDC)" in 1983 with the aim of development and use of biofertilizers, which functions through one national centre at Ghaziabad, 6 regional centres at Bengaluru, Bhubaneshwar, Hissar, Imphal, Jabalpur and Nagpur and 40 sub-centres for production of inoculants. Later on NBDC was renamed in 2003 as National Institute of Organic Farming (NIOF), and presently as National Centre for Organic Farming (NCOF) for production, promotion, market development and regulation of organic farming in the country. National Bank for Agriculture and Rural Development (NABARD) provides financial support for establishment of biofertilizer production units of 200 tonnes per annum capacity under the national project on organic farming (NPOF) as back-ended subsidy of 25%, restricted up to Rs. 40 lakh. Till 2000, biofertilizers were prepared and distributed to farmers mainly through agricultural research institutes, NCOF, agricultural universities, state departments of agriculture and cooperative sector units like IFFCO, RCFL, NFL & KRIBHCO which are located in different agroclimatic zones of India. However, recently the market is flooded with the biofertilizer products, both carrier-based as well as liquid formulations from various private manufacturers, with doubtful quality due to lack of enforcement of quality control measures by the government/local bodies in states, etc.

According to an estimate of the NCOF and the Biotech Consortium of India Ltd (BCIL) about 762, 372 tonnes of biofertilizers are required for Indian agriculture [7] (Fig. 1.2).

This estimated demand is based on the cultivated area and the treatment of the total seed sown at the rate of 200 g biofertilizer per 10 kg of seed. Although this assumption reflects only the macro-level requirement, if even 50% of the cultivated area is to be brought under biofertilizer application, there will be a wide gap between the actual production and the demand, keeping in view the steady increase in demand, particularly due to expansion of certified organic area.

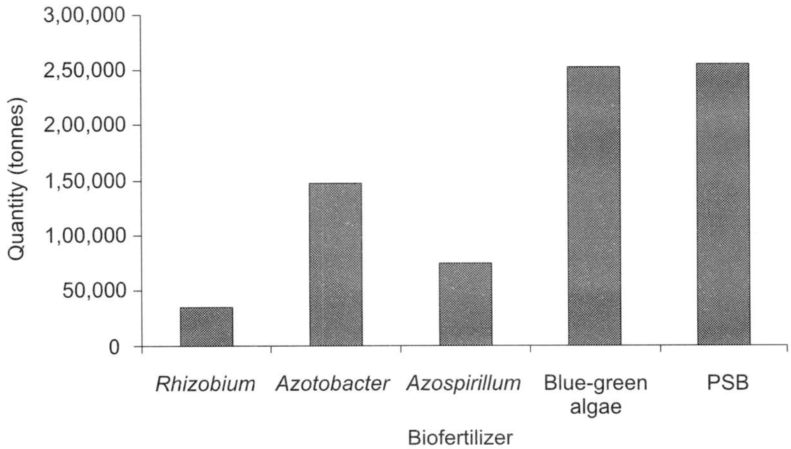

Fig. 1.2 Estimated potential demand for commonly used biofertilizers in India.

There are more than 150 producers of biofertilizers with an annual production of 46836.82 tonnes during 2012-13[7] (Fig. 1.3).

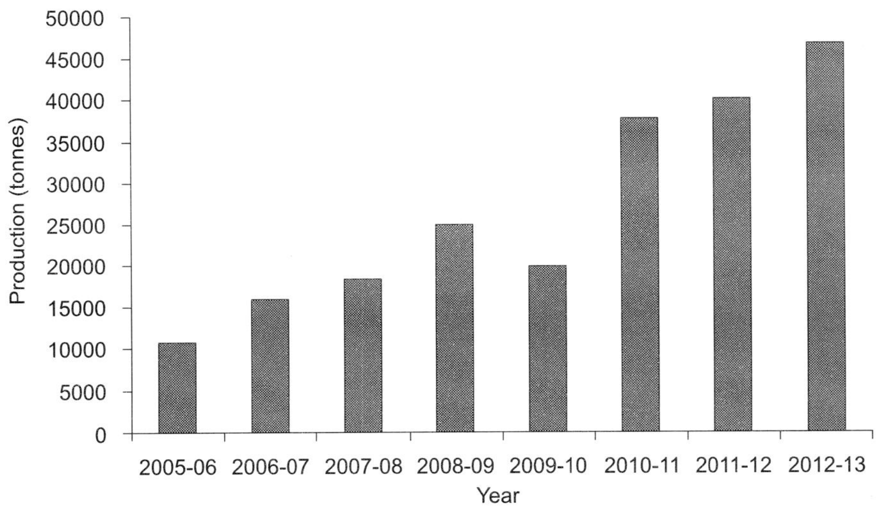

Fig. 1.3 Annual production of biofertilizers during recent years.

Among different biofertilizers, PSB constitutes around 50% of total biofertilizer production (Fig. 1.4). A lot of regional differences in the biofertilizer production is also observed, as south Indian states produce 54% of total biofertilizers followed by western (34%) and north Indian states (7%).

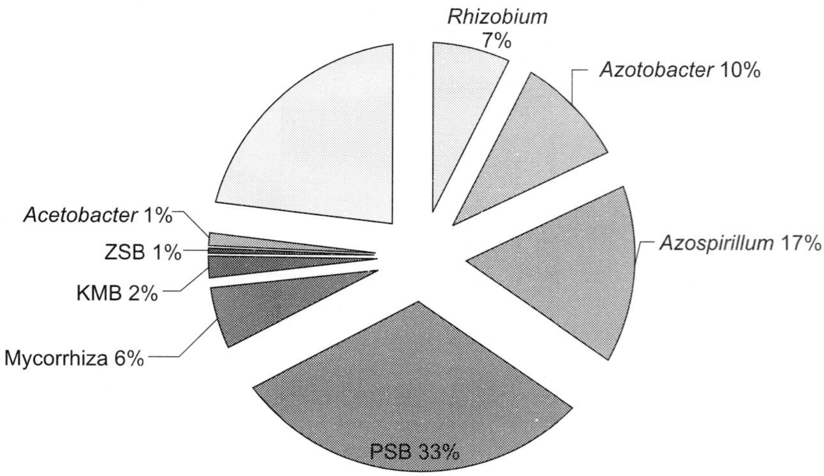

Fig. 1.4 Share of different biofertilizers to total production (2012-13).

1.4 QUALITY CONTROL OF BIOFERTILIZERS IN INDIA

Quality control is of prime importance in ensuring the benefits of biofertilizers to the farmers. The concept of quality is dynamic and the quality control parameters and critical levels need to be updated from time to time in the light of relevant scientific information (Fig. 1.5). Many studies suggest that the biofertilizers being sold in the market are of inferior quality and contaminated with low count of microorganisms which results in poor and uneven performance.

For example, in a survey of *Rhizobium* carried out by ICRISAT, 90% of samples from India were found to have a rhizobia count lower than that required for effective performance. For effective management of quality control regime, 5 important biofertilizers (*Rhizobium, Azotobacter, Azospirillum,* PSB and Mycorrhiza) have been declared as a part of an Essential Commodity and Fertilizer control Order (FCO), 2006, which was promulgated under Section 3 of Essential Commodities Act, 1955 since March 2006 to ensure the quality of these biofertilizers (Annexure 2). Recent amendments to FCO (1985) also included ZSB, KMB, *Acetobacter* and consortium biofertilizers and specifications have been published by Ministry of Agriculture [3]. These standards (Annexure 1) are formulated by subject specific technical experts comprising of all stakeholders - producers, consumers, technologists, government bodies and research and testing organizations.

Salient points of this amendment are:

1. No person can carry out the business of manufacturing biofertilizers, except under and in accordance with the terms and conditions of certificate of manufacture granted to him under the act.

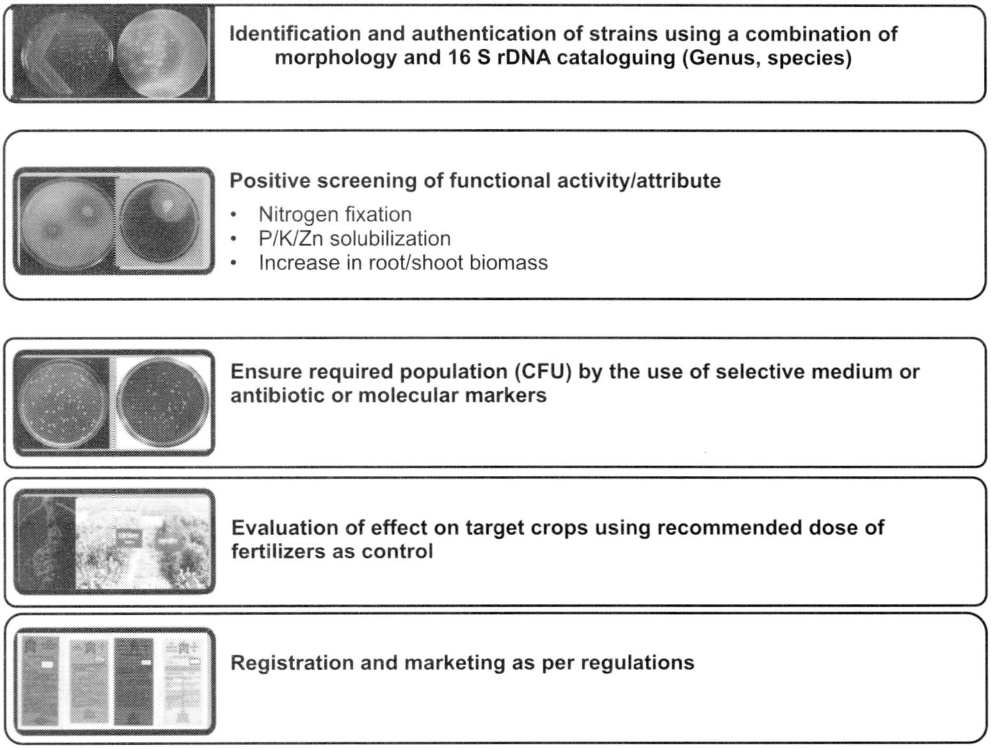

Fig. 1.5 Key steps in production of quality biofertilizers.

2. No person can manufacture any biofertilizer, unless it conforms to the standards set out in the order. Certificate of manufacture has to be obtained from registering authority
3. Every container of biofertilizer has to be labelled as "Biofertilizer"
4. Certificate of manufacture, shall, unless suspended or cancelled, be valid for a period of three years.
5. The State Governments can appoint Registering or Notifying Authority.
6. The Government can appoint inspectors of biofertilizers for the enforcement of the order
7. Withdrawal of Samples: Samples in their original unopened packets should be drawn and sent to the laboratory to prevent possible contamination of the samples during handling and help in revealing the true condition of the material.
8. Analysis of samples: A biofertilizer sample drawn by the inspector shall be analyzed at the National Centre of Organic Farming, Ghaziabad or Regional Centres of Organic Farming at Bengaluru, Bhubaneshwar, Hissar, Imphal, Jabalpur and Nagpur or any other laboratory notified by central or state government.
9. Time limit for analysis and communication of results: Drawn sample along with a memorandum in form K-1 is to be sent to the laboratory for analysis within a period of seven days from the date of its withdrawal.

10. The laboratory shall analyze the sample and forward the analysis report within 30 days from the date of receipt of the sample to the authority specified in the same memorandum.
11. The authority to whom the analysis report is sent, shall communicate the result of the analysis to the dealer/manufacturer, etc. from whom the sample was drawn within 15 days from the date of receipt of the analysis report of the laboratory.
12. Carrier-based biofertilizer (*Rhizobium, Azotobacter, Azospirillum* and PSB) should be packed in packaging material of low density polyethylene/polypropylene bags of minimum 75-100 micron thickness.
13. Instructions for packing and marking are also given to ensure consumer protection from substandard products. Each packet should be marked legibly to give the following information:
    (a) Name of the product
    (b) Crops for which intended, printed on a coloured ink background
    (c) Name and address of the manufacturer
    (d) Type of the carrier
    (e) Batch or code number
    (f) Date of manufacture, printed on a coloured ink background
    (g) Date of expiry (agreed between the manufacturer and the purchaser, subject to minimum 6 months from the date of manufacture), printed on a coloured ink background
    (h) Net quantity and the area meant for
    (i) Storage instructions worded as under:
    'STORE IN COOL PLACE AWAY FROM DIRECT SUN AND HEAT'
    (j) Any other information

Directions for the use of product should be printed briefly on the packet or preferably as separate pamphlet.

1.5 PROCEDURES FOR QUALITY TESTING OF BIOFERTILIZERS

The various steps involved in testing of quality are given below. The details of media employed are given as Annexure I.

1.5.1 Withdrawal of Samples

Sampling should be carried out by authorized and trained person from a protected place not exposed to dampness, air, light or dust. Samples in their original unopened packets should be drawn and sent to the laboratory to prevent possible contamination of sample during handling and to help in revealing the true condition of the material.

Desired number of packets/samples depending on the size of the lot should be selected randomly (As per IS 4905). Minimum of 3 samples per batch of 5000 packets should be taken. Out of 3 samples, one sample shall be sent for referee analysis and one will be kept under safe custody for reanalysis if required. One sample should be handed over to manufacturer or importer or dealer for reference. Sealed samples should be sent to in-charge of the laboratory notified by the State Government or to National Centre for Organic Farming or to any of its Regional Centres for analysis.

1.5.2 Methods of Analysis of Biofertilizers

Base material: The carrier material such as peat, lignite, peat soil, humus, talc or similar material, favouring the growth of the organism should be used for all bacterial inoculants. The carrier material should be neutralized (if required) with calcium carbonate and sterilized to obtain a fine powder capable of passing through 150 to 212 micron IS sieve. For mycorrhizal inoculants, the base material may be fine powder/tablets/granules/root biomass mixed with growing substrate, 90% of which should pass through 250 micron IS sieve.

Determination of pH: Add 20 g of the product into 50 mL of distilled water and place on a rotary shaker for 2 hours. Filter with Whatman No. 1 filter paper under vacuum using a Buchner funnel. Determine pH of the filtrate using a calibrated pH metre.

Moisture content: Heat 10 gm of sample for 12-16 hours in hot air oven at 100-105°C. Cool in a desiccator and weigh. The loss in weight represents the moisture. Calculate the moisture percentage on dry weight basis.

Viable count:

- For all bacterial inoculants, dispense 30 g of inoculants to 270 mL of sterile distilled water and shake for 10 min on a reciprocal shaker.
- Make serial dilutions up to 10^{-9} by suspending 10 mL aliquot of previous dilution to 90 mL of water. Except for *Azospirillum*, take 0.1 mL aliquots of 10^{-5} to 10^{-9} dilutions using sterile pipettes and spread plate on respective medium as given in Annexure II.
- Incubate the plates at 28 ± 2°C for 3 to 6 days.
- Count the colonies with help of magnifying glass or automatic colony counter.
- The colony morphology should be checked carefully specially for *Rhizobium* and *Azotobacter*. *Rhizobium* colonies are white, translucent, glistening and elevated colonies. Discard red colonies which are of contaminants usually. *Azotobacter* colonies are gummy, raised with or without striations, viscous and often sticky with very light brown to black pigmentation. For, PSB and ZSB count the colonies with solubilization zone in specified medium (Annexure II).
- For *Azospirillum*, pipette out 1 mL aliquots of 10^{-4} to 10^{-8} dilution and deliver it to screw cap tubes or test tubes containing N-free semi solid Nfb media for MPN (Annexure II).
- Label the tubes and incubate at 36 ± 1°C for 3-4 days in vertical position in a test tubes stand without disturbing them during the entire period of incubation. Examine each tube for colour change (to blue) and pellicle formation. Score for growth/pellicle in individual tube as +ve or –ve.

To calculate the MPN in the original sample:

1. Select as p1 the number of positive tubes in the least concentrated dilution in which the greatest number of tubes is positive, and let p2 and p3 represent the numbers of positive tubes in the next two higher dilutions.
    - From the table of MPN for 10-fold dilution and 5 tubes per dilution (Cochran, 1950), find the row of numbers in MPN table (Annexure II) which p1 and p2 correspond to the values observed experimentally.
2. Follow the row of numbers across the table to the column headed by the observed value of p3. The value at the point of intersection is the most probable number of *Azospirillum* in the quantity of the original sample represented in the inoculum added in the second dilution (p2). Multiply this value by the appropriate dilution factor to obtain the MPN for the original sample.

Example: Suppose following observations were made:

Dilution	+ve tubes
10^{-3}	5
10^{-4}	3
10^{-5}	1
10^{-6}	0

In this series: p1 = 5, p2 = 3, p3 = 1. For this combination of p1, p2 and p3, the MPN table gives 1.1 as the most probable number of *Azospirillum* in the quantity of inoculum applied in the 10^{-4} (p2) dilution. Multiplying this value with dilution factor 10^5 gives 1.1×10^5 as the MPN for the original sample.

For enumeration of mycorrhizal spores, the product should be mixed in a substantial volume of water and the soil solution decanted through a set of sieves (30, 60, 200, 250 and 350 micron mesh). Roots and coarse debris should be collected on a coarse (60-ISS) sieve, while spores can be captured on one or more finer sieves. Vigorous washing with water is necessary to free spores from aggregates of clay or organic materials, and the sieved material is collected in jars. The number of spores are calculated by transferring a known quantity (1-5 mL) from the above volume into eelworm counting plate or nematode dish under stereo microscope. The number of spores are expressed as per g of soil.

1.5.3 Efficiency Testing of Various Biofertilizers

Rhizobium:

1. Prepare the nutrient solution for plant test of pH 7.2 (Annexure II).
2. Surface sterilize the seeds with 95 per cent alcohol followed by washing in freshly prepared chlorine water (for 15 to 20 min) or 0.1 per cent mercuric chloride solution for 3 min in a suitable container. In case of seeds with tough seed coat, concentrated sulphuric acid may be used as a surface sterilant for 20 to 30 min.

3. The earthen pots are filled with soil (2 parts soil and 1 part washed coarse sand) (pH 6 to 7) and autoclaved for 2 h at 120°C.
4. After two days incubation at room temperature, autoclaving is repeated to ensure complete sterility of soil.
5. Surface sterilized seeds are inoculated with water slurry of the inoculant taken from a culture packet (15 to 100 g seeds per gram of inoculants is undertaken, depending on the size of the seed), and sowing is done with the seeds along with uninoculated seeds as control. Also a set of pots with ammonium nitrate at the rate of 100 kg N/ha as control is used after incubating them in a pot-culture house during appropriate seasons for appropriate plants, taking care to separate the inoculated pots from the control pots. If growth rooms or cabinets having facilities to adjust temperature and light are available, the pots may be incubated in crop specific controlled environmental conditions.
6. Sterilize the nutrient solution at 120°C for 20 min and irrigate each pot once to the moisture holding capacity of soil. Subsequently, water the seedling periodically with sterilized water without splashing of water from inoculated pots to uninoculated ones. Maintain 4-6 replicates for statistical analysis.
7. After two to three weeks of growth, thin down the number of plants in each pot to four uniform plants. At the end of 6 to 8 weeks, take one set of pots from both the control and inoculated series and separate the plants carefully from the soil under slow running water. Obtain data on the number, colour (effective nodules are pink or red) and mass of nodules.
8. At the end of 6 to 8 weeks, harvest the shoot system, dry at 60°C for 2 days and determine dry mass. For the above purpose, maintain adequate replications of pots (4 to 16).
9. Record the nodulation data regarding formation of pink colour of nodules as revealed visually when nodules are cut open by razor blade. After computing the data, based on the dry mass of plants and nodulation data, decide the effectiveness of culture. If good effective pink nodulation is obtained in inoculated plants together with local absence or sometimes presence of stray nodules in controls, and if there is a 50 per cent increase in the dry mass of plants over the uninoculated control without nitrate, it may be concluded that the culture is of the requisite quality.

Azotobacter

- Select pure culture of *Azotobacter* and plate on the medium given in Annexure I.
- For testing nitrogen fixation efficiency, inoculate culture in 50 mL sterilized medium in 250 mL conical flasks and incubate for 12 days at 28°C.
- Following incubation, test the purity of grown culture by streaking on fresh medium.
- Concentrate the flask contents over water bath (50 to 60°C) to dryness and take it as a sample.
- Take 0.25 g sample and determine total nitrogen using Kjeldahl Method.

- For determination of glucose/sucrose, dissolve 1 g sample in 250 mL of distilled water and shake for 5 minutes. Centrifuge at 15000 rpm for 5 minutes and discard pellet.
- Estimate glucose and sucrose in the supernatant in triplicate. The difference between the two provides the data of actual amount of glucose consumed. Calculate the amount of nitrogen fixed per gram of sucrose consumed.

Azospirillum: As per FCO (1985), formation of white pellicle in semisolid nitrogen free BTB medium is taken as efficiency character but it does not tell about nutrient supplying capacity of culture. BIS standard (IS 14806:2000) recommends plant growth test to check efficacy of *Azospirillum* inoculants.

Grow inoculated and control plants in pots using same medium (soil + sand) and nutrient solution as described for *Rhizobium* for 3 months. At the end of three months, harvest the shoot system. Dry at 60°C for 48 hours and determine the dry mass. Better vigour of roots in inoculated seeds, if established from the data when compared to the control, is the confirmation of effectiveness of *Azospirillum*. If there is 10 per cent increase in the plants over the dry mass of uninoculated control without nitrate, it may be concluded that the culture is of required quality.

PSB: Spread plate appropriate dilutions of PSB on ≥ 3 mm Pikovskya's medium plate and check for zone of solubilization after incubation for two weeks. A good quality strain should give at least 10 mm solubilization zone. When testing in broth, PSB should have phosphate solubilizing capacity in the range of minimum 30 per cent when tested spectrophotometrically. For spectrophotometric quantitative determination of solubilized P, inoculate Pikovskaya broth (Annexure I) with culture and incubate for 12 days. Following incubation, centrifuge the content and determine available P using ascorbic acid method [14]. Record the available P as % of total P added in the form of insoluble P source.

Mycorrhizal inoculants: The bioassay is used to determine the number of infective propagules present in the product.

- Mix 100 g test sample in a pot and plant 10-12 pregerminated seeds of crop and grow for 14 days.
- Harvest the pots and recover roots (fine roots can be rescued using sieve) completely.
- Chop the roots equally 1 cm in length.
- Clear the roots in KOH solution and stain the root pieces.
- Count the number of infection points/entry points formed on randomly picked 100 segments.
- Calculate the average number of entry points formed in 1 cm segment.
- Calculate the total number of infection points/infective propagules (IP) by multiplying the average number of entry points formed in 1 cm segment by the total root length.
- Extrapolate the IP present as numbers per gram of substrate/inocula.

KMB and ZSB

- Spread plate appropriate dilutions of KMB and ZSB on ≥ 3 mm thick plate of respective medium (Annexure 2) and check for zone of solubilization after incubation for two weeks at 28 ± 2°C. A good quality strain should give at least 10 mm solubilization zone.

Acetobacter

As per FCO (1985), the formation of yellowish pellicle in semisolid nitrogen free medium is taken as efficiency character.

Consortial formulations

The efficiency of individual member of consortium is checked by using the methods described above.

1.5.4 BIS Certification

Under the Bureau of Indian Standards (BIS) product certification marks scheme licences are granted to manufacturers for affixing BIS Mark on their products conforming to BIS. Regular and surprise surveillance inspections of the manufacturer's works, testing of samples from the production line as well as from the open market and a provision for free replacement of certified goods found to be defective are all highlights of BIS certification marks scheme designed to protect the interests of both consumer and manufacturer. The Bureau of Indian Standards has brought out standards for the following inoculants:

1. *Rhizobium* inoculants (IS 8268:2001)
2. *Azotobacter* inoculants (IS:9138-1979)
3. *Azospirillum* inoculants (IS 14806:2000)
4. Phosphate solubilizing bacteria (IS 14807:2000)
5. Phosphate solubilizing fungal inoculants - *Aspergillus awamori* (IS 15849:2009)

1.6 HURDLES TO BE OVERCOME AND FUTURE OUTLOOK

The problems associated with the adoption of the technology by the farmers, due to advocacy of different methods by agencies regarding methods of inoculation are among those voiced by most users. But the most significant matter of concern is the absence of immediate visual differences in the crop growth, as observed with the use of inorganic fertilizers. Despite the efforts of government and public organizations, there are still several gaps which need to be filled, such as:

- The standards prescribed as quality requirements of inoculants are mainly in terms of the number of viable cells in the carrier material, stipulated expiry period, freedom from contamination with other microorganisms, etc. The standards prescribe test procedures for assessing the quality besides sampling procedure. But these standards are not mandatory and moreover, the cost of getting the certification is high so manufacturers are reluctant for go for certification.

- Another important drawback is the lack of guidelines for the newly developed biofertilizers (biofilmed biofertilizers, P/K/Zn solubilizing fungi) and BGA biofertilizers.
- Adequate microbiological lab and qualified/trained microbiologist in units, who can ensure aseptic conditions, cleanliness and contamination free production lines and housing. Unsatisfactory storing conditions, particularly during the distribution period, can be harmful.
- Registration for manufacturing unit is difficult to get, only few reference labs are available, and there is less availability of area specific strains; FCO is not effectively implemented across the country resulting in marketing of spurious products. Financial constraints include the non-availability of sufficient funds and problems in getting bank loans with less return by sale of products in smaller production units.
- There is a need to develop facilities to maintain the germplasm collection of agriculturally important microbes at focal points in the country, so that these can serve as source as well as depositories for useful microbes. This will also help in generating region specific strains or abiotic/biotic stress tolerant strains. IARI has been at the forefront in developing high quality biofertilizers, whose demand is increasing every year. Through its entrepreneurship programmes and Annual Krishi Mela, IARI strives to meet the ever increasing demands of farmers for the quality bioinoculants.

CONCLUSION

With a view to bridge the gap between nutrient removal and replenishment to a limited extent, biofertilizers have assumed importance as among the most cost effective inputs in Indian agriculture, which are also environment-friendly. Despite the fact that biofertilizers are useful, at present many farmers are not able to practice crop inoculation with biofertilizers due to various reasons – including timely availability, poor quality material available locally or lack of awareness in interior parts of the country and poor economic and educational background. With the increasing problems of environmental pollution, global warming and depleting quality of soil, there is no alternative other than biofertilizers to supplement inorganic fertilizers. Although the increasing area under organic farming is heartwarming, providing an optimistic view regarding the prospects of the quality inoculants production, the role of researchers becomes all the more a responsible one, as they need to constantly strive to improve and innovate greener farming practices. This will help in improving the soil health and sustaining the crop yields in the years ahead.

REFERENCES

1. Anonymous (2014). "*Economic Survey of India*". Government of India, New Delhi.
2. Singh, S., Prasanna, R. and Lata and Saxena, A. (2012). "Quality biofertilizers for sustainable agriculture and better soil health", *Indian Farming,* 62, 4–6.

3. Singh, S., Singh, R.J., Kumar, K., Singh, B. and Shukla, L. (2013). "Biofertilizers and green manuring for sustainable agriculture". In: Kumar S., Prasad B (eds.) *Modern Technologies for Sustainable Agriculture*. New India Publishing Agency, New Delhi (India), pp. 129–150.
4. Tilak, K., et al. (2005). "Diversity of plant growth and soil health supporting bacteria", *Curr. Sci.,* 89, 136–150.
5. Kaushik, B.D. (2005). *"Advances in Microbiology at I.A.R.I. 1961-2004"*. Indian Agricultural Research Institute, New Delhi (India).
6. Venkataraman, G.S. (1972). *"Algal biofertilizers and rice cultivation"*. Today & Tommorrow's Printers & Publishers.
7. Dukare, A.S., Prasanna, R., Dubey, S.C., Nain, L., Chaudhary, V., Singh, R. and Saxena, A.K. (2011). "Evaluating novel microbe amended composts as biocontrol agents in tomato", *Crop Protect,* 30, 436–442.
8. Prasanna, R., et al. (2014). "Prospecting cyanobacteria fortified composts as plant growth promoting and biocontrol agents in cotton", *Exp. Agr.,* 1–24.
9. Prasanna, R., et al. (2013). "Evaluating the establishment and agronomic proficiency of cyanobacterial consortia as organic options in wheat-rice cropping sequence", *Exp. Agr.,* 49, 416–434.
10. Prasanna, R., Rana, A., Chaudhary, V., Joshi, M. and Nain, L. (2012). "Cyanobacteria-PGPR interactions for effective nutrient and pest management strategies in agriculture.". In: Satyanarayana T., Johri B., Prakash A. (eds) *Microorganisms in Sustainable Agriculture and Biotechnology*. Springer Science + Business Media B.V., pp. 173–195.
11. Prasanna, R., et al. (2014). "Evaluating the efficacy of cyanobacterial formulations and biofilmed inoculants for leguminous crops", *Arch. Agron. Soil Sci,* 60, 349–366.
12. Pabby, A., Prasanna, R. and Singh, P. (2004). "Biological Significance of *Azolla* and its Utilization in Agriculture", *Proc. Natl. Acad. Sci. Part B,* 70, 299–333.
13. Chen, Y., Rekha, P., Arun, A., Shen, F., Lai, W.-A. and Young, C. (2006). "Phosphate solubilizing bacteria from subtropical soil and their tricalcium phosphate solubilizing abilities", *Appl. Soil Ecol.,* 34, 33–41.
14. Aleksandrov, V., Blagodyr, R. and Ilev, I. (1967). "Liberation of phosphoric acid from apatite by silicate bacteria", *Mikrobiolohichnyĭ Zhurnal (Kiev),* 29, 111–114.
15. Barker, W., Welch, S., Chu, S. and Banfield, J. (1998). "Experimental observations of the effects of bacteria on aluminosilicate weathering", *Amr. Mineral.,* 83, 1551–1563.
16. Bennett, P., Choi, W. and Rogera, J. (1998). "Microbial destruction of feldspars", *Min. Mangmt.,* 8, 149–150.
17. Balogh-Brunstad, Z., Keller, C., Gill, R., Bormann, B. and Li, C. (2008). "The effect of bacteria and fungi on chemical weathering and chemical denudation fluxes in pine growth experiments", *Biogeochemistry,* 88, 153–167.

18. Iqbal, U., Jamil, N., Ali, I. and Hasnain, S. (2010). "Effect of zinc-phosphate-solubilizing bacterial isolates on growth of *Vigna radiata*", *Ann. Microbiol.*, 60, 243–248.
19. Ramesh, A., Sharma, S.K., Sharma, M.P., Yadav, N. and Joshi, O.P. (2014). "Inoculation of zinc solubilizing *Bacillus aryabhattai* strains for improved growth, mobilization and biofortification of zinc in soybean and wheat cultivated in Vertisols of central India", *Appl. Soil Ecol.*, 73, 87–96.
20. Wang, H., Dong, Q., Zhou, J. and Xiang, X. (2013). "Zinc phosphate dissolution by bacteria isolated from an oligotrophic karst cave in central China", *Front. Earth Sci.*, 7, 375–383.
21. Crane, F., Sun, I., Clark, M., Grebing, C. and Löw, H. (1985). "Transplasma-membrane redox systems in growth and development", *BBA Bioener.*, 811, 233–264.
22. Hughes, M.N. and Poole, R.K. (1991). "Metal speciation and microbial growth—the hard (and soft) facts", *J. Gen. Microbiol.*, 137, 725–734.
23. Wakatsuki, T. (1995). "Metal oxidoreduction by microbial cells", *J. Ind. Microbiol.*, 14, 169–177.
24. Rodrìguez, H. and Fraga, R. (1999). "Phosphate solubilizing bacteria and their role in plant growth promotion", *Biotech. Adv.*, 17, 319–339.
25. Sarathambal, C., Thangaraju, M., Paulraj, C. and Gomathy, M. (2010). "Assessing the Zinc solubilization ability of *Gluconacetobacter diazotrophicus* in maize rhizosphere using labelled ^{65}Zn compounds", *Indian J. Microbiol.*, 50, 103–109.
26. Nain, L., et al. (2010). "Evaluation of synergistic effects of bacterial and cyanobacterial strains as biofertilizers for wheat", *Plant Soil*, 331, 217–230.
27. Rana, A., Joshi, M., Prasanna, R., Shivay, Y.S. and Nain, L. (2012). "Biofortification of wheat through inoculation of plant growth promoting rhizobacteria and cyanobacteria", *Eur. J. Soil Biol.*, 50 (S1), 118–126.
28. Ophir, T. and Gutnick, D.L. (1994). "A role of exopolysaccharides in the protection of microorganisms from desiccation", *Appl. Environ. Microbiol.*, 60, 740–745.
29. Prasanna, R., Pattnaik, S., Sugitha, T.K., Nain, L. and Saxena, A. (2011). "Development of cyanobacterium-based biofilms and their in vitro evaluation for agriculturally useful traits", *Folia Microbiol.*, 56, 49–58.
30. Triveni, S., Prasanna, R., Shukla, L. and Saxena, A.K. (2013). "Evaluating the biochemical traits of novel *Trichoderma*-based biofilms for use as plant growth-promoting inoculants", *Ann. Microbiol.*, 63, 1147–1156.
31. Chandra, K. (ed.) (2013). National Project on Organic Farming – Annual Report 2012-13. NCOF, Ghaziabad (India).
32. Anonymous (2014). "*Fertilizer Control Order Amendment* (1985)". Governement of India, New Delhi.
33. Murphy, J. and Riley, J. (1962). "A modified single solution method for the determination of phosphate in natural waters", *Anal. Chim. Acta*, 27, 31–36.

ANNEXURE 1

Specifications for different biofertilizers for obtaining certification.

Biofertilizers	*Viable cell/g or mL before expiry*	*pH*	*Efficiency characters*
Rhizobium CB	5×10^7	6.5-7.5	Should show effective nodulation on all the species indicated
Rhizobium Liquid	1×10^8	6.5-7.5	
Azotobacter CB	5×10^7	6.5-7.5	Able to fix minimum of 10 mg N/g of sucrose utilized
Azotobacter Liquid	1×10^8	6.5-7.5	
Azospirillum CB	5×10^7	6.5-7.5	Formation of white pellicle in semisolid N-free bromothymol blue media
Azospirillum Liquid	1×10^8	6.5-7.5	
PSB CB	5×10^7	6.5-7.5	35-50 % phosphate solubilizing efficiency, 10 mm solubilization zone
PSB Liquid	1×10^8	6.5-7.5	
KMB CB	5×10^7	6.5-7.5	Minimum 10 mm solubilization zone in prescribed media of at least 3 mm thickness
KMB Liquid	1×10^8	5.0-7.5	
ZSB CB	5×10^7	6.5-7.5	Minimum 10 mm solubilization zone in prescribed media of at least 3 mm thickness
ZSB Liquid	1×10^8	5.0-7.5	
Acetobacter CB	5×10^7	5.5-6.0	Formation of yellowish pellicle in N free semi-solid medium
Acetobacter Liquid	1×10^8	3.5-6.0	
Consortia* CB	1×10^7 of each microbe/Total minimum viable count of 5×10^7	6.5-7.5	Same as for *Azotobacter, Rhizobium, Azospirillum*, KMB, PSB, ZSB depending on the members of consortium
Consortia* Liquid	1×10^8 CFU each microbe/Total minimum viable count of 5×10^8	5.0-7.0	
Mycorrhiza	100/gm of finished product	6.0-7.5	80 infection points in test roots/g of mycorrhizal inoculum used

*Mixture of any 2 or maximum three from following microorganisms: *Azotobacter, Rhizobium, Azospirillum*, KMB, PSB, ZSB

Carrier: For bacterial biofertilizers, the carrier may be peat, lignite, peat soil, humus, talc. For mycorrhizal inoculants, the base material may be fine powder/tablets/granules/root biomass mixed with growing substrate.

Moisture content for carrier based bacterial formulation should be 30-40%. For mycorrhizal inoculants, moisture content should be 8-12%.

Contamination tolerance: There should not be any contamination at 10^5 dilution for all, except for consortium for which it is 10^4 and for consortium, there should not be any contamination at any dilution

Expiry: 6 months for carrier based formulations and 12 months for liquid formulations

Source: The Fertilizer (Control) Order, 1985.

ANNEXURE 2

1. **Congo red medium for analysis of viable count and contaminants in *Rhizobium* formulations.**

K_2HPO_4	0.5 g
$MgSO_4.7H_2O$	0.2 g
NaCl	0.1 g
Yeast extract	1.0 g
Mannitol	10.0 g
Agar	20.0 g
Distilled water	1000 mL
Congo red	0.025 g
pH	7.0

2. **Medium for analysis of viable count and contaminants in *Azotobacter* formulations.**

Sucrose	20.0 g
Ferric sulphate $Fe_2(SO_4)_3$	0.1 g
Dibasic potassium phosphate (K_2HPO_4)	1.0 g
Magnesium sulphate ($MgSO_4$, $7H_2O$)	0.5 g
Sodium chloride (NaCl)	0.5 g
Calcium carbonate ($CaCO_3$)	2.0 g
Sodium molybdate (Na_2MoO_4)	0.005 g
Agar	20 g
Distilled water	1000 mL
pH	6.8 to 7.2

3. **Semi solid N-free Malate medium for MPN of *Azospirillum* formulations.**

D-Malic acid	5.0 g
K_2HPO_4	0.5 g
KOH	4.0 g
$MgSO_4$	0.1 g
NaCl	0.02 g
$CaCl_2$	0.01 g
Trace element Soln.	2.0 ml
Fe -EDTA (1.64% Soln.)	4.0 mL
Vitamin soln.	1.0 mL

0.5% Alc. BTB	2.0 mL
Agar	1.75 g (semi-solid)/15.0 g (solid)
pH	6.8-7.0

Trace element solution (g/litre)

Na_2MoO_4 $2H_2O$	0.2 g
$MnSO_4.H_2O$	0.235 g
H_3BO_3	0.28 g
$CuSO_4.5H_2O$	0.00 g
$ZnSO_4.7H_2O$	0.024 g
Distilled water	1000 mL

Use 2 ml of this solution in one litre of Nfb media

Vitamin solution (g/litre)

Biotin	0.01 g
Pyridoxin	0.02 g
Distilled water	1000 mL

Use one mL of this sol. in one litre of Nfb media

4. **Pikovskaya agar for analysis of viable count, contaminants and efficiency testing of PSB.**

Glucose	10.0 g
Tri-calcium phosphate	5.0 g
Ammonium sulphate	0.5 g
Magnesium sulphate	0.1 g
Sodium Chloride	0.2 g
Yeast extracts	0.5 g
Manganese sulphate	Trace
Ferrous sulphate	Trace
Distilled water	1000 mL
Agar	15.0 g
pH	7 ± 0.2

5. **Medium for analysis of viable count and contaminants in KMB formulations.**

Manitol	15.0
Yeast extract	3.0
Peptone	2.0
Trace element solution	1 mL

Agar	18.5
Distilled water	1000 mL
Trace element solution	
Sodium molybdate	0.20
Boric acid	0.28
Manganese sulphate	0.23
Copper sulphate	0.01
Zinc sulphate	0.03
Distilled water	1000 mL

6. Medium for efficiency testing of KMB.

Glucose	5.0 g
Manganese sulphate	0.005 g
Ferric chloride	0.1 g
Calcium carbonate	2.0 g
Potassium mineral (Mica powder)	2.0 g
Calcium phosphate	2.0 g
Agar	18.5
Distilled water	1000 mL
pH	7.0± 0.2

7. Medium for analysis of viable count, contaminants and efficiency testing of ZSB.

Zinc oxide	1.0
Ammonium sulphate	0.5
Potassium chloride	0.2
Yeast extract	0.5
Ferrous sulphate	0.01
Manganese sulphate	0.01
Dipotassium hydrogen phosphate	0.5
Distilled water	1000 mL
pH	7.0 ± 0.2

8. Medium for analysis of viable count and contaminants in *Acetobacter* formulations.

Sucrose	100.0 g
Ferric chloride	0.01 g

Dibasic potassium phosphate (K_2HPO_4)	0.4 g
Monopotassium phosphate (KH_2PO_4)	0.6 g
Magnesium sulphate ($MgSO_4.7H_2O$)	0.2 g
Sodium Chloride (NaCl)	0.5 g
Calcium Chloride ($CaCl_2$)	0.02 g
Sodium Molybdate (Na_2MoO_4)	0.02 g
BTB (0.5% in 0.2 m KOH)	5.0 mL
Agar	18.5 g
Distilled water	1000 mL
pH	5.5

9. Plant nutrient solution for plant test

S. No	*Composition*	*Conc.*	*g/L*
(a)	Potassium chloride	0.001 M	0.0745
(b)	Di -potassium hydrogen phosphate (K_2HPO_4)	0.001 M	0.175
(c)	Calcium sulphate ($CaSO_4$ $2H_2O$)	0.002 M	0.344
(d)	Magnesium sulphate ($MgSO_4$ $7H_2O$)	0.001 M	0.246
(e)	Trace elements solution		
	1. Copper sulphate ($CuSO_4$ $5H_2O$)	0.01 mg/kg	0.78
	2. Zinc Sulphate ($ZnSO_4$ $7H_2O$)	0.25 mg/kg	2.22
	3. Ammonium molybdate ($(NH_4)_6Mo_7O_{24}$ $4H_2O$)	0.0025 mg/kg	0.01
	4. Magnesium sulphate ($MgSO_4$ $7H_2O$)	0.25 mg/kg	2.03
	5. Boric acid (H_3BO_4)	0.125 mg/kg	1.43
	Prepare the solution no (e) consisting of trace elements in one liter of stock solution and add to final nutrient solution at the rate of 0.5 mL per liter.		
(f)	Iron solution:		g/100 mL
	Ferrous sulphate		5
	Citric acid		5
	Water		100 mL
	Prepare the solution no. (f) as 100 mL of stock solution and add final nutrient solution at the rate of 0.5 mL per liter.		

10. Most Probable Numbers for use with 10 fold dilution and 5 tubes per dilution (Cochran, 1950).

P1	P2	Most probable number for indicated values of P3					
		0	1	2	3	4	5
0	0	–	0.018	0.036	0.054	0.072	0.090
0	1	0.018	0.036	0.055	0.073	0.091	0.11
0	2	0.037	0.055	0.074	0.092	0.11	0.13
0	3	0.056	0.074	0.093	0.11	0.13	0.15
0	4	0.075	0.094	0.11	0.13	0.15	0.17
0	5	0.094	0.11	0.13	0.15	0.17	0.19
1	0	0.020	0.040	0.060	0.080	0.10	0.12
1	1	0.040	0.061	0.081	0.10	0.12	0.14
1	2	0.061	0.082	0.10	0.12	0.16	0.17
1	3	0.089	0.10	0.13	0.16	0.17	0.19
1	4	0.11	0.13	0.15	0.17	0.19	0.22
1	5	0.13	0.15	0.17	0.19	0.22	0.24
2	0	0.046	0.068	0.091	0.12	0.14	0.16
2	1	0.068	0.092	0.12	0.14	0.17	0.19
2	2	0.093	0.12	0.14	0.17	0.19	0.22
2	3	0.12	0.14	0.17	0.20	0.22	0.25
2	4	0.15	0.17	0.20	0.23	0.25	0.28
2	5	0.17	0.20	0.23	0.26	0.29	0.32
3	0	0.078	0.11	0.13	0.16	0.20	0.23
3	1	0.11	0.14	0.17	0.20	0.23	0.27
3	2	0.14	0.17	0.20	0.24	0.27	0.31
3	3	0.17	0.21	0.24	0.28	0.31	0.35
3	4	0.21	0.24	0.28	0.32	0.36	0.40
3	5	0.25	0.29	0.32	0.37	0.41	0.45
4	0	0.13	0.17	0.21	0.25	0.30	0.36
4	1	0.17	0.21	0.26	0.31	0.36	0.42
4	2	0.22	0.26	0.32	0.38	0.44	0.50
4	3	0.27	0.33	0.39	0.45	0.52	0.59
4	4	0.34	0.40	0.47	0.54	0.62	0.69
4	5	0.41	0.48	0.56	0.64	0.72	0.81
5	0	0.23	0.31	0.43	0.58	0.76	0.95
5	1	0.33	0.46	0.64	0.84	1.1	1.3
5	2	0.49	0.70	0.95	1.2	1.5	1.8
5	3	0.79	1.1	1.4	1.8	2.1	2.5
5	4	1.3	1.7	2.2	2.8	3.5	4.3
5	5	2.4	3.5	5.4	9.2	16.0	–

2

Status of the Biofertilization Technology and Biosafety of Biofertilizers in Egypt

Youssef Yanni[1], Frank Dazzo[2] and Abu-Bakr Gomaa[3,4]

[1]*Sakha Agricultural Research Station, Kafr El-Sheikh – 33717, Egypt*
[2]*Dept. of Microbiology and Molecular Genetics, Michigan State University, USA*
[3]*Biological Sciences Dept., Faculty of Science, King Abdulaziz University, Kingdom of Saudi Arabia*
[4]*Agricultural Microbiology Department, National Research Centre, Cairo, Egypt*
E-mail: *yanni244@yahoo.com*

ABSTRACT

Biofertilization was taken just as a tool to reduce fertilizers input while other benefits were not considered properly, leading to a situation in which less educated/informed farmers were mostly unaware about biofertilization especially when chemical fertilizers are available in the market. In Egypt, biofertilization started more than 50 years ago by legume seed inoculation with rhizobia and rice fields with cyanobacteria, and later with *Azospirilla* for different crops. Recent emerging issues include stimulation of cereals performance by certain strains of cereal root inhabitants belonging to *Rhizobium* sp. Constraints facing the technology involve: (1) lack of fixed data describing the magnitude of adverse effects of excessive application of chemical fertilizers and pesticides on man, animal, plant, environment and national economy; (2) unsatisfactory information about potential of indigenous candidates; (3) most of the agronomic studies for evaluating the biofertilization lay in a "black box" category of research, in which, the positive results are deduced from indirect evidence; (4) unsatisfactory extension programmes and inadequate training of both the farmers and the extension experts; (5) conflicts between microbiologists and agronomists about what is better, synthetic fertilizers or biofertilizers; and, most importantly; (6) less adoption of tests for verification of the biosafety of biofertilizers before released. The latter is given special attention in this context. Future high priority topics may involve: (1) necessity of testing biosafety before release of biofertilizer preparations; (2) development of rapid assays that can predict performance of formulations in field level; (3) evaluation of compounds that promote cell survival on inoculated seed; (4) determination of effects of amendments, chemical fertilizers and pesticides; (5) development of new extension strategies and initiating innovate informative methods statistically sound in design; and (6) adoption of molecular biology methods to identify desirable characteristics, manipulates and introduces more efficient strains for specific inoculation purposes, (7) explore new associations and capitalize on their diversity, PGP activities, ecology (survival, colonization, endophytic state, biogeography, dispersal, etc., and 8) strengthen the electronic connections between researchers and extension experts working on informatics concerning active, effective and modern research and extension service programmes.

Keywords: Biofertilizers, Biosafety of Preparations, Current Situation, High Priorities.

2.1 INTRODUCTION

Implementation of microbial biofertilizers in crop production can manage nutrient dynamics and pathogenesis to more environmental biosafety, improved crop productivity and low inputs of hazardous chemicals. While inoculant-manufacturing industries and marketing of inoculant preparations are well-established in Europe, North America, and Australia, only some farmers in developing countries practice biofertilization of their fields, and when used, it is just used as a tool to reduce synthetic fertilizer inputs while many other benefits are unknown, not considered properly or taken seriously. This leads to a situation in which less-educated farmers are mostly careless about utilizing microbial biofertilizers, especially when chemical fertilizers are available in the local market and/or can be easy imported. In Egypt and many other parts of the world, field inoculation trials on many legume crops and cereals showed that increases up to 33 - 50% of the yield can be achieved by proper adoption of the biofertilization technology. In Egypt, application of rhizobial and cyanobacterial inocula is a relatively old practice, adopted since the 1960s. Utilization of other preparations containing *Azospirillum* and phosphate-dissolvers began later in the 1980s. More recent emerging issues in the biofertilization technology included stimulation of rice and wheat performances by inoculation with certain strains of root inhabitants belonging to *Rhizobium* [1, 2, 3].

Constraints facing the technology are mostly similar in both developed and developing countries. The most important ones are: (1) lack of information about indigenous beneficial candidates that can be included in biofertilizer preparations; (2) shortage of data describing the magnitude of adverse effects of chemical fertilizers on man, animal, plant, environment and national economies, and how biofertilization can address them; (3) mistakes of seed or field inoculation procedures; (4) lack of proper field evaluation to test the contribution of biofertilizers; (5) the translational field inoculation work run in a "black box" category of research in which the positive results are concluded from indirect evidence, mostly using comparisons between inoculated and non-inoculated experimental field plots; (6) unsatisfactory extension programmes for both the farmers and the extension specialists; (7) inadequate training of junior researchers and lack of experience on advanced lab facilities; and (8) conflicts between microbiologists and agronomists about which is better: adoption of biofertilizers or complete dependence on chemically formulated preparations. This list of mismanagements of the technology led to a situation in which less-educated farmers have been mostly careless regarding utilization of microbial inoculants in their fields. Catastrophically, when biofertilizers were adopted, they were and still are taken lightly, just considered a tool to reduce inputs of macro-nutrients like nitrogen and phosphorus while their many other benefits are still not taken into consideration.

Recent emerging plans for the future include stimulation of cereal performance, specially rice and wheat by certain strains of root inhabitants belonging to isolates of *Rhizobium* sp. Measures for testing the biosafety of this new type of cereal biofertilization along with biosafety of the cyanobacterial biofertilizers are currently attracting special attention not only for the traditional practice of biofertilization of lowland rice but also for upland rice and other crops containing wheat [4], leguminous crops [5], vegetables [6], and control of soil properties [7].

2.2 CURRENT STATUS OF CROP PRODUCTION AND POTENTIAL OF ADOPTION OF BIOFERTILIZATION

The area of agricultural land in Egypt is confined mainly to the Nile delta and the Nile valley that extends to > 1000 km south of the delta, plus a few oases and some arable land in Sinai. The total cultivated area is 3.1 million ha. Except for some rain-fed areas on the Mediterranean coast, the entire crop area is irrigated. Over the past four decades, an area of 0.462 million ha of newly reclaimed land has been added to the agricultural area. The landholdings are fragmented, with an average farm area of 1.05 ha. The land is cropped twice annually (summer and winter cropping seasons) with a total area of about 4.956 million ha, representing a cropping ratio of about 2:1. Egypt has an arid climate with an annual average rainfall ranging from 60 to 190 mm along the Mediterranean coast to 25 to 60 mm in the Nile delta, and less than 25 mm in Upper Egypt and adjacent areas. The climate is generally very uniform with good sunshine. In addition, the Nile is an exceptional source of water, and soil near the Nile is generally of excellent quality for farming due to the accumulation of clay and silt from the annual Nile flood sediments over thousands of years. However, most of the yearly supplements of clay and silt were affected due to precipitation in the Naser Lake of 500 km^2 created since establishment of the High dam located about 950 km south of Cairo. For many social and economic reasons, this loss of part of the annual contribution of clay and silt could not be avoided. Nothing is free! However, Egypt's total agricultural crop production has increased by around 24% in the past decade. During the same period, the rate of population growth has increased at a percentage moderately higher than that percentage increase in crop production.

For cereals in Egypt, rice is one of the major summer crops, grown on nearly 0.546 million ha. It is considered the second most important cereal crop after wheat, which is the major winter cereal grain crop and the third major crop in terms of area planted (about 0.252 million ha). Maize is the third most important crop (0.132 million ha), but at least 50 per cent of its production is used for livestock and poultry feed. For fibre crops, cotton has traditionally been the most important fibre crop in Egypt and the leading agricultural export crop due to its fine and unique quality. Unfortunately, its importance is dramatically affected owing to an increase in its production in many countries and also to introduction of synthetic fibres which replaced part of the Egyptian cotton in textiles. Sugar cane is the main sugar crop in Upper Egypt. About 90 per cent of its yield is used for sugar extraction. Sugar beets are also grown in large areas of saline and less fertile areas in the Nile delta, contributing to the sugar industry. Food legumes include numerous crops that are used for human consumption, such as broad beans, soybeans, pea, common bean, lentil, lupine, chickpea and fenugreek. Egyptian berseem clover is the major winter forage crop cultivated all over the country. It is the most widely grown field crop and occupies an area that totals 0.504 million ha. Clover is rotated with rice in 60% of the rice cultivation area in and around the Nile delta. Citrus, primarily oranges that represent 85 per cent of the total citrus production area, makes up 50 per cent of Egypt's total fruit production. The fruit-planted area has expanded over the last four decades to reach about 88,200 ha. Other subtropical fruits also grown in Egypt include grapes, stone fruits and pome fruits. For vegetables, tomatoes is grown annually in three seasons – winter, summer and autumn – on about 3 per cent of Egypt's total planted area. Potatoes are the second most important vegetable after tomatoes, both in terms of cash value and total tonnage produced.

2.3 CURRENT STATUS OF THE BIOFERTILIZATION TECHNOLOGY IN EGYPT

A large area available for agriculture is cultivated with summer and winter legumes and cereal crops, in which the biofertilization technology can be successfully adopted to assist their production. Considering that the N-demand for these crops ranges between 175 and 300 kg N/ha, the amount of fertilizer N added to the soil to meet this demand ranges from 420 to 720 thousand tonnes per year, equivalent to 1.98-3.39 million tonnes of ammonium sulphate (21% N), 2.46-4.22 million tonnes of calcium nitrate (14.3% N), 0.9 - 1.48 million tonnes of urea (46% N) or 1.20-2.06 million tonnes of ammonium nitrate (35% N). Unfortunately, large amounts of these chemicals find their way to groundwater, drainage water runoff, and eventually to drinking water. About 50 years ago, attempts were made to reduce application of nitrogen fertilizers by performing large-scale seed or soil inoculation of legume crops by rhizobia [(http://www.ctahr.hawaii.edu/bnf/Downloads/Training/BNF%20 technology/Legumes. PDF), [8.9.10]. Several field application trials demonstrated that biofertilization can increase crop production while decreasing the need of high inputs of chemical fertilizers, thereby reducing the demands for high energy-consuming chemical industries, and helping to restore ecosystem health [11]. Although legume inoculation with rhizobia and rice inoculation with cyanobacteria (http://www.ias.ac.in/resonance/Volumes/09/06/0006-0010.pdf) and *Azospirillum* and *Pseudomonas* (http://www.sciencedirect.com/science/article/pii/S0929139312000157), for instance, are an accepted and mature agricultural practice, unfortunately the utilization and transfer of these technologies are still going slowly due to many reasons; some of them are presented elsewhere in the context. However, an increase in productivity per unit area is a result of integrated crop management practices that are sometimes not followed correctly and in some cases completely ignored. The importance of these urgently needed practices include: (1) new high-yielding and pest-resistant varieties; (2) better fertilization practices including starter chemical nitrogen and nodulation with indigenous or inoculated symbiotic N_2-fixers, phosphate fertilizers and/or inoculation with phosphate-dissolving bacteria; (3) foliar or soil application of trace elements and pesticides including herbicides, insecticides and fungicides; (4) proper water management including surface, sprinkled and foliar irrigation accompanied with efficient drainage systems; (5) the interactions between application of biofertilizers and the dynamics of soil structural changes; (6) water content and diffusional constraints of nutrients compartmentalization in inoculated plants compared to their non-inoculated counterparts; and (7) coupling of N-cycling processes and linkage of microbial community structure.

The research and application programmes dealing with the biofertilization technology in Egypt include studies on various cultivars of legume fodder crops including new varieties of alfalfa, berseem clover, guar and fodder lupine. These new cultivars have been tested for responses to inoculation with different strains of N_2-fixing bacteria. In addition, new soybean varieties that are resistant to the cotton leaf worm have been widely released to minimize use of insecticides and reduce production input costs. These have been tested for successful nodulation and efficiencies in N_2-fixation. Decayed organic matter derived from residues of legumes, cereals and other field vegetables, fibre crops, oil crops like sunflower, and fruit crops is being used for production of compost and compost tea to reduce application of chemical fertilizers, maintain soil fertility and control environmental pollution while targeting environment-friendly practices that can initiate/maintain agricultural sustainability. Special emphasis is directed to production of new fodders from combinations of different agricultural wastes and recycling of other crop residues (mainly rice straw) through production of on-farm organic manure using microbial preparations. An advanced research programme was initiated for production of microbial preparations including symbiotic N_2 fixers for legumes and *Azospirillum* for cereals and to assess the beneficial endophytic association between rhizobia and cereals [1, 2, 3], taking into account the biosafety of the biofertilizers that will be used in this new approach in the cereals biofertilization technology. This includes not only biosafety of the microbial candidates but also quality control during different stages of production of the biofertilizer preparation, effects of storage and transportation and proper methods of field application. However, until now, less than 5% of the area of legumes annually receives biofertilizer preparations — an undesired situation, but it creates high potential and active BNF programmes that would help to establish new production strategies for major field crops. International collaborations on BNF in Egypt started in early 1980s. A strong collaboration was established with the NifTal project targeting a stronger BNF research/technology programmes. This involved inoculation of different legume crops with locally isolated cultures of symbiotic N_2 fixers, a field-testing programme to compare effects of inoculation with indigenous vs. imported rhizobial strains, and field management that supported maximum benefits from field inoculation. In addition, a strong research and application programme with the Indian Agricultural Research Institute (IARI), New Delhi, resulted in setting up of mass production of cyanobacterial inocula for rice fields using the open-air soil culture method [12]. Other activities included enhancement of production of major cereal and legume crops through multidisciplinary research and application and bringing together experts of different areas in agricultural research with extension specialists. This resulted in the development of inocula, now available in the market: rhizobia for legumes, cyanobacteria for rice and azospirilla for legumes and cereals. Most of the completed research has shown a positive response of crops to inoculation using the proper strain(s) [1, 2, 3, 9, 14, 15], and confirmed the importance of nodulated legumes in maintaining soil fertility through agronomic rotations involving both cereal and legume crops. The potential benefit of using efficient BNF inocula is high as more than ninety per cent of the total area can be targeted for efficient extension programmes on biofertilization. Economists have estimated that this practice can save the country millions of USDs per year through diminished N-fertilizers application [10], regardless of several additional indirect benefits due to decreasing environmental pollution, enhanced human and animal health and increasing work abilities.

Although N cycling in Egyptian rice farming systems has received considerable attention over the years, an active research programme was established addressing emerging issues such as stimulation of rice growth by *Rhizobium leguminosarum* by *trifolii*, the new beneficial rice/microbe association that was first described by Yanni and coauthors [1], and then, independently, validated by independent research teams in many other agro-ecosystems worldwide. Similar associations containing the *Rhizobium leguminosarum* bv. *trifolii* were later introduced to the scientific community. For instance, although tens of strains of this species isolated from within surface-sterilized rice roots were effective N_2-fixing symbionts in root nodules of their natural host, clover, they were found to be active stimulators of growth and performance of cereals that rotate with the corresponding legumes in the crop rotations like those of corn-beans (Mexico), wheat-lentil (Morocco), wheat-clover (Canada & Egypt), rice-clover (Egypt), barley-peas (Canada), and sorghum and millet - soybeans (Kenya) plus others. Egypt initiated programmes to establish research institution network that worked in the areas of advanced biotechnology and genetic engineering, including activities related to BNF [16]. Another major collaborative effort was launched in 1996 sponsored by the US-Egypt Science and Technology Joint Fund in which major priority areas were identified and more than forty-four collaborative research projects including BNF research and technology transfer were setup. It is clear then that BNF is still considered a promising research area in the focus of agricultural sustainability for enhancement of crop production without additional chemical fertilizer inputs and environmental pollution through excessive catastrophic consequences of their use [16, 17].

Inoculation with some (not all) of the newly discovered candidates sharing the cereal/ rhizobia associations were tested for biosafety and found to have potentials to adversely affect the growth of wild rice-aeshcynomene (Senegal) and rice-clover in Egypt. This phenomenon attracted our attention to biofertilizer biosafety issues early in the last decade. So, this chapter is also concerned with the measures undertaken to assess the biosafety of the preparations that are in use or on their way to become part of the biofertilization technologies in the Egypt agriculture landscape.

2.4 BIOSAFETY TESTS OF BIOFERTILIZER PREPARATIONS IN EGYPT

Most of the microbial candidates locally used for biofertilization of legume and cereal crops are isolated, demonstrated to be effective, efficient and more adapted to the Egypt agroecosystem than their imported counterparts. Recommendations for use of these candidates in production of biofertilizer inoculants are based mostly on their efficiencies in biological dinitrogen fixation, their ability to increase phosphorus availability through dissolution of insoluble phosphates, and other plant growth-promoting activities.

2.4.1 Clover/Rice and Wheat Agronomic Rotations

Figure 2.1 illustrates the three-dimensional life cycle of the clover endosymbiont *Rhizobium leguminosarum* bv. *trifolii* in Egypt agro-ecosystem in which rice is traditionally followed by clover in the same crop rotation [1]. Root-adapted rhizobia indigenous to soil in the Nile delta were found to be active endophytic plant-growth promoting rhizobacteria (PGPR) within rice roots, but some of the isolated strains were subsequently found to be Nod$^+$ Fix$^-$ (borderline pathogens) on clover. By performing many biosafety tests, the association between rice and clover *Rhizobium* was found to include all the possible types of association (beneficial, neutral, and deleterious). Some of the clover rhizobia strains were found to be active clover root nodulators and N_2-fixers while being capable of stimulating rice growth in both gnotobiotic lab culture tests, PGPR$^+$ [1,18] and later

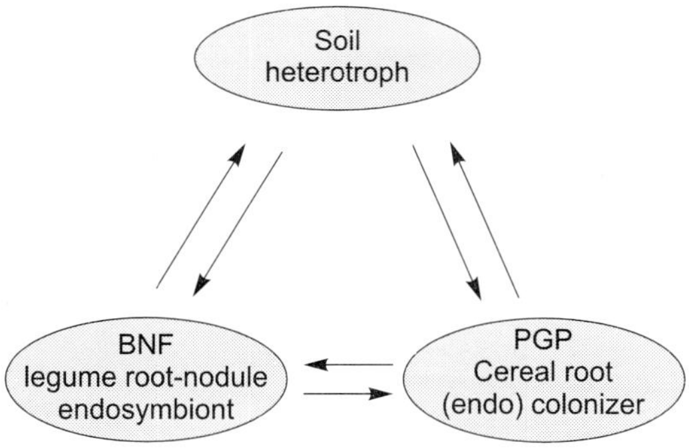

Fig. 2.1 *Rhizobium* life cycle in legume-cereal rotations.

in translational field inoculation studies in the Egypt Nile delta. Some of the strains in the gnotobiotic tests were found to be Nod$^+$Fix$^+$ on clover but inhibitors to rice growth, PGPR$^-$. Some others were found to inhibit clover growth (toxi) while they expressed PGPR$^+$ on rice. Table 2.1 clearly indicates these various interrelationships between the *Rhizobium leguminosarum* bv. *trifolii* strains, the clover legume host, and the rice plant. We concluded that utilization of inocula as biofertilizers requires their careful evaluation, including biosafety studies to assess their effects on other legume and cereal crops that share the same crop rotation with the natural legume host, the clover. Figure 2.2 illustrates examples of undesirable characteristics of certain rice endophytic strains of *Rhizobium leguminosarum* bv. *trifolii* (Rlt) isolated in the Nile delta recognized through PGP bioassays under gnotobiotic conditions. Recognition of the adverse effects of the inoculant strain on non-target plant host avoids potentially catastrophic consequences in future field applications. In this biosafety test example, both the endophytic rhizobial strains E3 and E13 cannot be used for biofertilization of clover or rice in a rotation containing the two crops.

Table 2.1 Examples of undesirable characteristics of certain rice endophytic strains of *Rhizobium leguminosarum* bv. *trifolii* (Rlt) isolated in the Nile delta recognized through PGP bioassays under gnotobiotic conditions.

Rlt strain	*Response on clover*	*Response on rice*
E 11	Nod$^+$ Fix$^+$ (Clover nodulator and growth promoter)	PGP+ (Growth promoter)
E 3	Nod$^+$ Fix$^+$ (Clover nodulator and growth promoter)	PGP$^-$ (Growth inhibitor)
E 13	Nod$^+$Fix$^-$ (Clover nodulator but pathogenic – ultimately lethal)	PGP$^+$ (Growth promoter)

Next, we evaluated the symbiotic performance of 8 isolates of the wheat-adapted rhizobial endophytes representing the range of strain diversity in this collection on berseem clover plants in enclosed gnotobiotic cultures grown in nitrogen-free growth media in growth chambers. Evaluation of the Nod/Fix symbiotic phenotypes indicated that both effective and ineffective isolates were included in the culture collection (Fig. 2.2).

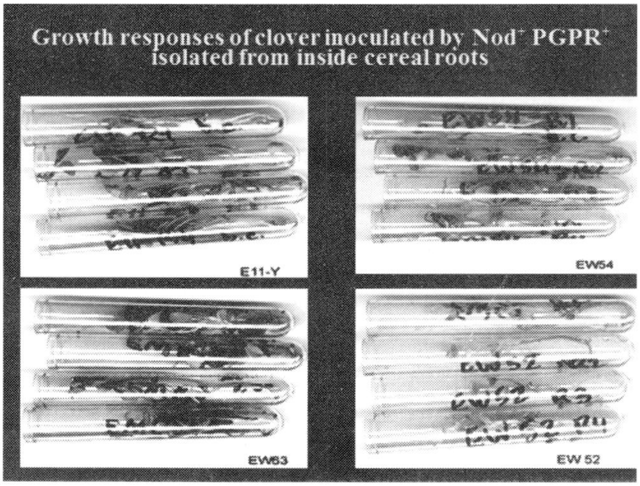

Fig. 2.2 Gnotobiotic growth responses of berseem clover to inoculation with rhizobial endophyte strains isolated from rice (E11) and wheat (EW52, EW54, and EW63).

We also assessed the ability of some isolates of wheat-adapted rhizobia to promote wheat growth in gnotobiotic culture (Fig. 2.3). The experimental design of this bioassay was similar to the one used in our earlier rice studies under growth chamber conditions, except that the containers for wheat growth did not include a top layer of water-submerged sand. The growth response of 4 currently economically important wheat cultivars (2 from Egypt, 2 from USA farming systems) revealed the range of wheat growth responses to the various strains of wheat-adapted rhizobial endophytes, and showed strain-variety specificity in the rhizobia-wheat association. They helped us in selecting the strains of wheat-adapted rhizobial endophytes that could be used in the field inoculation trials.

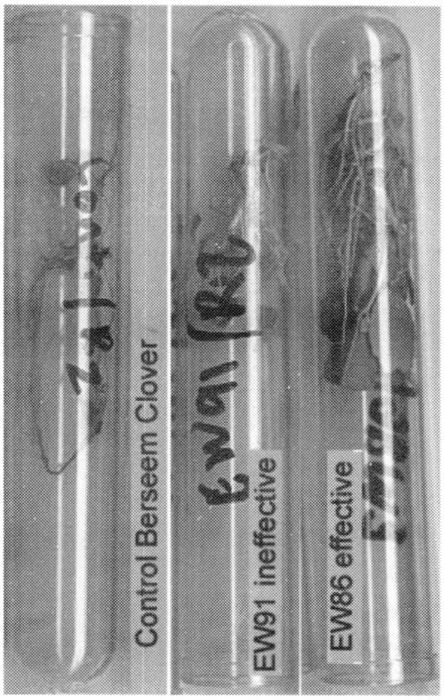

Typical ineffective and effective symbiotic responses of berseem clover inoculated with 2 wheat endophytes of *R.l. trifolii*. Note the uninoculated control plant for comparison.

Fig. 2.3 Gnotobiotic growth responses of berseem clover. Plants were uninoculated (left) or inoculated with wheat endophyte strains of *R. leguminosarum* bv. *trifolii* (middle, ineffective strain EW91; right effective strain EW86).

2.4.2 Another Track for Testing Rhizobial Strains for Cereal Biofertilization

Wheat develops a natural, beneficial association with mycorrhizal fungi. A biosafety concern we had was whether biofertilization with these growth-promoting rhizobia may interfere with that fungus-wheat root association. For this test, wheat roots were sampled from 7 field testing sites receiving 1/3rd, 2/3rds, and the full recommended fertilizer-N doses (180 kg N/ha), cleared with 10% KOH, and stained with 0.05% trypan blue [12]. After lactophenol decolorization, roots were cut into 1 cm segments and quantitatively examined for mycorrhizal infection using brightfield microscopy. Microscopic examination of the inoculated and uninoculated roots indicated that they were all infected with endomycorrhizal fungi. Examples of mycorrhizal infections are shown in the composite image of Fig. 2.4. Evaluation of the intraradical spores, vesicles and hyphae of these endomycorrhizal micrographs indicate morphologies of the endomycorrhizal fungal symbionts that resembled Glomus intraradices (= *Rhizophagus irregularis*) or other related members of the Glomaceae. The intensity of infection in the sampled root segments varied from 34% to 92%, depending on the level of fertilizer N, inoculation, and field testing site. The means of mycorrhizal infection with vs. without rhizobial infection at the 7 sites were 73.2% vs. 70.4% at the 1/3rd dose, 83.1% vs. 74.1% at the 2/3rd dose, and 71.7% vs. 60.1% at the full recommended dose of fertilizer N.

These differences in quantitative levels of mycorrhizal infection of wheat roots were not statistically significant (df. 6, p = 0.385, 0.169 and 0.203, respectively), indicating that there was no statistically significant difference in overall level of mycorrhizal infection due to rhizobial inoculation. These results indicate that our wheat biofertilizer strains successfully pass biosafety tests indicating that the wheat plants would likely continue to gain benefit from this natural, plant-fungal symbiosis.

2.4.3 Screening of Toxicity of Cyanobacterial Isolates for Rice Biofertilization

Inoculation of Egypt rice fields with dinitrogen-fixing cyanobacteria [13] was adopted on an experimental scale for about 20 years from 1970 to 1990. Large-scale production of inocula containing active contributors that biologically fix nitrogen is now an important practice for enhancing rice performance [19, 20]. Increases in rice yield up to 15-25% were achieved by substitution of 1/3 to 1/2 of the recommended rate of fertilizer nitrogen by inoculation with cyanobacteria (8, 20, 21]. Cyanobacterial contributions other than merely biologically fixed-N are also reported. They included antagonism against aquatic macrophytes in rice fields [22, 23, 24]), increasing availability of native phosphorus [25], decreasing sulfide injury [26], aiding soil aggregation [27], and stabilization of the N-status of the ecosystem to be sufficient for healthy plant growth without increasing susceptibility to pests like blast fungus, stem-borers or leaf-miners [19, 20, 21, 25, 28].

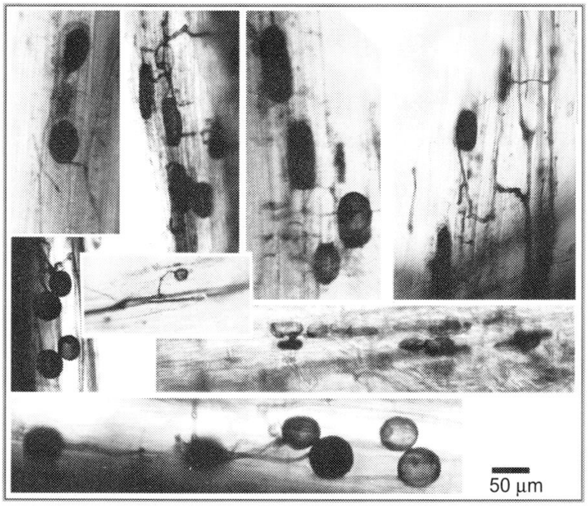

Fig. 2.4 Examples of endomycorrhizal fungal infection of wheat roots inoculated with rhizobia and grown under field conditions.

Although this technology became a common practice in many developing countries, the ability of cyanobacterial inoculants to produce dangerous toxins seems nearly out of consideration of the inocula producers. Since isolates in an inoculum normally will find their way to soil and water resources that are frequently used by marginal farmers for animal and human consumption, this necessitates important biosafety testing of the cyanobacterial isolates used for production of inocula to assess their potential (possible potential and potency) to produce different types of toxins. Occurrence of toxic cyanobacteria in water blooms has been documented in several areas of the world since the first published reports in the middle to late 1800s [29, 30, 31, 32]. Gorham and Carmichael [33] found that species and strains belonging to *Anabaena, Aphanizomenon, Microcystis, Nodularia* and *Oscillatoria* are the main toxic bloom-formers in mesotrophic to eutrophic fresh and brackish waters. Wild and domestic animal poisonings plus some human intoxication attributed to the presence of toxic blooms have been found with increasing frequency in several areas of the world [34, 35].

2.4.3.1 Methodology

We initiated a programme for examining samples collected all over the Nile delta to shed light on the abundance and distribution of toxic cyanobacteria in this area. Isolation, purification and identification of some 75 cyanobacterial isolates collected from the Egypt Nile delta revealed that the collection is dominated by two main categories having distinguished cultural and morphological characteristics. For the first category, 21 day-old cultures on solid medium showed localized spreading growth of colonies, fibrous in appearance, on the agar surface. Colonies have a glossy lustre, are opaque and medium green, but the surrounding media are not coloured (lack water-soluble pigments). Eighteen day-old cultures in liquid medium showed green homogenous growth without colouring the medium itself. Microscopic examination revealed that trichomes have no ramifications. They are uniseriate, single wavy and show no tapering. No sheath or motility of the trichomes is observed. Vegetative cells are short angular, 4-4.5 micrometres wide, and 2-3.5 micrometres long. Most apical cells are pointed. Heterocysts are mostly intercalary and single. They are long barrel-shaped, 5-6 μm wide and 6.5-7.5 μm long. No spores were observed. Hormogonia have the same width as the filaments. Comparison of this group of isolates with previously isolated ones [36] suggested that they belong to the genus *Anabaena*. The second group of isolates was most commonly obtained from rice fields located in the middle Nile delta. Twenty days-old cultures on solid medium showed colonies with an average of 1-1.5 mm diameter and low convex elevation at their edge. Their margins were rough. The colonies had dull lustre, opaque and dark-green colour. Their growth media was not coloured. Sixteen day-old cultures in liquid medium had aggregative sedimentary types of growth and an incohesive patchy growth at the bottom of the flasks. The cultures showed medium green colour; the patchy growth was medium yellowish-green. The medium itself was not coloured. Microscopic observation revealed trichomes having no ramifications. They were uniseriate, single, aggregated and exhibited neither polarity nor tapering. No sheath or trichome motility was observed. There were three sizes and shapes of cells; (a) barrel-shaped cells, 5 - 6 μm wide and 5.5 - 7 μm long; (b) granular, ellipsoidal cells, 5 μm in width and 5.5-7.5 μm long; (c) yellowish-brown round cells, 8 - 9 μm in diameter. Few heterocysts were observed. They were single and both intercalary and terminal. These criteria closely resemble those of the genus *Nostoc* (most likely *Nostoc muscorum*).

2.4.3.2 Biosafety test for production of hepatotoxic microcystins

The enzyme-linked immunosorbent assay test (ELISA) revealed that most of the isolates exhibit varying capabilities of producing hepatotoxic microcystins. Figure 2.5

indicates that about 58 of the 75 isolates (about 77%) produced up to 500 ng hepatotoxic microcystins/L of their cell-free culture media (A). About 23% of them (17 out of 75) produced more than 500 ng/L having varied toxin-producing abilities. 10.7% produced concentrations from 501 to 1000 ng/L (B), and 6.7% produced 1001 to 1500 ng/L (C). Extracts of two isolates contained up to 2000 ng/L (D), one registered 2800 ng/L (E) and another scored 4240 ng/L (F) [37, 38].

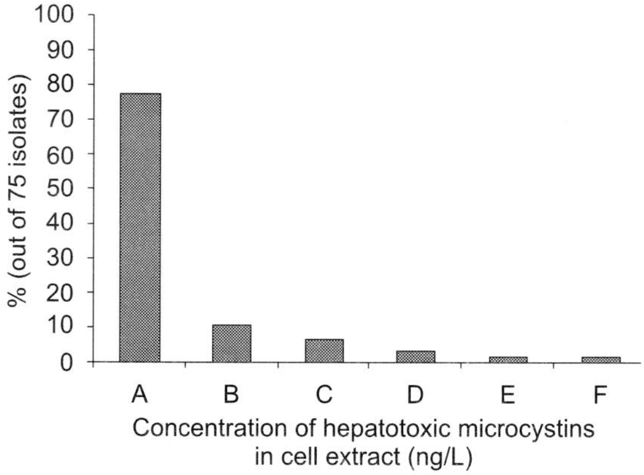

Fig. 2.5 Screening of cyanobacterial isolates belonging to *Anabaena* and *Nostoc* for production of hepatotoxic microcystins.

As aforementioned, 23% of the Egyptian isolates were found to be active producers of hepatotoxic microcystins that are capable of inducing acute liver toxicosis. Recently, these cyclic peptide hepatotoxins (which are called microcystin and nodularin) have been found to be potent liver tumour promoters through their inhibition of type I and type 2A protein phosphatases. This raises the possibility that they could be involved in liver tumour promotion in humans through consumption of water or other sources contaminated with toxic cyanobacterial types. However, the presence of these toxic types may have a relation with liver failure and other liver diseases recognized by epidemiological evidence between marginal farmers of the Egypt Nile delta. Nevertheless, when about one-quarter of the collected isolates tested in this work were found to be hepatotoxic microcystins producers, the situation cannot be taken lightly. It is, therefore, important to have these toxins monitored and regulated when present in human drinking supplies. In addition, it is very important to test the biofertilizer candidate cyanobacteria for their toxin production in order to identify which ones are toxic cyanobacterial forms that should absolutely be avoided as a biofertilizer for rice fields.

Fortunately, this study revealed that the isolates of cyanobacteria that are currently being used as a biofertilizer for rice fields in Egypt are very low hepatotoxic microcystins producers. Future work worldwide for using cyanobacteria in biofertilizers preparations must involve screening of the cyanobacterial isolates on several mammalian and microbial cell lines for cytotoxic activities.

2.5 MISIDENTIFICATION OF CANDIDATES TO BE USED IN BIOFERTILIZATION

We have an experience with this issue. Briefly, we performed two cycles of isolations of endophytic rhizobacteria that were targeted for isolation of *Azospirillum* using the common N-free malate semi-solid medium designed for isolation of that soil microorganism. In both cycles, various isolates that matched the phenotypic description of *Azospirillum* were obtained in pure culture. They were capable of microaerophilic growth in the N-free malate semi-solid medium, formed characteristic pink pigmented colonies on beef extract peptone agar plates, and had Gram-negative rod morphology. However, in both isolation cycles, when the 16S rDNA sequences of many of the isolates were determined and compared to established databases, their taxonomic status was indicated as species of *Pseudomonas,* not *Azospirillum.* Since this unexpected result repeated in the 2nd cycle of isolations, we had no doubt of its accuracy. These surprising results cast doubt about the use of "standard" procedures in the microbiology literature to isolate *Azospirillum.* Although this was our unique experience, it reinforces the necessity of advanced molecular genetics methods to accurately identify the candidates that will be used for production of biofertilizer preparations. However, several recent research works referred to some *Pseudomonas* species as PGPR. A lot of these research reports can be accessed in the proceedings of the "8th International Workshop on Plant Growth-Promoting Rhizobacteria, Portland USA, May 17–22, 2009 (http://capps.wsu.edu/pgpr/Abstract_Booklet.pdf).

2.6 BIOSAFETY OF COMPOSTED MATERIALS FOR USE AS BIOFERTILIZERS

Recommendations for use of microbial candidates for production of compost and compost tea from farm residues is a long precarious story. The practice has depended only on testing the candidates in the starter used as inocula in preparations for biosafty while completely ignoring the biosafety and quality of the produced composted material to assure that it is safe before use in the open fields. However, recent studies have revealed that some of these microbes enriched in the composts may adversely affect humans, animals, plants and the environment.

Conclusively, some of the microbial candidates used or planned to be used as biofertilizers in Egypt, if not tested, may adversely affect humans, animals, plants and/or the environment.

2.7 REASONS FOR LOW ADOPTION OF THE BIOFERTILIZATION TECHNOLOGY

Reasons for the current low adoption rates of the biofertilization technology in some developed countries and most of the developing countries are of two main categories:

2.7.1 Reasons for Which Discriminating Issues can be Found Through Improvement of the Technology

- Agriculture uses few - if any – modern production inputs.
- Deficiency in the extension and technology transfer programmes.
- Poor performance of some biofertilizer preparations. In one study, half of the inoculants from 12 countries had less than 10 viable cells/g representing a population density where yield response is compromised [39]. The farmer then no longer trusts the technology for sound and low cost fertilization management.
- Lack of advanced facilities for fast and effective quality control during and after manufacture, and for monitoring the candidate dynamics after field application.
- Inaccurate seed inoculation procedures; faulty preparation of the soil like in case of lack of appropriate water content in the seed bed; and incorrect times, rates and methods of application of supplementary chemical fertilizers and pesticides needed as starter doses to support successful biofertilization.
- Lack of information about the potential of indigenous strains, their biodiversity, infectivity, efficiencies and competitiveness to inoculated strains. Most of the agronomic studies to evaluate the efficacy of field inoculation lie in a "black box" category of research, recording positive results from indirect evidence. In this case, competitiveness of the inoculant strains against their indigenous counterparts needs to be explored.

2.7.2 Reasons That Require Policy Interventions

- There are no fixed data describing the magnitude of adverse effects related to excessive use of chemical fertilizers on man, animal, plant and environment and the overall effects on national economy.
- Extension programmes for biofertilization are mostly taken lightly, especially where poorly-educated farmers can access large amounts of locally produced or imported chemical fertilizers in the market. The farmers mentality prefer chemical fertilizers as they induce rapid changes in plant features and growth responses. The plant mostly turns to the delighting dark green color and turn healthy within 2-3 days after chemical fertilization, comparing to application of the biofertilizers which show similar changes slowly and on a long run period. The farmers do not properly consider the potential adverse effects of chemical fertilization.
- Neither the less-educated farmers nor agronomists and policy makers care enough about environmental pollution caused by excessive use of chemical fertilizers. This needs some governmental regulations and interventions.

2.8 HIGH PRIORITY TOPICS AND FUTURE AREAS OF MULTINATIONAL COLLABORATION

- Develop rapid assays that can predict performance of microbial formulations, especially in the field.
- Evaluate compounds that promote cell survival on inoculated seed.

- Determine effects of agrochemicals on performance of the microbial formulations.
- Develop new extension strategies and initiate informative methods that are statistically sound in design.
- Adopt molecular biological methods to identify, manipulate and introduce more efficient strains for specific inoculation purposes. They must be critically examined regarding their biosafety before use in biofertilizer preparations.
- Explore new plant/microbe associations and capitalize on their diversity, plant growth promotion activities (PGP) and ecology (survival, colonization, endophytic state, biogeography, dispersal, etc.).
- Strengthen the electronic connections between researchers, extension experts and policy makers working on informatics concerning active, effective, efficient and modern research, application and extension programmes that put agricultural sustainability and environmental biosafety as urgent future priorities for workers through extension programmes based on feed/feedback based systems.

CONCLUSION

Many constraints are facing the biofertilization technology in Egypt and elsewhere. Included among them are deficiencies of biofertilizer candidates, insufficient data on adverse effects of agrochemicals, environmental factors, biosafety of biofertilizers, lack of sufficient field evaluations, unsatisfactory extension programs, inadequate training, and farmers' carelessness about the necessity of biofertilization to control various adverse consequences on human health, environment and work abilities. Current recommendations for use of biofertilizer candidates are based on their plant growth-promoting efficiencies regardless of their biosafety for use in open fields. Improvement of the technology needs, among others, accurate evaluation of microbiological and agronomical factors that support successful biofertilization, effective and efficient extension and technology transfer programs, and novel governmental regulations and interventions to diminish excessive use of chemical fertilizers. High priority topics and necessary future plans include: initiation of rapid assays that can predict the performance of microbial formulations in fields, evaluation of compounds that can promote cell survival on inoculated seeds, assessment of effects of agrochemicals, development of extension strategies and informative methods that are statistically sound in design, adoption of molecular biological methods to identify, manipulate, and introduce more efficient strains after critical examination of their biosafety, capitalization of their diversity, biogeography, colonization and survival in preparations and dispersal in fields. Strengthen electronic connections between researchers, extension experts and policy makers must be adopted. In conclusion, considering agricultural sustainability, environmental soundness and biosafety of crop production practices by using biosafe biofertilizers have currently become a matter of "to be or not to be" for humanity!

REFERENCES

1. Yanni, Y.G., Rizk, R.Y., Conch, V., Squartini, A., Philip-Hollingsworth, S., Mateos, P.F., Orgambide, C., De Bruijn, F., Stolfzfus, J., Buckley, D., Schmidt, T., Ladha, J.K. and Dazzo, F.B. (1997). "Natural endophytic association between *Rhizobium leguminosarum* bv. *trifolii* and rice roots and assessment of its potential to promote rice growth". *Plant and Soil,* 194, 99-114.
2. Yanni, Y.G., Rizk, R.Y., Abd El-Fattah, F.K., Squartini, A., Conch, V., Giacomini, A., de Bruijn, F., Rademaker, J., Maya-Flores, J., Ostrom, P., Vega-Hernandez, M., Hollingsworth, R.I., Martinez- Molina, E., Mateos, P., Velazquez, E., Wopereis, J., Triplett, E., Umali-Garcia, M., Anarna, J.A., Rolfe, B.C., Ladha, J.K., Hill, J., Mujoo, R., Ng, P.K. and Dazzo, F.B. (2001). "The beneficial plant growth promoting association of *Rhizobium leguminosarum* bv. *trifolii* with rice roots". *Aust. J. Plant Physiol.,* 28, 845–870.
3. Yanni, Y.G. and Dazzo, F.B. (2010). "Enhancement of rice production using endophytic strains of *Rhizobium leguminosarum* bv. *trifolii* in extensive field inoculation trials within the Egypt Nile delta". *Plant and Soil,* 336, 129–142.
4. Karthikeyan, N., Prasanna, R., Nain, L. and Kaushik, B.D. (2007). "Evaluating the potential of plant growth promoting cyanobacteria as inoculants for wheat". *Eur. J. Soil Biol.* 43, 23–30.
5. Prasannaa, R., Trivenia, S., Bidyarania, N., Babua, S., Yadava, K., Adaka, A., Khetarpalb, S., Palb, M., Shivayc and Saxenaa, A.K. (2014). "Evaluating the efficacy of cyanobacterial formulations and biofilmed inoculants for leguminous crops. DOI:10.1080/03650340.2013.792407, *Arch. of Agron. and Soil Sci.,* 60, 349–366
6. Talita, C.H. and Bittencourt-Oliveira, M.D. (2012). "Bioaccumulation of microcystins in lettuce". *J. of Phy.,* 48, 1535–1537.
7. Shahane Amit, A., Singh, Y.V., Kumar, D., Prasanna, R., Chakraborty, D. and Kumar, A. (2013). "Influence of cyanobacterial inoculants and planting methods on nutrient uptake of rice (Oryza sativa) and soil chemical properties", *Indian J. of Agron.,* 58, 607–610.
8. Yanni, Y.G. and Hegazy, M.H. (1990). "Efficiency of algalization of rice in the Nile Delta with different amounts of nitrogen and potassium fertilization". *World J. of Microb. and Biot.,* 6, 395–399.
9. Yanni, Y.G. (1998). "Efficiency of biofertilization management of rice and soybean in the Nile Delta with application of pesticides". Pages 643-648 in Proceedings of the 1998 Brighton Conference on Pests & Diseases, Brighton, UK, 16–19 November 1998.
10. Abdel-Aziz, R.A. (2001). "Production and application technology of rhizobial inoculants". Page 30 in Proceedings of the First US-Egypt Workshop on Microbial Ecology (Dazzo F, Badr El-Din S, Tiedje J and Eissa S, eds.), 6–10 May 2001, National Research Center, Cairo, Egypt.
11. Yanni, Y.G. and Abd El-Fattah, F.K. (1999). "Towards integrated biofertilization management with free living and associative dinitrogen fixers for enhancing rice performance in the Nile Delta". Symbiosis, 27, 319–331.

12. Phillips, J.M. and Hayman, D.S. (1970). Improved procedures for clearing roots and staining parasitic and vesicular-arbuscular mycorrhizal fungi for rapid assessment of infection. Trans. of the Brit. Mycol. Soc., 55, 158–161.
13. Venkataraman, G.S. (1966). Algalization. *Phykos*, 5, 164–174.
14. Yanni, Y.G. (1990). "Response of symbiotic interrelationship between soybean and the indigenous or inoculated microsymbiont, *Bradyrhizobinm japonicum*, to soil application of manganese and molybdenum". *World J. Microb. Biot.*, 5, 289–294.
15. Yanni, Y.G. (1992a). "Contribution of inoculation with azolla combined with nitrogen, phosphorus and zinc to rice in the Nile Delta". *World J. Microb. Biot.*, 8, 579–584.
16. Moawad, H. (2001). "Overview of microbial ecology research in Egypt". Pages 22-28 in *Proceedings of the First US-Egypt Workshop on Microbial Ecology* (Dazzo F, Badr Eldin, Tiedje J and Eissa, eds.), 6-10 May 2001, National Research Center, Cairo, Egypt.
17. Tiedje, J.M. (2001). "Status and directions in U.S. microbial ecology research". Pages 19-21 in *Proceedings of the First US-Egypt Workshop on Microbial Ecology*, (Dazzo F, Badr Eldin, Tiedje J and Eissa, eds.), 6-10 May 2001, National Research Center, Cairo, Egypt.
18. Yanni, Y.G. and Dazzo, F.B. (2001). "Necessity for testing the biosafety of biofertilizers used in the Egyptian farming system". Page 32 in *Proceedings of the First US-Egypt Workshop on Microbial Ecology* (Dazzo F, Badr El-Din 5, Tiedje J and Eissa 5, eds.), 6-10 May 2001, National Research Center, Cairo, Egypt.
19. Yanni, Y.G. and Sehly, M.R. (1991). "Rice performance and natural infection with blast (*Pyricularia oryzae* Cav.) under different algalization techniques and rates of fertilizer nitrogen". *World J. Microb. Biot.*, 7, 43–47.
20. Yanni, Y.G. and Osman, Z.H. (1990). "Contributions of algalization to rice growth yield, N attributes and incidence of infestation with the blast fungus *Pyricularia oryzae* under different fungicidal treatments". *World J. Microb. Biot.*, 6, 371–376.
21. Yanni, Y.G. and Abd El-Rahman, A.A. (1993). "Assessing phosphorus fertilization of rice in the Nile Delta involving nitrogen and cyanobacteria". *Soil Biol. Biochem.*, 25, 289–293.
22. Subramanyan, R., Relwani, L.L. and Manna, G.B. (1965). "Fertility build-up of rice soils by blue-green algae". *Proceedings of the Indian Academy of Sciences*, Series B 62, 252–277.
23. Yanni, Y.G., Salem, K.G. and Shalan, S.N. (1988). "Response of transplanted rice to algalization under different treatments of herbicides". In: *Proceedings of the Second Conference of the Agricultural Development Research*, Ain-Shams University, Cairo, 216–226.
24. Yanni, Y.G. and Abdallah, F.E. (1990). "Role of algalization in rice growth, yield and incidence of infestation with the stem borer *Chilo agamemnon* Bles. and the leaf miner *Hydrellia prosternalis* Deeming in the Nile Delta". *World J. Microb. Biot.*, 6, 383–389.
25. Arora, S. (1969). "The role of algae in availability of phosphorus in paddy fields". *Riso*, 18, 135–138.

26. Jacq, V. and Roger, P.A. (1977). "Diminution des fonts de semis dues a' la sulfatoreduction, par un pretraitement des grains de riz avec des cyanophycees". *Cabiers ORSTOM Serie Biologie,* 12, 101–107.
27. Roychodhury, R., Krishnamurti, G.S.R. and Venkataraman, G.S. (1980). "Effect of algal inoculation on soil aggregation in rice soils". *Phykos* 19, 224–227.
28. Yanni, Y.G., Abdallah, F.E., Zidan, M.I. and Foda, M.E. (1993). "Rice yield and resistance to infestation with the stem borer *Chilo agamemnon* as affected by algalization under different agrochemical conditions". In: *Nitrogen-Fixation with Non-Legumes,* N.A. Hegazi, M. Fayez & M. Monib, eds. (American Univ. Press, Cairo), 143–144.
29. Francis, G. (1878). "Poisonous Australian lake". *Nature,* 18, 11–12.
30. Skulberg, O.M., Codd, G.A. and Carmichael, W.W. (1984). "Toxic blue-green algal blooms in Europe: a growing problem". *Ambio,* 13, 244–247.
31. Codd, G.A. and Poon, G.K. (1988). "Cyanobacteria toxins". In: *Biochemistry of the Algae and Cyanobacteria,* L.J. Roger & J.G. Gallon, eds. (Oxford Science Publishers, Oxford), 283–296.
32. Carmichael, W.W. (1989). "Freshwater cyanobacteria (blue-green algae) toxins". In: *Natural Toxins: Characterization, Pharmacology and Therapeutics,* C.L. Ownby and G.V. Odell, eds. (Pergamon Press, Elmsford, New York), 3–16.
33. Gorham, P.R. and Carmichael, W.W. (1988). "Hazards of freshwater blue-greens (cyanobacteria)". In: *Algae and Human Affairs,* C.A. Lembi & J.R. Waaland, eds. Cambridge University Press, Cambridge, 403–431.
34. Carmichael, W.W. (1987). "*The Water Environment: Algal Toxins and Health*". W.W. Carmichael (Ed.) Plenum Press, New York, 1–13.
35. Mahmood, N.A., Carmichael, W.W. and Pfahler, D. (1988). "Anticholinesterase poisoning in dogs from a cyanobacterial (blue-green alga) bloom dominated by *Anabaena flosaquae*". *Am. J. Vet. Res.,* 49, 500–503.
36. Roger, P.A. and Ardales, S. (1991). The IRRI blue-green algae collection and computerized information on the strains available for distribution. *J. Appl. Phycol.,* 3, 375–376.
37. Beasley, V.R., Cook, W.O., Dahlem, A.M., Hooser, S.B., Lovell, R.A. and Valentine, W.M. (1989). "Intoxication in livestock and water flow". *Clinical Toxicology-Veterinary Clinics of North America. Food Animal Practice,* 5, 345–361
38. Yanni, Y.G. and Carmichael, W. (1998). "Screening of cyanobacteria isolated from soil, rice fields and water resources of the Nile Delta for production of cyanotoxins". In: *Harmful Algae* (Reguera B, Blanco J, Fernandez ML and Wyatt T, Eds.). Xunta de Galicia and Intergovernmental Oceanographic Commission of UNESCO, 493–494.
39. Bottomley, P.J. and Myrold, D.D. (2001). "The microbial ecology of terrestrial N cycling examined at different spatial scales". Page33 in *Proceedings of the First US-Egypt Workshop on Microbial Ecology* (Dazzo F, Badr El-Din S, Tiedje J and Eissa S. eds.), 6-10 May 2001, National Research Center, Cairo, Egypt.

3

Environment-friendly Phosphorus Biofertilizer as an Alternative to Chemical Fertilizers

Dinesh Chandra*, Rashmi Srivastava and A.K. Sharma

Department of Biological Sciences, G.B. Pant University of Agriculture and Technology
Pantnagar – 263 145, Uttarakhand, India
***E-mail:** dchandra.009@gmail.com*

ABSTRACT

Plant growth promoting rhizobacteria (PGPR) are a group of soil microorganisms which can enhance plant growth parameters and can be used as biofertilizers. The aim of this chapter is to emphasize on the importance and use of microbial inoculants possessing P-solubilizing activities as an environment-friendly alternative to chemical phosphorus (P) fertilizers in agricultural soils. The reckless and unbalanced use of chemical fertilizers and pesticides in agricultural soil has resulted not only in the deterioration of soil health but also created some major environmental and health related issues such as soil and water pollution. Phosphorus is an essential macronutrient required for every aspect of cell biology from energy metabolism to the structure of the genetic material, and its deficiency is a severe constraint to crop production. Soil microorganisms that solubilize the least mobile P play very important role in soil biology, plant health and subsequent enhancement of crop yield. To avoid the phosphorus deficiency, phosphate solubilizing bacteria (PSB) play an important role in supplying phosphate to plants in a more environment-friendly and sustainable manner. Therefore, PSB can be used in an ideal cropping system with a lesser impact on the environment through decreased application of chemical fertilizers.

Keywords: Biofertilizers, PSB, Rhizobacteria, Microbial Inoculants, Phosphorus.

3.1 INTRODUCTION

Soil sustains an immense diversity of prokaryotic and eukaryotic organisms; it is the natural growth media for living plants and microbes. Rhizosphere is niche for a group of soil microorganisms inhabiting around the plant root zone. Basically, rhizosphere is comprised of three distinct components: the rhizosphere soil, the rhizoplane, and the root itself. The term "rhizosphere" was coined by German agronomist Lorenz Hiltner in 1904 and is defined as biologically active zone of soil in the immediate vicinity of plant roots, an extremely important area for root metabolism, where microbes flourish and accelerate by root exudates and rhizodeposits [1]. The soil particles strongly adhere to plant root surfaces called rhizoplane [2], and the root itself is a part of the system, because certain endophytic microorganisms are able to colonize inner root tissues [3]; the stimulation of microbial growth by root exudates is called rhizosphere effect [4]; and it is a dynamic process initiated by root exudates and release of organic nutrients and is influenced by host factors (stage of development, species and age), and soil factors (moisture level, fertility, physical properties and environmental factors such as light and temperature).

Generally, rhizosphere is a dynamic environment, which harbours diverse group of microbes in which the effects of the plant on soil microorganisms and effects of microorganisms on plants are interacting and interdependent [5]. The plant-associated bacteria that are able to colonize and persist on roots are called rhizobacteria [6]. The Rhizosphere harbours numerous microorganisms such as algae, bacteria, protozoa and fungi which coexist in the rhizospheric region; bacterial population are most abundant among them. These microbial communities in nature tend to live in relative harmony where all microbial populations generally balance each other out in their quest for food and space [7].

3.2 PLANT GROWTH PROMOTING RHIZOBACTERIA (PGPR)

The term PGPR was coined by Kloepper and Schroth in late 1970s to define microbial population in the rhizosphere which is naturally occurring, actively colonize plant roots, and shows plant growth promotion activity. They are the potential tools for sustainable agriculture and trend for the future. The plant roots basically provide niche for the proliferation of soil bacteria that thrive on root lysates and root exudates. The population densities of bacteria in rhizosphere is 100-fold higher than in bulk soil because of the high nutrient levels in rhizoshere; and up to 15% of plant roots are covered by micro-colonies of a variety of bacterial strains [8-10]. Bacterial species belonging to genera *Pseudomonas, Bacillus, Rhizobium, Azospirillum, Arthrobacter, Burkholderia, Serratia, Flavobacterium, Enterobacter, Erwinia, Acinetobacter, Alcaligenes,* etc. are well known for their growth promotion effects on plants [11, 12].

PGPR can be divided into two groups on the basis of their relationship with the host plants:

1. Symbiotic rhizobacteria, which may invade the interior of cells and survive within the cell (also called intracellular PGPR (iPGPR), e.g., nodule bacteria (*Rhizobium*); and
2. Free-living or non-symbiotic rhizobacteria that exist outside plant cells (called extracellular PGPR (ePGPR), e.g., *Bacillus, Pseudomonas, Burkholderia,* and *Azotobacter*) [13, 14].

3.2.1 Characteristics of Ideal PGPR

1. Enhancement of plant growth
2. Safe to environment
3. Compatible with other rhizobacteria
4. High rhizosphere competence
5. High competitive saprophytic ability
6. Ease for mass multiplication
7. Broad spectrum of action against pathogens
8. Tolerant to drought, heat, oxidizing agent and UV radiations.
9. Excellent and reliable pest control. [15].

3.2.2 Root Colonization

Root colonization generally refers to microbial attachment on roots. It is an important factor in plant pathogenesis of soil-borne microorganisms as well as in beneficial interactions used for phytoremediation, microbiological control, phytostimulation and biofertilization [16]. A two phase process was hypothesized by [17] for microbial root aggregation, in phase I: the bacterial cells on the seeds attached to the growing root tips and passively transported into the soil. During root growth some cells remain associated with the terminal tip while others were left on older portions of the root. In phase II: the bacterial cells deposited along the root multiplied and formed colonies in nutrient-rich sites, compete with indigenous microflora and avoided displacement. The concept of phase I and II is meant to define more easily two different stages, rather than a strict temporal sequence of events because both phases occur simultaneously on different parts of the roots [18]. Basically, root colonization is influenced mainly by two factors:

1. Abiotic factor (temperature, soil pH, growth substrate and soil humidity)
2. Biotic factor (genetic traits of the host plant and the colonizing organism)

During extended period root systems are characterized by strong microbial competition and exert plant growth promoting traits, so soil bacteria need to be rhizosphere competent [19].

3.3 DIVERSITY IN RHIZOSPHERIC BACTERIA

Microorganisms exist in every conceivable place on earth and soil may harbour up to 10 billion microorganisms per gram. It is estimated that 1 g of soil may contain about 4,000 different bacterial genomic units based on DNA–DNA reassociation. It is supposed that up to one billion bacterial species actually exist in the earth environment and yet only about 5,000 species have been described [20, 21]. Moreover, only about 1% of the soil bacterial population can be cultured by standard laboratory practices. The tropics are considered to be richer in microbial diversity than boreal or temperate environments. The determination of bacterial diversity is essential for the systematic study of bacterial community analysis and biogeography. The soil microbial communities exhibit the greatest portion of biodiversity known in the world so far [22]. The rhizosphere is a narrow zone of soil that is influenced by root secretions such as organic acid, amino acid, carbohydrates, protein, etc. and can contain up to 10^{11} microbial cells per gram root [23] and more than 30,000 prokaryotic species [24]. The microbial population in soil is diverse containing an enormous diversity of microorganisms (Table 3.1).

Table 3.1 The average number of organisms in 1 cubic centimetre of agricultural soil [25]

Microbial Diversity	*Average numbers*
Bacteria	9,00,00,000
Actinomycetes	40,00,000
Fungi	2,00,000
Algae	30,000
Protozoa	5000
Nematodes	30
Earthworms	< 1

The estimated presence of approximately 6000 different bacterial genomes per gram of soil by taking the genome size of *Escherichia coli* (candidate microorganism) as a unit suggests that microbial biomass is large in temperate grassland soil in which the bacterial and fungal biomass amounted to 1–2 and 2–5 t ha^{-1}, respectively [26, 27].

3.4 APPLICATION OF PGPR IN AGRICULTURE

Environment-friendly biotechnological approaches offer alternatives to chemical fertilizers [28]. The excessive use and increasing costs of chemical fertilizers creates a negative impact on agricultural practices, therefore, PGPB is being considered as an alternative way of reducing the use of chemicals in agriculture [29-32]. During the past couple of decades, the use of plant growth promoting rhizobacteria (PGPR) for sustainable agriculture has increased tremendously in various parts of the world. Treatment with PGPR has increased the germination percentage, seedling vigour, emergence, shoot growth, root growth, total biomass of the plants, seed weight, early flowering, increased grain, fodder, fruit yields, etc. Significant increases in growth and yield of agronomically important crops in response to inoculation with PGPR have been repeatedly reported [33-35]. The use of PGPR with the aim of improving nutrient availability for plants is an important practice in agriculture production [36]. According to their mode of action, PGPR have been divided into two broad groups:

1. PGPR that directly affect plant growth, seed emergence or improve crop yields; and
2. biocontrol PGPR that indirectly benefit the plant growth [37].

The direct mechanism of plant growth promotion by PGPR includes increased nitrogen fixation, production of phytohormone and solubilization and uptake of nutrients. In contrast, indirect mechanism of growth promotion includes induced systemic resistance (ISR), antibiotic protection against pathogens, production of HCN, siderophore production, and synthesis of antifungal enzymes or lytic enzymes [38-40]. The work of Glick and coworkers on PGPR that produce 1-aminocyclopropane-1-carboxylate deaminase (ACC deaminase) and Vanderleyden and coworkers on indole-3-acetic acid (IAA) by *Azospirillum brasilense* in the promotion of plant growth are examples of the mode of action of a PGPR on genetic level.

3.4.1 Biological Nitrogen Fixation (BNF)

Nitrogen is one of the most common nutrients required for optimal plant growth and productivity. It is a key element for all cellular forms of life and is necessary for the synthesis of nucleic acids (DNA and RNA), proteins and other organic nitrogenous compounds. More than 78% N_2 is present in the atmosphere, yet it remains unavailable to growing plants. Hence, it needs to be converted into ammonia, an available form, that can be used by plants and eukaryotes. The atmospheric N_2 is converted into plant utilizable forms by three different processes:

1. The conversion of atmospheric N_2 into oxides of nitrogen.
2. The industrial N_2 fixation that involves the use of catalysts and high temperature to convert N_2 into ammonia.
3. Biological N_2 fixation which converts nitrogen to ammonia by using complex enzyme known as nitrogenase [41].

The key enzyme in biological process of nitrogen fixation is an enzyme complex called nitrogenase — a two component metalloenzyme consisting of:

1. Dinitrogenase reductase (also called Fe protein), which is a dimer of identical 30-kd subunits bridged by a 4Fe-4S cluster; and
2. Dinitrogenase which is a tetramer with two copies of different subunits.

Based on metal cofactor three different N_2-fixing systems have been identified:

1. Mo — dinitrogenase
2. V — nitrogenase and
3. Fe — nitrogenase [42].

Biological N_2 fixation (BNF) is restricted only to bacteria and archaea (e.g., *Methanococcus volate*). Till now no plant species (eukaryotic organisms) is able to fix atmospheric N_2 into ammonia and use it directly for its growth. It is one of the major mechanisms through which plants get benefited from PGPR and enrich the soil in form of ammonia [43]. The nitrogen fixing prokaryotes are called diazotrophs. The biological nitrogen fixation is very important in enhancing the soil fertility. The most studied PGPR in N_2 fixation is rhizobia (including *Bradyrhizobium, Sinorhizobium, Mesorhizobium, Allorhizobium, Azorhizobium,* and *Rhizobium*) for their ability to fix N_2 in their legume hosts [44-48]. According to an estimate, BNF imparts 180×10^6 metric tons per year globally, out of which symbiotic nitrogen fixation contributes 83% (e.g., *Rhizobium* in legumes, *Frankia* in non-leguminous plants) and the remaining part comes from free living or non-symbiotic nitrogen fixation (e.g., *Acetobacter diazotrophicus, Cyanobacteria, Azoarcus, Azospirillum, Azotobacter,* etc [49].

One of the benefits that diazotrophic microorganisms provide to plants is fixed nitrogen in exchange for fixed carbon released as root exudates [50].

A large number of PGPR strains have been identified for their ability to fix atmospheric N_2, however, enhanced plant growth through this type of mechanism is rarely credited to biological nitrogen fixation, but there is little evidence that PGPR with the ability to fix N_2 stimulate the growth of a specific host plant using nitrogenase activity. Such PGPR include *Azoarcus* sp., *Beijerinckia* sp., *Klebsiella pneumonia, Pantoea agglomerans* and *Rhizobium* sp. [48, 51-55].

3.4.2 Plant Growth Hormones and Biologically Active Substances

Plants synthesize a variety of chemical compounds called phytohormones that are actively involved in the regulation of plant growth and development [56]. Basically, plant growth hormone or plant growth regulator (PGR) is an organic compound that affects the physiological processes of plants even in extremely low concentration. There are mainly five classes of PGRs: auxins, gibberellins, cytokinins, ethylene, and abscisic acid. The soil microbiota, particularly the rhizosphere region, is a potential source of these PGRs [57-60]. Although numerous strains of PGPR have been reported to produce phytohormones such as auxins, ethylene and cytokinins but little is known regarding the production of gibberellins [10]. Auxin is an important phytohormone which promotes cell enlargement, cell division, root initiation and changes the plant morphological functions to uptake more nutrients from the soil. It is well-known that 96% symbiotic nitrogen-fixing rhizobia produce IAA [61]. In plant, the role of cytokinin is cell division, tissue expansion and cell enlargement. Microbial production of plant growth regulator seems to be important for improvement of plant growth and development in response to seed or root inoculation.

The stress hormone ethylene is known to be an inhibitor of plant growth, but ethylene at low concentrations may promote growth in several plant species including *Arabidopsis* [62]. The moderate levels of ethylene inhibit root and shoot elongation and elevated levels may induce senescence and organ abscission [63]. 1-aminocyclopropane-1-carboxylic acid (ACC) is immediate precursor of ethylene biosynthesis and it is secreted in root exudates by plants. PGPR that possess the enzyme ACC deaminase can utilize ACC as nitrogen source by degrading it, due to which the levels of ACC in plant roots can lower down. This leads to improved root growth by lowering the levels of ACC [10, 64]. However, bacteria lacking ACC deaminase could also promote plant growth by employing such mechanisms which are not well understood. It is believed that under such conditions bacterial cells secrete certain compounds that act as elicitors of plant growth. Plant roots perceive and recognize such elicitors in the ways similar to the recognition of elicitors from plant pathogens.

3.4.3 Solubilization of Phosphates

Phosphorus is an essential macronutrient required in every aspect of cell biology from energy metabolism to the structure of the genetic material, and in soil it is present at levels of 400–1200 mg kg^{-1} [65]. It is the most important key element in the nutrition of plants, next to nitrogen (N), and its cycle in biosphere can be described as open or sedimentary because there is no interchange with the atmosphere [66]. It is the most common limiting factor for biological growth and productivity of terrestrial plants and is taken by the plants from soil as phosphate anions. Depending on the quality of the soil, reactive phosphate anions are trapped by precipitation with cations like Mg^{2+}, Ca^{2+}, Al^{3+} and Fe^{3+}. Usually, a large proportion of phosphorus is insoluble in soil, but the amount accessible to plants is usually a little of total P unless it is solubilized by microbes [67]. The ability of different microbial strains to solubilize unavailable or insoluble inorganic phosphate compounds such as rock phosphate, hydroxyapatite, dicalcium and tricalcium phosphate (TCP) through the production of organic acids (such as gluconic acid, 2-ketogluconic acid, lactic acid, isovaleric acid, isobutyric acid, acetic acid, citric acid, propionic acid, etc.) and phosphatases helps in increasing plant growth and productivity, and these bacteria are known as phosphate solubilizing bacteria (PSB). Plants can only absorb P in two soluble forms, the monobasic [phosphoric acid (HPO_4^{2-})] and the diabasic [dihydrogen phosphate (H_2PO4^-)] ion forms [68].

Example of phosphate solubilizing bacteria (PSB)

Alcaligenes, Acinetobacter, Arthrobacter, Azospirillum, Burkholderia, Bacillus, Enterobacter, Erwinia, Flavobacterium, Pseudomonas, Rhizobium and *Serratia*, etc.

It has been reported that a large number of *Bradyrhizobium* strains are able to solubilize inorganic phosphate. The PSB are common in rhizospheres and their ability to solubilize phosphates by no means indicates that a rhizospheric bacterium will function as a PGPR, however, a study revealed that only two of the five rhizospheric isolates positive for P solubilization actually promote seedling growth. Similarly, not all P solubilizing PGPR increase plant growth by increasing P availability to the hosts. The *Bacillus* sp. and *Xanthomonas maltophilia* isolates from Canola (*Brassica napus* L.) rhizosphere had positive effects on plant growth, but no effects on phosphorus content of the host plants. [69-71].

3.4.4 Sequestration of Iron by Siderophores

For proper growth and development of plant adequate supply of mineral nutrients is inevitable. The most common PGPR activity is to increase the accessibility of nutrients in the rhizosphere [72, 73]. Siderophores are iron binding proteins of low molecular weight compound and have a high binding affinity with ferric ion (Fe^{3+}) however, iron in the soil is unavailable for direct assimilation by microbes because of its inaccessible form (Fe^{3+}) which is abundant in soil and sparsely soluble. Iron is one of the bulk minerals present in ample amount on earth, yet it is inaccessible in the soil for the plants. To overcome this problem, PGPR secrete siderophores that improve plant growth and development by increasing the accessibility of iron in the soil surrounding the roots [74]. Numerous studies reported siderophore production in root nodulating bacteria such as *Rhizobium* spp. and *Bradyrhizobium* [61, 75, 76]. However, plant growth promoting rhizobacteria are actively involved in the solubilization of important minerals such as phosphorus, iron, thereby enhancing the availability of these essential nutrients to plants [50]. Plants such as oats, sorghum, cotton, peanut, sunflower and cucumber demonstrate the ability to use microbial siderophores as sole source of iron.

Siderophores are complexing agents that have a high affinity for iron and are produced by almost all microorganisms in response to iron deficiency. Thus, siderophores act as solubilizing agents for iron from minerals or organic compounds under conditions of iron limitation. There are approximately 500 known siderophores, with the majority of them being used by a wide range of microorganisms and plants and some of them being exclusively used by the microbial species and strains that produce them [77]. In a long list of siderophores compound, only few of them are structurally characterized which are produced by root nodulating rhizobia. Some are listed in Table 3.2.

Table 3.2 Siderophore compound produced by root nodulating rhizobia under iron restricted condition

Siderophore compound	*Sources*	*References*
Rhizobactin, Dihydroxamate rhizobactin	*Sinorhizobium meliloti*	[78, 79]
Catechol	*Bradyrhizobium* (cowpea)	[80]
Anthranilate, Vicibactin, (trihydroxamate)	*Rhizobium leguminosarum* biovar *viciae*	[81, 82]
Citrate	*Bradyrhizobium japonicum*	[83]
Catechol	*Bradyrhizobium* (Peanut) *Rhizobium leguminosarum* *Rhizobium leguminosarum* biovar *trifolii*	[84-86]

Plants growing on the soils enriched with nutrients may still exhibit nutritional deficiencies due to inaccessibility of mineral nutrients. The positive role of PGPR in stimulating the plant growth by improving solubilization of essential mineral nutrients and their uptake by the plants has been well-documented in the literature [50, 87-89].

3.4.5 HCN Production

Many rhizobacteria produce HCN which plays a role in biological control of pathogens [90]. The cyanide ion is released as HCN and metabolized to a lesser degree into other compounds. First, it inhibits the electron transport chain and as a consequence the energy supply to the cell is disrupted, leading to the death of the organisms. It inhibits proper functioning of enzymes, natural receptors and also inhibits the action of cytochrome oxidase. PGPR produce HCN to control growth of different types of pathogens [91]. The production of cynogens by PGPR as a biocontrol agent (BCA) is one of the reasons to inhibit the growth of different pathogens. It has been observed that 4 out of 32 rhizobial strains produce HCN [92] and less than 1% rhizobia isolated from tomato rhizosphere were found to be positive for HCN production [93]. Strains of *Rhizobium, Bradyrhizobium* and *Pseudomnas fluorescens* have been reported to produce HCN and some toxin molecules which inhibit the growth of soil-borne pathogens as well as fungi [54, 94]. In-vitro experiment confirmed that *Rhizobium* strain NBRI 19513 completely inhibited the growth of pathogens such as *Rhizoctonia bataticola, Fusarium oxysporum,* and *Pythium* sp., and *Bradyrhizobium elkanii* accumulated rhizobiotoxine in culture and as well as in nodules. The *Bradyrhizobium* strains AHR-2amp+, AHR-5amp+ and AHR-6amp+ showed antagonistic activity against *Macrophomina phaseolina* [76, 95-97].

3.4.6 Antibiosis

Antibiosis is a biological interaction between two or more organisms that is harmful to at least one of them. A numbers of antibiotic compounds used as biocontrol mechanism of PGPR have been identified such as 2,4-diacetylphloroglucinol (DAPG), amphisin, hydrogen cyanide, oomycin A, phenazine, pyoluteorin, pyrrolnitrin, tensin, tropolone, and cyclic lipopeptides produced by pseudomonads [98-100] and zwittermicin A, oligomycin A, kanosamine, and xanthobaccin produced by *Bacillus*, *Streptomyces*, and *Stenotrophomonas* spp. [101-104].

3.4.7 Induced Systemic Resistance (ISR)

The enhanced defensive ability of plants, when appropriately stimulated by non-pathogenic rhizobacteria, has been shown to suppress disease by inducing a resistance mechanism in the plant. This phenomenon is called induced systemic resistance [105]. When cucumber plant was inoculated with rhizobacterial strains, it protected the leaves against anthracnose caused by colletotrichum orbiculare [106]. At first ISR was described in carnation, a plant that was systemically protected by the *Pseudomonas fluorescens* strain WCS417r against *F. oxysporum* f. sp. *dianthi* [107]. Recent work on the broad spectrum of PGPR-mediated induced systemic resistance against different pathogens (such as bacteria, fungi, viruses, nematodes and insects etc.) in different crops has gained importance worldwide.

Research carried out in the past showed that antagonism is an alternative mechanism to ISR by which most bacterial biological control agents (BCA) of plants pathogens functions partially or completely for achieving biological control of plant disease. According to Beattie (2006), bacteria that reduce the incidence or severity of plant diseases are often referred to as biocontrol agents, whereas those that exhibit antagonistic activity toward a pathogen are defined as antagonists. The difference lies in mechanism of biological control between antagonism and ISR which are summarized as follows:

1. The process of antagonism depends on direct functions of the BCA such as production of antibiotics, siderophores, HCN and nutrients competitions, while ISR is based on plant defense mechanisms that are activated by inducing agents as a consequence increased activity of chitinases, β-1,3- glucanases, peroxidases, pathogenesis related (PR) proteins, accumulation of antimicrobial low molecular weight substances, (e.g., phytoalexins) and formation of protective polymers like lignin, callose and hydroxyproline-rich glycoprotein which works against a wide range of pathogens.
2. Antagonism is not active against diverse pathogens, while ISR is active against a broad spectrum of pathogens that can be controlled with a single inducing agent.
3. Antagonistic mechanism of biological control is generally not systemic, while ISR protects the plants systemically following induction with an inducing agent to a single part of the plant.

There are two different forms of induced resistance – systemic acquired resistance (SAR) and induced systemic resistance (ISR). Phenotypically, ISR is similar to SAR in that it acts ambiguously against taxonomically different pathogens [108, 109] and both require the functioning of NPR1 protein and can be associated with an accumulation of PR-proteins and phytoalexins, as a consequence, alterations in cell wall composition [110]. In addition to the PR-proteins, the plants produce other enzymes of defence, including peroxidases (PO), phenylalanine ammonia-lyase (PAL), and polyphenoloxidase (PPO). PO and PPO are catalysts in the formation of lignin. PAL and other enzymes are involved in the formation of phytoalexins.

The interactions between plants and pathogens can lead to either a successful infection (compatible response) or resistance (incompatible response). In incompatible interactions, infection by viruses, bacteria or fungi will elicit a set of localized responses in and around the infected host cells. Due to this there will be an oxidative burst which can lead to cell death. Thus, the pathogen may be trapped in dead cells and appears to be prevented from spreading from the site of initial infection. The volatile organic compounds (VOC) secreted by *Bacillus subtilis* GBO3 and *B. amyloliquefaciens* IN937a play an important role in activation of ISR pathway in *Arabidopsis thaliana* seedlings challenged with the soft rot pathogen *Erwinia carotovora* subsp. *carotovora* [111-113].

3.5 DIVERSITY AND PHOSPHATE-SOLUBILIZING BACTERIA AS PLANT GROWTH PROMOTERS

PSM is mainly concentrated in the rhizosphere and they are metabolically more active than from other sources [114]. Typically, one gram of fertile soil contains 10^1 to 10^{10} bacteria and their live weight may exceed 2000 kg ha^{-1}. Soil bacteria are in cocci (sphere, 0.5 μm), bacilli (rod, 0.5–0.3 μm) or spiral (1-100 μm) shapes. Bacilli are common in soil whereas spirilli are very rare in natural environments [115]. The PSB are ubiquitous with variation in forms and population in different soils. Population of PSB depends on different soil properties (physical and chemical properties, organic matter, and P content) and cultural activities [116]. Larger populations of PSB are found in agricultural and rangeland soils [117]. Numerous microbial species which exhibit P solubilisation ability include bacteria, fungi, actinomycetes and even algae.

Table 3.3 Bacterial genera showing P-solubilizing ability

Bacteria (P- solubilizing ability)	*References*
Pseudomonas, Bacillus, Chryseobacterium, Arthrobacter, Phyllobacterium, Gordonia, Serratia, Rhodococcus, Delftia sp.,	[118-120]
Enterobacter, Pantoea, and Klebsiella	[121, 122]
Vibrio proteolyticus, Xanthobacter agilis	[114]
Rhizobium leguminosarum bv. *trifolii*	[123]
Azotobacter	[124]
Xanthomonas	[71]
Micrococcus roseus, Micrococcus luteus, Citrobacter freundii, Escherichia coli	[125]

Phosphorus is a light-bearing, nonmetallic chemical element that is constituent of nucleic acids, phospholipids and ATP, and therefore, plants cannot grow without a reliable supply of this nutrient. Phosphorus plays an important role in all major metabolic processes such as photosynthesis, respiration, energy transfer, signal transduction and macromolecular biosynthesis in plant, and it has been estimated that accumulated phosphates in agricultural soils, if it were available to plants, would be sufficient to sustain maximum crop yields worldwide for about 100 years [126-129]. The term microphos is generally used for the viable microbial preparations possessing P-solubilizing activity. The realization of all these potential problems associated with chemical P fertilizers together with the enormous cost involved in their manufacture has led to the search for environmentally compatible and economically feasible alternative strategies for improving crop production in low or P-deficient soils [130]. Phosphate-solubilizing microbes (PSM) exhibiting greater solubilization of insoluble phosphates qualitatively and quantitatively *in vitro* conditions are selected for field trials prior to production as biofertilizer.

An important feature of phosphorus biogeochemistry is that only 1% of the total soil P (400–4,000 kg P/ha in the top 30 cm of soil) is incorporated into living plant biomass during each growing season (10–30 kg P/ha); these figures suggest its low accessibility to plant uptake [131, 132]. Furthermore, P is a finite resource and based on its current rate of use, it has been estimated that the world's known reserves of high quality rock P may be depleted within the current century [133]. A number of different sources which may affect the P-solubilizing ability of microbes include C, N, concentration of iron and temperature. The ammonium salts are found to be best source of N followed by asparagine, sodium nitrate, potassium nitrate, urea and calcium nitrate [134]. Pikovskaya (1948) suggested that microbes could dissolve unavailable forms of soil P and play an important role in providing P to plants. Numerous methods and media for P- solubilisation have been used such as Pikovskaya agar, bromophenol blue dye method and National Botanical Research Institute P (NBRIP) medium [135-137].

The P solubilizing activity is determined by the formation of clear halo around bacterial colonies and it is not necessary to consider the production of a halo on a solid agar medium is the sole test for P solubilization. An additional test to be performed when bacterial colonies grow without a halo after several replacements of the medium then test the desirable strains in liquid media to assay P dissolution and the few isolates that are obtained after such rigorous selection should be again tested for abundant production of organic acids and isolates satisfy these criteria should be tested on a model plant as the ultimate test for potential P solubilization [138].

Various P-solubilizing bacteria have been isolated from stressed environments for example, the halophilic bacteria *Kushneria sinocarni* isolated from the sediment of Daqiao saltern on the eastern coast of China, which may be useful in salt affected agricultural soils [139]. Bacteria are more effective in P solubilization than fungi [140] and number of PSB among total PSM in north Iranian soil was around 88% [141]. Among the whole microbial population in soil, phosphorus solubilizing bacteria (PSB) and phosphorus solubilizing fungi (PSF), respectively constitute 1 to 50% and 0.1 to 0.5% and evidence of naturally occurring rhizospheric phosphorus solubilizing microorganism (PSM) dates back to 1903 [121, 142].

Microbes having phosphate-solubilizing activity not only improve the availability of organic (soluble) P to plants but also facilitate the biomass, yield and other growth contributing parameters of different agronomically important crops by other mechanisms [143, 144]. In a follow-up study, cold tolerant phosphate-solubilizing *Serratia marcescens* with inherent plant growth facilitating activities, such as, IAA, HCN and siderophore production, substantially increased dry matter accumulation and nutrient uptake of wheat (*Triticum aestivum*) seedlings grown in cold temperatures [145].

3.6 BIOAUGMENTATION AND BIOSTIMULATION APPROACHES USING MICROBIAL INOCULANTS FOR SOIL QUALITY

Microbial inoculants and biofertilizers are an important component of organic farming accounting for about 65% of the nitrogen supply to crops worldwide. Biofertilizers have emerged as an important component of the integrated soil nutrient supply system and hold great promise to improve agricultural productivity. Biofertilizers contain different types of microorganisms which have an ability to convert nutritionally important elements from inaccessible to accessible form through biological processes in soil. The importance of chemical fertilizers in agricultural practices cannot be overemphasized due to their excessive costs, therefore this needs to be substituted with cheaper available alternatives such as beneficial microbial inoculants, biofertilizers, and organic amendments to improve soil quality, fertility, biology, and agricultural productivity [146-148].

The organic materials digested by microbes enhance plant growth and improve soil structure and nutrient status of soil. Biofertilizers have following advantages over chemically synthesized pesticides and fertilizers:

(a) Greater relative safety
(b) Potentially reduced environmental damage and human health risk
(c) Much more targeted activity
(d) Effectiveness in small quantities
(e) Capacity for self-multiplication while being controlled by the plant as well as by the indigenous microbial populations
(f) Faster decomposition than conventional chemical pesticides
(g) Ability to be used in conventional or integrated pest management systems [149].

Microbial functions in ecosystems are as diverse as the microbes themselves. Microorganisms play an essential role in the functioning and sustaining of all natural ecosystems including biogeochemical cycling of nutrients. Improved plant productivity, disease suppression and accelerated nutrient availability and assimilation are the beneficial effects of plant-microbe interaction in agricultural areas. A seasonal fluctuation in environmental conditions such as temperature, water content and nutrient availability alters the soil microflora and their activity.

Plant-associated bacteria may be endophytic, phyllospheric and rhizospheric bacteria. Among these, endophytes are bacteria or fungi that colonize healthy plant tissue inter or intracellularly without causing any apparent symptoms of disease [150]. The soil organic matter and the associated biological agents are major contributing factors to nutrient cycling in the biosphere, and their presence plays very important role in plant growth, impacts soil quality (soil structure, resistance to erosion) and also acts as a major carbon reservoir of the biosphere-atmosphere system.

Quorum sensing based behaviour of plant-microbe interactions includes phyto-stimulation, biofertilization, rhizoremediation, biocontrol and in all these processes the small diffusible extracellular signalling molecules play very important role in mediating cell-cell communication. For Gram-negative bacteria the signaling molecules are named as autoinducers, usually acylated homoserine lactones (AHLs) and the Gram-positive bacteria use peptide as signalling molecules. The general biochemical properties of soil such as microbial biomass, dehydrogenase activity and nitrogen mineralization potential and specific biochemical parameters are the activities of hydrolytic enzymes, such as phosphatase, urease, and β-glucosidase. These are the general indicators of soil quality related to the biocycling of elements such as C, N, P, and S [151]. A variety of compounds are present within the soil organic matter, so a great diversity of enzymes exists in soil which efficiently helps in degradation of different compounds.

3.7 BIOTECHNOLOGICAL POTENTIALS OF PHOSPHATE SOLUBILIZING MICROORGANISMS

Soil has an abundance of mineral phosphates, but it is deficient in the soluble orthophosphate (Pi) essential for the growth and development of most plants and microorganisms [152]. Microflora which efficiently solubilize the mineral phosphate are includes *Bacillus* spp., *Pseudomonas* spp., *Rhizobium, Azotobacter, Azospirillum,* etc. These play important role in the supply of nutrients to the plants and therefore help in plant growth promoting activities. Numerous reports revealed plant growth promotion by rhizospheric microbes that have the ability to solubilize inorganic and organic P from soil after their inoculation in soil or plant seeds [153]. The mechanism for mineral phosphate solubilization is the production of organic acids, and acid phosphatases play a major role in the mineralization of organic phosphorus in soil. Gluconic acid and 2-ketogluconic acid seem to be the most frequent agents of mineral phosphate solubilization [154]. Also, other organic acids such as lactic acid, citric acid and tartaric acid are identified in strains with phosphate solubilizing ability. Therefore the use of PGPR inoculants as biofertilizers provides a promising alternative to chemical fertilizers.

The application of phosphate solubilizing bacterial preparation such as poly-myxobacterin in pond aquaculture resulted in increased level of mineral phosphate content in water by developing natural forage base and increased fish productivity. It was observed that in aquatic ecosystem there are optimal conditions for growth of *Paenibacillus polymyxa* KB. The polymyxobacterin treatment contributes to an increase in mineral phosphorus in pond water by 1.3–3.7 times compared to the control and does not exceed the normative values (less than 0.5 mg P/dm^3), which in turn stimulates phytoplankton vegetation and zooplankton development. After polymyxobacterin application due to development of natural forage organism, fish productivity of nursery and fattening pond increased than the control [155].

3.8 GENETICS OF PHOSPHATE SOLUBILIZING BACTERIA

Mainly two components of P in soil such as organic and inorganic phosphates and a greater per cent of phosphorus is present in insoluble forms and therefore, not accessible for plant nutrition. The apatite, hydroxyapatite and oxyapatite are important mineral forms of phosphorus that constitute the biggest reservoirs of element in soil because under appropriate conditions they can be solubilized and become available for plants and microorganisms. [65].

3.8.1 Genetics of Mineral Phosphate Solubilization (MPS)

The genetic basis of phosphate solubilization is not well understood, but studies at the molecular levels revealed that how the PSM brings out the solubilization of insoluble P into soluble is debatable [156]. The production of organic acids from mineral phosphate with the help of enzymes such as acid phosphatases and the gene involved in MPS synthesis might have an effect on phosphate solubilization. P solubilizing gene in Gram-negative bacteria *Erwinia herbicola* has been cloned that is involved in MPS by screening the antibiotic-resistant recombinants from a genomic library in a medium containing hydroxyapatite as the source of P [157]. The expression of P solubilizing gene allowed the production of gluconic acid and mineral phosphate solubilization activity in *E.coli* HB101, and its sequence analysis revealed its probable involvement in the synthesis of the enzyme pyrroloquinolinequinone (PQQ) synthase, which directs the synthesis of PQQ, a co-factor necessary for the formation of the holoenzyme (GDH)-PQQ.

Similarly, a mineral phosphate solubilization gene *gabY* isolated from *Pseudomonas cepacia,* whose expression allowed the induction of mineral phosphate solubilization via gluconic acid production in *E. coli* JM109, showed no evident homology with the cloned PQQ synthase gene, but it did with a permease membrane protein and the napA phosphatase gene from the soil bacterium *Morganella morganii* was transferred to *Burkholderia cepacia* IS-16 (a strain used as a biofertilizer), using the broad-host range vector pRK293. As a consequence an increased level of extracellular phosphatase activity of the recombinant strain was achieved [158-161]. The information about the genetic and biochemical mechanisms involved in the synthesis of the GDH-PQQ holoenzyme is insufficient, and variations between constitutive and inducible phenotypes are reported in several bacterial species, and it has been shown that the glucose, gluconate, mannitol, and glycerol are the possible inducers of the holoenzyme activity of GDH-PQQ [162, 163]. However, the manipulation of these genes through recombinant DNA technology and molecular biotechnology followed by their expression in selected rhizobacterial strains open an emerging area for obtaining PSM strains with enhanced phosphate solubilizing capacity and thus, a more effective use of these microbial strains as agricultural inoculants.

There are several advantages of developing genetically-modified PSM over transgenic plants:

(1) With current technologies, it is far easier to modify a bacterium than complex higher organisms.

(2) Several plant growth-promoting traits can be combined in a single organism.

(3) Instead of engineering crop by crop, a single engineered inoculant can be used for several crops (especially when using a non-specific genus like *Azospirillum* [156].

Table 3.4 Cloning of genes involved in mineral phosphate solubilization (MPS)

Microorganism	*Gene or plasmid*	*Features*	*References*
Erwinia herbicola	mps	Produces gluconic acid and solubilizes mineral P in *E. coli* HB101	[157]
Pseudomonas Cepacia	gabY	Produces gluconic acid and solubilizes mineral P in *E. coli* JM109. No homology with PQQ genes	[159]
Enterobacter agglomerans	pKKY	Solubilizes P in *E. coli* JM109 Does not lower pH	[164]
Rahnella aquatilis	pK1M10	Solubilizes P and produces gluconic acid in *E. coli* DH5α	[165]
Serratia marcescens	pKG3791	Produces gluconic acid and solubilizes mineral P	[166]

3.8.2 Genetics of Organic Phosphate Mineralization

The second major component of soil P is organic matter including phosphomonoesters, phosphodiesters (phospholipids and nucleic acids) and phosphotriesters and may constitute upto 30–50% of the total phosphorus in soils, although it may range from as low as 5% to as high as 95% [167]. The organic phosphate in soil is mostly in the form of inositol phosphate accounting up to 50% of total organic P (most stable form of P), and it is synthesized by microorganisms and plants [168-170]. Phosphorus can be released from organic compounds in soil by three groups of enzymes:

(i) nonspecific phosphatases, which perform dephosphorylation of phosphoester or phosphoanhydride bonds in organic matter;

(ii) phytases, which specifically cause P release from phytic acid; and

(iii) phosphonatases and C–P Lyases, enzymes that perform C–P cleavage in organophosphonates [156].

Particularly, most of the alkaline phosphatases are found in family *Enterobacteriaceae,* in which the phosphatase activity is widely investigated (e.g., *Escherichia coli and Salmonella typhimurium*) and its regulatory mechanism is still lacking. The production of phosphatases is controlled by complex regulatory mechanisms, so that the enzyme activity is noticeable only under certain set of environmental conditions [171]. The regulatory mechanism of phosphatases is best studied in the alkaline phosphatase (gene *phoA*) of *E. coli,* based on inorganic phosphate (Pi) concentration dependent manner which is induced when the Pi concentration decreases from 100 mM to 0.16 mM. *Pseudomonas fluorescens* MF3, *Providencia stuartii,* and *Providencia rettgeri* also produce alkaline phosphatase activity, which is repressed by phosphate [172-174]. The mechanism involves a Pi transport operon as a regulatory element as well as the sensor-activator operon, and the genes controlled by Pi and activated by PhoB constitute the PHO regulon [175].

Alkaline phosphatase of *Morganella morganii* is another example of Pi repressible bacterial phosphatase. It is produced under conditions of low-Pi availability; its regulation and the molecular mass of its polypeptide is similar to that of *E. coli,* and the cleavage of the C-P bond from organophosphonates by phosphonoacetaldehyde hydrolase is also inducible only under conditions of low phosphate availability [176, 177]. The production of Pi starvation based induction of PhoN enzyme (class A acid phosphatase) in *Salmonella enteric* serovar *typhimurium* is under the control of *phoP-phoQ,* two component regulatory system which promotes transcription of *phoN* and other PhoP activated genes under low environmental Mg^{2+}concentrations [178, 179].

High-level phosphate-irrepressible production of acid phosphatase activity (HPAP phenotype) in some species of *Enterobacteriaceae* family,*Providencia stuartii* and *Morganella morganii* produces a major phosphate-irrepressible class A acid phosphatase (PhoC), which is associated with the HPAP phenotype and a minor phosphate-irrepressible class B acid phosphatase (NapA). The regulatory mechanism of class B enzyme is almost similar to that of class B phosphatase of *E. coli.* The *phoD* gene of *Zymomonas mobilis* is another example of Pi irrepressible phosphatase which is expressed constitutively [176, 180-183].

CONCLUSION

Phosphorus solubilizing bacteria (PSB) play an important role in phosphorus nutrition by enhancing its accessibility to plants through solubilization and mineralization of inorganic and organic phosphates in soil. The insufficient phosphate availability limits the growth and yield of plants and therefore, lesser solubility of hydroxyapatite, calcium triphosphate and aluminium phosphate resulted into low accessibility of phosphate in agricultural soil as well as is the plants. Phosphorus is an important limiting factor in agriculture production and considering the negative effects of chemical P fertilizers, microbial intervention of PSM seems to be an effective way to solve the phosphorus availability in soil. PGPR with phosphate solubilization, pesticide degradation potential and ability to excrete phytohormones and antimicrobial metabolites may be used as potent biological inoculants for sustainable agriculture. Now it is evident that biological methods demonstrate a number of economic and environmental advantages over chemical-based P fertilizers applications. Numerous evidences indicate that PSB pose wide variety of functions which facilitate their implications in large number of agro-, bio-, and environmental technological strategies.

FUTURE PROSPECTS

The emphasis on the use of PSB, to better exploit soil P resources as alternative sustainable strategy for management of soil to optimize P bioavailability has considerable promise for the future and is one of the newly emerging areas for meeting agricultural challenges in varied environmental conditions. The biotechnological and molecular approaches of PSB provides better opportunity to develop environment-friendly phosphorus biofertilizer to be used as alternatives to chemical fertilizers that could lead to more successful in plant-microbe interaction. Therefore, more research is needed to investigate the Phosphate Solubilizing Bacteria (PSB) as bioinoculants to enhance the sustainable management of soil and agriculture that could help in economic stability and environment friendliness.

REFERENCES

1. Smalla, K., Sessitsch, A. and Hartmann, A. (2006). "The rhizosphere: soil compartment influenced by the root", *FEMS Microbiol. Ecol.*, 56(2): 165.
2. Kennedy, A.C. (2005). "Rhizosphere, In: *Principles and applications of soil microbiology*", Sylvia, D.M., Fuhrmann, J.J., Hartel, P.G. and Zuberer, D.A. (eds.)., 2nd ed. Pearson, Prentice Hall, New Jersey, pp. 242–262.
3. Bowen, G.D. and Rovira, A.D. (1999). "The rhizosphere and its management to improve plant growth", *Adv. Agron.*, 66, 1–102.
4. Morgan, J.A.W. and Whipps, J.M. (2001). "Methodological approaches to the study of rhizosphere carbon flow and microbial population dynamics", In: Pinton, A., Varanini, Z. and Nannipieri, P., (eds.)., *"The rhizosphere, Biochemistry and Organic Substances at the Soil-Plant Interface"*, New York, USA: Marcel Dekker, 373–409.
5. Mukerji, K.G., Manoharachary, C. and Singh, J. (2006). "Microbial activity in the rhizosphere (Soil Biology)", Springer. 16(7), p. 349.
6. Kloepper, J.W., Zablowicz, R.M., Tipping, B. and Lifshitz, R. (1991). "Plant growth mediated by bacterial rhizosphere colonizers", In: Keister D.L and Gregan B. (eds.)., *"The Rhizosphere and Plant Growth."* BARC Symp., 14, 315-326.
7. Be´langer, R.R. and Avis, T.J. (2002). "Ecological processes and interactions occurring in leaf surface fungi", In: Lindow, S.E., Hecht-Poinar, E.I., Elliot, V.J. (eds.)., *Phyllosphere Microbiol.*, APS Press, St. Paul, MN, pp. 193–207.
8. Bayliss, C., Bent, E., Culham, D.E., Mac Lellan, S., Clarke, A.J., Brown, G.L. and Wood, J.M. (1997). "Bacterial genetic loci implicated in the *Pseudomonas putida* GR12-2R3-canola mutualism: identification of an exudate-inducible sugar transporter", *Can. J. Microbiol.*, 43, 809–818.
9. Penrose, D.M. and Glick, B.R. (2001). "Levels of ACC and related compounds in exudate and extracts of canola seeds treated with ACC deaminase containing plant growth-promoting bacteria", *Can. J. Microbiol.*, 47, 368–372.
10. Van Loon, L.C. (2007). "Plant responses to plant growth-promoting rhizobacteria", *Eur. J. Plant Pathol.*, 119, 243–254.

11. Tilak, K.V.B.R., Ranganayaki, N., Pal, K.K., De, R., Saxena, A.K., Nautiyal, C.S. Mittal, S., Tripathi, A.K. and Johri, B.N. (2005). "Diversity of plant growth and soil health supporting bacteria", *Curr. Sci.,* 89, 136–150.
12. Egamberdiyeva, D. (2007). "The effect of plant growth promoting bacteria on growth and nutrient uptake of maize in two different soils", *Appl. Soil Ecol.,* 36, 184–189.
13. Khan, A.G. (2005). "Role of soil microbes in the rhizospheres of plants growing on trace metal contaminated soils in phytoremediation", *J. Trace Elem, Med. Biol.,* 18, 355–364.
14. Babalola, O.O. and Akindolire, A.M. (2011). "Identification of native rhizobacteria peculiar to selected food crops in Mmabatho Municipality of South Africa", *Biol. Agric. Hort.,* 27(3–4), 294–309.
15. Jeyarajan, R. and Nakkeeran, S. (2000). "Exploitation of microorganisms and viruses as biocontrol agents for crop disease management", In: *Biocontrol Potential and their Exploitation in Sustainable Agriculture,* (ed. Upadhyay et al.,) Kluwer Academic/ Plenum Publishers, USA. pp. 95–116.
16. Chin-A-Woeng, T.F.C. and Lugtenberg B.J.J. (2004). "Root colonization following seed inoculation", Plant Surface Microbiology pp. 13–33.
17. Howie, W.J., Cook, R.J. and Weller, D.M. (1987). "Effects of soil matric potential and cell motility on wheat root colonization by fluorescent pseudomonads suppressive to take-all. *Phytopathol.,* 77: 286–292.
18. Weller, D.M. and Thomashow, L.S. (1994). F.O'Gara, D.N. Dowling, B. Boesten, "Current Challenges in introducing beneficial microorganisms into the rhizosphere: Molecular ecology of rhizosphere microorganisms: biotechnology and the release of GMOs", pp. 1–18.
19. Whipps, J.M. (2001). "Microbial interactions and biocontrol in the rhizosphere", *J. Experimental Botany.,* 52, 487–511.
20. Hunter-Cevera, J.C. (1998). "The value of microbial diversity", *Curr. Opin. Microbiol.,* 1, 278–285.
21. Curtis, T.P. and Sloan, W.T. (2004). "Prokaryotic diversity and its limits: microbial community structure in nature and implications for microbial ecology", *Curr. Opin. Microbiol.,* 7, 221–226.
22. Curtis, T.P., Sloan, W.T. and Scannell, J.W. (2002). "Estimating prokaryotic diversity and its limits", Proceedings of the National Academy of Sciences of the United States of America., 99, 10494–10499.
23. Egamberdieva, D., Kamilova, F., Validov, S., Gafurova, L., Kucharova, Z. and Lugtenberg, B. (2008). "High incidence of plant growth stimulating bacteria associated with the rhizosphere of wheat grown on salinated soil in Uzbekistan", *Environ. Microbiol.* 10, 1–9.
24. Mendes, R., Kruijt, M., de Bruijn, I., Dekkers, E., van der Voort, M., Schneider, J. H. and Raaijmakers, J.M. (2011). "Deciphering the rhizosphere microbiome for disease-suppressive bacteria", Sci., 332(6033): 1097–1100.

25. Alexander, M. (1991). "*Introduction to Soil Microbiology*", 2nd ed. Malabar, F.L, Krieger Publishing Company.
26. Torsvik, V.L., Sorheim, R. and Goksoyr, J. (1996). "Total bacterial diversity in soil and sediment communities - a review", *J. Industrial Microbiol.*, 17, 170–178.
27. Killham, K. (1994). "*Soil Ecology*", Cambridge University Press, Cambridge. pp. 242.
28. Dobbelaere, S., Vanderleyden, J. and Okon, Y. (2003). "Plant growth promoting effects of diazotrophs in the rhizosphere", *Crit. Rev. Plant Sci.* 22, 107–149.
29. De Weger, L.A., Vander A.J., Bij, L.C., Dekkers, M., Simons, Wijffelman, C.A. and Lugtenberg, B.J.J. (1995). "Colonization of the rhizosphere of crop plants by plant-beneficial Pseudomonads", *FEMS Microbiol. Ecol.*, 17, 221–228.
30. Gerhardson, B. (2002). "Biological substitutes for pesticides", *Trends Biotechnol.*, 20, 338–343.
31. Postma, J., Montanari, M., and van den Boogert, P.H.J.F. (2003). "Microbial enrichment to enhance the disease suppressive activity of compost", *Eur. J. Soil Biol.*, 39: 157–160.
32. Welbaum, G., Sturz, A.V., Dong, Z. and Nowak, J. (2004). "Fertilizing soil microorganisms to improve productivity of agroecosystems", *Crit. Rev. Plant Sci.*, 23, 175–193.
33. Kloepper, J.W., Schroth, M.N. and Miller, T.D. (1980). "Effects of rhizosphere colonization by plant growth-promoting rhizobacteria on potato plant development and yield", *Phytopathol.*, 70, 1078–1082.
34. Figueiredo, M.V.B., Burity, H.A., Martinez, C.R. and Chanway, C.P. (2008a). "Alleviation of water stress effects in common bean (*Phaseolus vulgaris L.*) by co-inoculation with *Paenibacillus* x *Rhizobium tropic*", *Appl. Soil. Ecol.*, 40, 182–188.
35. Araujo, F.F. (2008). "Seed inoculation with *Bacillus subtilis* formulated with oyster meal and growth of corn, soybean and cotton", *Ciência Agrotecnologia Lavras* 32, 456–462.
36. Freitas, A.D.S., Vieira, C.L., Santos, C.E.R.S., Stamford, N.P. and Lyra, M.C.C.P. (2007). "Caracterizac¸a˜o de rizo´bios isolados de Jacatupe´ cultivado em solo salino no Estado de Pernanbuco", *Brasil. Bragantia.*, 66, 497–504.
37. Glick, B.R., Patten, C.L., Holgin, G. and Penrose, D.M. (1999). "*Biochemical and Genetic Mechanisms used by Plant Growth Promoting Bacteria*", Imperial College Press, London, p. 267.
38. Burdman, S., Jurkevitch, E. and Okon, Y. (2000). "Recent advances in the use of plant growth promoting rhizobacteria (PGPR) in agriculture, In: *Microbial Interactions in Agriculture and Forestry*", Subba Rao, N.S. and Dommergues, Y.R. (eds.)., Science Publishers, Enfield, USA, Vol II, pp. 229–250.
39. Dobbelaere, S. and Okon, Y. (2007). "The plant growth promoting effects and plant responses", In: *Nitrogen Fixation: Origins, Applications and Research Progress,* Newton, W. (ed.), Vol V: "Associative and endophytic nitrogen-fixing bacteria and cyanobacterial associations."

40. Lucy, M., Reed, E. and Glick, B.R. (2004). "Applications of free living plant growth-promoting rhizobacteria", *A. Van Leeuw.*, 86, 1–25.
41. Kim, J. and Rees, D.C. (1994). "Nitrogenase and biological nitrogen fixation", *Biochem.*, 33, 389–397.
42. Dean, D.R. and Jacobson, M.R. (1992). "Biochemical genetics of nitrogenase". In: Stacey, G., Burris, R.H. and Evans, H.J. (eds.)., *Biological Nitrogen Fixation.* Chapman and Hall, New York, NY, pp. 763–834.
43. Young, P. (1992). "Phylogenetic classification of nitrogen-fixing organisms", In: Stacey, G., Burris R.H., Evans, H.J. (eds.)., *Biological Nitrogen Fixation.* Chapman and Hall Inc, New York, pp. 43–86.
44. Hansen, A.P. (1994). "Symbiotic N_2 fixation of crop legumes: Achievement and perspectives", Hohenheim Tropical Agricultural Series. Margraf Verlag, pp. 31–36
45. Gualtieri, G. and Bisseling, T. (2000). "The evolution of nodulation", *Plant Mol. Biol.*, 42, 181–194.
46. Schultze, M. and Kondorosi, A. (1998). "Regulation of symbiotic root nodule development", *Annu. Rev. Genet.*, 32, 33–57.
47. Sessitsch, A., Howieson, J.G., Perret, X., Antoun, H. and Martínez-Romero E. (2002). "Advances in *Rhizobium* Research", *Crit. Rev. Plant Sci.*, 21(4): 323–378.
48. Vessey, J.K. (2003). "Plant growth promoting rhizobacteria as biofertilizers", *Plant and soil.*, 255, 571–586.
49. Graham, P.H. (1988). "Principles and Application of Soil Microbiology", 40, 322–345
50. Glick, B.R. (1995). "The enhancement of plant growth promotion by free living bacteria", *Can. J. Microbiol.*, 41, 109–117.
51. Hurek, T., Reinhold-Hurek, B., Van Montagu, M. and Kellenberger, E. (1994). "Root colonization and systemic spreading of *Azoarcus* sp. strain BH72 in grasses", *J. Bacteriol.*, 176, 1913–1923.
52. Baldani, J.I., Caruso, L., Baldani, V.L.D., Goi, S.R. and Dobereiner, J. (1997). "Recent advances in BNF with non-legume plants", *Soil Biol. Biochem.*, 29, 911–922.
53. Riggs, P.J., Chelius, M.K., Iniguez, A.L., Kaeppler, S.M. and Triplett, E.W. (2001). "Enhanced maize productivity with diazotrophic bacteria", *Aust. J. Plant Physiol.* 28, 829–836.
54. Antoun, H., Beauchamp, J., Goussard, N., Chabot, R. and Lalande, R. (1998). "Potential of *Rhizobium* and *Bradyrhizobium* species as plant growth-promoting rhizobacteria on non-legumes: Effect on radishes (*Raphanus sativus L.*)", *Plant Soil.*, 204, 57–67.
55. Yanni, Y.G., Rizk, Y., Abd-El, F.F.K., Squartini, A., Corich, V., Giacomini, A., De Bruijn, F., Rademaker, J., Maya, J.F., Ostrom, P., Veja, H.M., Hollingsworth, R.I., Martinez, E.M., Mateos, P., Velazquez, E., Wopereis, J., Triplett, E., Umali, G.M., Anarna, J.A., Rolfe, B.G., Ladha, J.K., Hill, J., Mujoo, R., Ng, P.K. and Dazzo, F.B. (2001). "The beneficial plant growth-promoting association of *Rhizobium leguminosarum* bv. *trifolii* with rice roots", *Aust. J. Biol. Sci.*, 28(9): 845–870.

56. Santner, A., Calderon-Villalobos, L. and Estelle, M. (2009). "Plant hormones are versatile chemical regulators of plant growth", *Nat. Chem. Biol.*, 5, 301-307.
57. Frankenberger, J.W.T. and Arshad, M. (1995). "Microbial synthesis of auxins. In: Frankenberger W.T., Arshad, M., (eds.)., "*Phytohormones in Soils*". Marcel Dekker Inc, New York, pp. 35–71.
58. Frankenberger, J.W.T. and Arshad, M. (1998). "Plant growth regulating substances in the rhizosphere: microbial production and functions", *Adv. Agron.*, 62, 146-151
59. Costacurta, A. and Vanderleyden, J. (1995). "Synthesis of phytohormones by plant associated bacteria", *Crit. Rev. Microbiol.*, 21, 1–18.
60. Patten, C.L. and Glick, B.R. (1996). "Bacterial biosynthesis of indole-3-acetic acid", *Can. J. Microbiol.*, 42, 207–220.
61. Arora, N.K., Kang, S.C. and Maheshwari, D.K. (2001b). "Isolation of siderophore producing strains of *Rhizobium meliloti* and their biocontrol potential against *Macrophomina phaseolina* that causes charcoal rot of groundnut", *Curr. Sci.* 81, 673–677.
62. Pierik, R., Tholen, D., Poorter, H., Visser, E.J.W. and Voesenek, L.A. (2006). "The Janus face of ethylene: growth inhibition and stimulation", *Trends Plant Sci.*, 11, 176–183.
63. Abeles, F.B., Morgan, P.W. and Saltveit, Jr. M.E. (1992). "Ethylene in plant biology, 2nd ed. San Diego, CA: Academic Press, Inc.
64. Glick, B.R. (2005). "Modulation of plant ethylene levels by the bacterial enzyme ACC deaminase", *FEMS Microbiol. Lett.*, 251, 1–7.
65. Fernandez, C., R., Novo, V. and Microbiana, E., Suelo (1988). "II. La Habana: Editorial Pueblo y Educación. Fibroblasts", *Life Sci* 52, 1909–1915.
66. Begon, M., Harper, J.L. and Townsend, C.R. (1990). "*Ecology: Individuals, Populations and Communities*", 2nd ed. Blackwell Scientific Publications, Oxford, USA.
67. Rovira, A.D. (1965). "Interaction between plants roots and soil microorganisms", *Ann. Rev. Microbiol.*, 19, 241–266.
68. Galleguillos, C., Aguirre, C., Barea, J.M. and Azcón, R. (2000). "Growth promoting effect of two *Sinorhizobium meliloti* strains (a wild type and its genetically modified derivative) on a non-legume plant species in specific interaction with two arbuscular mycorrhizal fungi", *Plant Sci.*, 159, 57–63.
69. Halder, A.K. and Chakraborty, P.K. (1993). "Solubilization of inorganic phosphate by *Rhizobium*", *Folia Microbiol.*, 38, 325–330.
70. Cattelan, A.J., Hartel, P.G. and Fuhrmann, J.J. (1999). "Screening of plant growth promoting rhizobacteria to promote early soybean growth", *Soil Sci. Soc. Am. J.*, 63: 1670–1680.
71. De Freitas, J.R., Banerjee, M.R. and Germida, J.J. (1997). "Phosphate-solubilizing rhizobacteria enhance the growth and yield but not phosphorus uptake of canola (*Brassica napus*)", *Biol. Fertil. Soils.*, 36, 842–855.

72. Obreht, Z., N. Kerby W., Gantar M. and Rowell, P. (1993). "Effect of root associated N_2-fixing cyanobacteria on the growth and nitrogen content of wheat (*Triticum vulgare*) seedlings", *Biol. Fert. Soils.*, 15: 68–72.

73. Gantar, M. and Elhai. J. (1999). "Colonization of wheat para-nodules by the N_2-fixing cyanobacterium *Nostoc* sp. strain 2S9B", *New Phytol.*, 141, 373–379.

74. Marschner, H. and Rohmeld, V. (1994). "Strategies of plants for acquisition of iron", *Plant Soil.*, 165, 261-274.

75. Carson, K.C., Meyer, J.M. and Dilworth, M.J. (2000). "Hydroxamate siderophore of root nodule bacteria", *Soil Biol. Biochem.*, 32: 11–21.

76. Deshwal, V.K., Dubey, R.C. and Maheshwari, D.K. (2003a). "Isolation of plant growth-promoting *Bradyrhizobium (Arachis)* sp. with biocontrol potential against *Macrophomina phaseolina* causing charcoal rot of peanut", *Curr. Sci.* 84(3): 443–448.

77. Crowley, D.E. (2007). "Microbial siderophores in the plant rhizosphere", In: Barton L.L., Abadia J (Eds.)., *Iron Nutrition in Plants and Rhizospheric Microorganisms, Dordrech,* Springer. pp. 169–198.

78. Smith, M.J., Schoolery, J.N., Schwyn, B., Holden, I. and Neilands, J.B. (1985). "Rhizobactin, a structurally novel siderophore from *Rhizobium meliloti*", *J. Am. Chem. Soc.*, 107, 1739–1743.

79. Persmark, M., Pittman, P., Buyer, J.S., Schwyn, B., Gill, P.R. Jr. and Neilands, J.B. (1993). "Isolation and structure of rhizobactin 1021, a siderophore from the alfalfa symbiont *Rhizobium meliloti* 1021", *J. Am. Chem. Soc.*, 115(10): 3950–3956.

80. Modi, M., Shah, K.S. and Modi, V.V. (1985). "Isolation and characterization of catechol like siderophores from cowpea *Rhizobium* RA-1", *Arch. Microbiol.*, 141, 156–158.

81. Rioux, C.R., Jordan, D.C. and Rattray, B.M. (1986a). "Iron requirement of *Rhizobium leguminosarum* and secretion of anthranilic acid during growth on an iron-deficient medium", *Arch. Biochem.*, 248(1): 175–182.

82. Dilworth, M.J., Carson, K.C., Giles, R.G.F., Byrne, L.T. and Glenn, A.R. (1998). "*Rhizobium leguminosarum* bv. *viciae* produces a novel cyclic trihydroxamate siderophore, vicibactin", *Microbiol.*, 144, 781–791.

83. Guerinot, M.L., Meidl, E.J. and Plessner, O. (1990). "Citrate as a siderophore in *Bradyrhizobium japonicum*", *J. Bacteriol.*, 172, 3298–3303.

84. Nambiar, P.T.C. and Sivaramkrishnan, S. (1987). "Detection and assay of siderophores of cowpea rhizobia *Bradyrhizobium* using radioactive Fe (^{59}Fe)", *Lett. Appl. Microbiol.*, 4, 37–40.

85. Patel, H.N., Chakraborty, R.N. and Desai, S.B. (1988). "Isolation and partial characterization of phenolate siderophore from *Rhizobium leguminosarum* IARI 102", *FEMS Microbiol Lett.*, 56, 131–134.

86. Skorupska, A., Derylo, M. and Lorkiewicz, Z. (1989). "Siderophore production and utilization by *Rhizobium trifolii*", *Biol. Metals.*, 2, 45–49.

87. Chabot, R., Antoun, H. and Cescas, M.P. (1996). "Growth promotion of maize and lettuce by phosphate solubilizing *Rhizobium leguminosarum biovar phaseoli*", *Plant Soil.*, 184, 311-321.

88. Biswas, J.C., Ladha, J.K. and Dazzo, F.B. (2000b). "Rhizobial inoculation influences seedling vigour and yield of rice", *Agron. J.*, 92, 880–886.
89. Dazzo, F.B., Yanni, Y.G., Rizk, R., de Bruijn, F.J., Rademaker, J., Squartini, A., Corich, V., Mateos, P. and Martı´nez-Molina, E. (2000). "Progress in multinational collaborative studies on the beneficial association between *Rhizobium leguminosarum bv. trifolii* and rice", In: "*The Quest for Nitrogen Fixation in Rice*", (Ladha J.K. and Reddy, P.M. et al., eds.); IRRI, Los Banos, Philippines: pp. 167–189.
90. De´fago, G., Haas, D., Berling, C.H., Burger, U., Keel, C., Voisard, C., Wirthner, P. and Wuthrich, B. (1990). "Suppression of black root rot of tobacco and other root diseases by strains of *Pseudomonas fluorescens*: potential applications and mechanisms", In: Hornby, D. (ed.) "*Biological Control of Soil-Borne Plant Pathogens*", CAB International, Wallingford, pp. 93–108.
91. Bagnasco, P., Fuente, L., De La., Gualtieri, G., Noya, F. and Arias, A. (1998). "Fluorescent *Pseudomonas* spp. as biocontrol agents against forage legume root pathogenic fungi", *Soil Biol. Biochem.*, 30, 1317-1322.
92. Beauchamp, C.J., Dion, P., Kloepper, J.W. and Antoun, H. (1991). "Physiological characterization of opine-utilizing rhizobacteria for traits related to plant growth-promoting activity", *Plant Soil.*, 132, 273–279.
93. Alvarez, M.A.B., Gagne, S. and Antoun, H. (1995). "Effect of compost on rhizosphere microflora of the potato and on the incidence of plant growth promoting rhizobacteria", *Appl. Environ. Microbiol.*, 61, 194–199.
94. Voisard, C., Keel, C., Haas, D. and Défago, G. (1989). "Cyanide production by *Pseudomonas fluorescens* helps suppress black root rot of tobacco under gnotobiotic conditions", *E.M.B.O. J.*, 8, 351–358.
95. Nautiyal, C.S. (1997). "Selection of chickpea rhizosphere competent *Pseudomonas fluorescens* NBRI 1303 antagonistic to *Fusarium oxysporum* f.sp. *ciceri*, *Rhizoctonia bataticola* and *Pythium* sp.", *Curr. Microbiol.*, 35, 52–58.
96. Devine, T.E., Kuykenendall, L.D. and O'Neill, J.J. (1988). "DNA homology group and the identity of *Bradyrhizobium* strains producing rhizobitoxine-induced foliar chlorosis on soybean", *Crop. Sci.*, 28, 939–941.
97. Kuykendall, L.D., Saxena, B., Devine, T.E. and Udell, S.E. (1992). "Genetic diversity in *Bradyrhiozbium japonicum* Jordan 1982 and a proposal for *Bradyrhizobium elkani* sp. *nov*", *Can. J. Microbiol.*, 38, 501–505.
98. Raaijmakers, J.M., Vlami, M. and De souza, J.T. (2002). "Antibiotic production by bacterial biocontrol agents", *A. Van Leeuw.*, 81, 537–547.
99. De Souza, J.T., de Boer, M., de Waard, P., van Beek, T.A. and Raaijmakers. J.M. (2003). "Biochemical, genetic and zoosporicidal properties of cyclic lipopeptide surfactants produced by *Pseudomonas fluorescens*", *Appl. Environ. Microbiol*, 69, 7161–7172.
100. Nielsen, T.H. and Sorensen, J. (2003). "Production of antifungal cyclic lipopeptides from *Pseudomonas fluorescens* strains in bulk soil and sugar beet rhizosphere", *Appl. Environ. Microbiol.*, 69, 861–868.

101. Milner, J.L., Raffel, S.J., Lethbridge, B.J. and Handelsman, J. (1995). "Culture conditions that influence accumulation of zwittermicin A by *Bacillus cereus* UW85", *Appl. Microbiol. Biotechnol.*, 43, 685–691.

102. Hashidoko, Y., Nakayama, T., Homma, Y. and Tahara, S. (1999). "Structure elucidation of xanthobaccin A, a new antibiotic produced from *Stenotrophomonas* sp. strain SB-K88", *Tetrahedron Lett.*, 40, 2957–2960.

103. Kim, B.S., Moon, S.S. and Hwang, B.K. (1999). "Isolation, identification and antifungal activity of a macrolide antibiotic, oligomycin A, produced by *Streptomyces libani*", *Can. J. Bot.*, 77, 850–858.

104. Nakayama, T., Homma, Y., Hashidoko, Y., Mizutani, J. and Tahara, S. (1999). "Possible role of xanthobaccins produced by *Stenotrophomonas* sp. strain SB-K88 in suppression of sugar beet damping-off disease", *Appl. Environ. Microbiol.*, 65, 4334–4339.

105. Van Loon, L.C., Bakker, P.A.H.M. and Pieterse, C.M.J. (1998). "Systemic resistance induced by rhizosphere bacteria", *Annu. Rev. Phytopathol.*, 36, 453–483.

106. Wei, G., Kloepper, J.W. and Tuzun, S. (1991). "Induction of systemic resistance of cucumber to *Colletotrichum orbiculare* by select strains of plant growth promoting rhizobacteria", *Phytopathol.*, 81, 1508–1512.

107. Van Peer, R., Niemann, G.J. and Schippers, B. (1991). "Induced resistance and phytoalexin accumulation in biological control of *Fusarium* wilt of carnation by *Pseudomonas* sp strain WCS 417r ", *Phytopathol.*, 81, 728–734.

108. Zehnder, G., Murphy, J., Sikora, E. and Kloepper, J. (2001). "Application of rhizobacteria for induced systemic resistance", *Eur. J. Plant Pathol.*, 107, 39–50.

109. Pieterse, C.M.J. and Van Loon, L.C. (2007). "Signalling cascades involved in induced resistance", In: Walters, D., Newton, A. and Lyon, G. (eds.)., *Induced Resistance for Plant Disease Control: A Sustainable Approach to Crop Protection*. Oxford: Blackwell Publishing, pp. 65–88.

110. Ramamoorthy, V., Viswanathan, R., Raguchander, T., Prakasam, V. and Samiyappan, R. (2001). "Induction of systemic resistance by plant growth promoting rhizobacteria in crop plants against pests and diseases", *Crop Protect.*, 20, 1–11.

111. Lamb, C. and Dixon, R.A. (1997). "The oxidative burst in plant disease resistance", *Annu. Rev. Plant Physiol. Plant Mol. Biol.*, 48, 251–275.

112. Kombrink, E. and Schmelzer, E. (2001). "The hypersensitive response and its role in local and systemic disease resistance", *Eur. J. Plant* Pathol., 107, 69–78.

113. Ryan, P.R., Delhaize, E. and Jones, D.L. (2001). Function and mechanism of organic anion exudation from plant roots", *Annu. Rev. Plant Physiol. Plant Mol. Biol.*, 52, 527–560.

114. Vazquez, P., Holguin, G., Puente, ELopez, M., Cortes, A. and Bashan, Y. (2000). "Phosphate solubilizing microorganisms associated with the rhizosphere of mangroves in a semi-arid coastal lagoon", *Biol. Fert. Soils.*, 30, 460–468.

115. Baudoin, E., Benizri, E. and Guckert, A. (2002). "Impact of growth stages on bacterial community structure along maize roots by metabolic and genetic fingerprinting", *Appl. Soil Ecol.*, 19, 135–145.

116. Kim, K.Y., Jordan, D. and McDonald, G.A. (1998). "Effect of phosphate-solubilizing bacteria and vesicular-arbuscular mycorrhizae on tomato growth and soil microbial activity", *Biol. Fert. Soils.*, 26:79–87.

117. Yahya, A. and Azawi, S.K.A. (1998). "Occurrence of phosphate solubilizing bacteria in some Iranian soils", *Plant Soil.*, 117, 135–141.

118. Igual, J.M., Valverde, A., Cervantes, E. and Velázquez, E. (2001). "Phosphate-solubilizing bacteria as inoculants for agriculture: use of updated molecular techniques in their study", *Agron.*, 21, 561–568.

119. Wani, P.A., Zaidi, A., Khan, A.A. and Khan, M.S. (2005). "Effect of phorate on phosphate solubilization and indole acetic acid (IAA) releasing potentials of rhizospheric microorganisms", *Ann. Plant Protect. Sci.*, 13, 139–144.

120. Chen, Y.P., Rekha, P.D., Arun, A.B., Shen, F.T., Lai, W.A. and Young, C.C. (2006). "Phosphate solubilizing bacteria from subtropical soil and their tricalcium phosphate solubilizing abilities", *Appl. Soil Ecol.*, 34: 33–41.

121. Chung, H., Park, M., Madhaiyan, M., Seshadri, S., Song, J., Cho, H. and Sa, T. (2005). "Isolation and characterization of phosphate solubilizing bacteria from the rhizosphere of crop plants of Korea", *Soil Biol. Biochem.*, 37, 1970–1974.

122. Whitelaw, M.A. (2000). "Growth promotion of plants inoculated with phosphate solubilizing fungi", *Adv. Agron.*, 69, 99–151.

123. Abril, A., Zurdo-Pineiro, J.L., Peix, A., Rivas, R. and Velazquez, E. (2007). "Solubilization of phosphate by a strain of *Rhizobium leguminosarum* bv. *trifolii* isolated from *Phaseolus vulgaris* in El Chaco Arido soil (Argentina) First International Meeting On Microbial Phosphate Solubilization, Developments in Plant and Soil Sciences", Springer, The Netherlands., 102: 135–138.

124. Kumar, V., Behl, R.K. and Narula, N. (2001). "Establishment of phosphate solubilizing strains of *Azotobacter chroococcum* in the rhizosphere and their effect on wheat cultivars under greenhouse conditions", *Microbiol. Res.*, 156, 87–93.

125. Ravikumar, S., Williams, G.P., Shanthy, S., Gracelin, N.A., Babu, S. and Parimala, P.S. (2007). "Effect of heavy metals (Hg and Zn) on the growth and phosphate solubilising activity in halophilic phosphobacteria isolated from Manakudi mangrove." *J. Environ. Biol.*, 28(1): 109–114.

126. Khan, M.S., Zaidi, A., Ahemad, M., Oves, M. and Wani, P.A. (2010). "Plant growth promotion by phosphate solubilizing fungi – current perspective", *Arch. Agron. Soil Sci.*, 56, 73–98.

127. Khan, A.A., Jilani, G., Akhtar, M.S., Naqvi, S.M.S. and Rasheed, M. (2009a). "Phosphorus solubilizing bacteria: occurrence, mechanisms and their role in crop production", *J. Agric. Biol. Sci.*, 1(1): 48–58.

128. Khan, M.S., Zaidi, A. and Wani, P.A. (2009b). "Role of phosphate solubilizing microorganisms in sustainable agriculture", In: Lictfouse E et al (eds.)., *Sustainable Agriculture*. Springer, p. 552.

129. Khan, M.S., Zaidi, A., Wani, P.A. and Oves, M. (2009c). " Role of plant growth promoting rhizobacteria in the remediation of metal contaminated soils", *Environ. Chem. Let*, 7, 1–19.

130. Zaidi, A., Khan, M.S., Ahemad, M., Oves, M. and Wani, P.A. (2009). "Recent advances in plant growth promotion by phosphate-solubilizing microbes", In: Khan, M.S *et al.* (eds.)., *Microbial Strategies for Crop Improvement.* Springer-Verlag, Berlin Heidelberg, pp. 23–50.

131. Blake, L., Mercik, S., Koerschens, M., Moskal, S., Poulton, P.R., Goulding, K.W.T., Weigel, A. and Powlson, D.S. (2000). "Phosphorus content in soil, uptake by plants and balance in three European long term field experiments", *Nutr. Cycl. Agroecosyst.*, 56, 263–275.

132. Quiquampoix, H. and Mousain, D. (2005). "Enzymatic hydrolysis of organic phosphorus", In: Turner, B.L., Frossardand, E., Baldwin, D.S. (eds.)., *Organic Phosphorus in the Environment*, Wallingford UK: CAB International. pp. 89–112.

133. Cordell, D., Drangert, J.O. and White, S. (2009). "The story of phosphorus: global food security and food for thought", *Glob. Environ. Chang.*, 19, 292–305.

134. Ahuja, A., Ghosh, S.B. and D'Souza, S.F. (2007). "Isolation of a starch utilizing, phosphate solubilizing fungus on buffered medium and its characterization", *Bioresour. Technol.*, 98, 3408–3411.

135. Pikovskaya, R.I. (1948). "Mobilization of phosphorus in soil in connection with vital activity of some microbial species", *Microbiol.*, 17, 362–370.

136. Gupta, R.R., Singal, R., Shanker, A., Kuhad, R.C. and Saxena, R.K. (1994). "A modified plate assay for secreening phosphate solubilizing microorganisms", *J. Gen. Appl. Microbiol.*, 40, 255–260.

137. Nautiyal, C.S. (1999). "An efficient microbiological growth medium for screening of phosphate solubilizing microorganisms", *FEMS Microbiol. Lett.*, 170, 265–270.

138. Bashan, Y., Kamnev, A.A. and de Bashan, L.E. (2013a). "A proposal for isolating and testing phosphate-solubilizing bacteria that enhance plant growth", *Biol. Fertil. Soils.*, 49, 1–2.

139. Zhu, F., Qu, L., Hong, X. and Sun, X. (2011). "Isolation and characterization of a phosphate-solubilizing halophilic bacterium *Kushneria* sp. YCWA18 from Daqiao Saltern on the coast of Yellow Sea of China", *Evid. Base Compl. Alternative Med.*, 615032:6.

140. Alam, S., Khalil, S., Ayub, N. and Rashid, M. (2002). "In vitro solubilization of inorganic phosphate by phosphate solubilizing microorganism (PSM) from maize rhizosphere", *Intl. J. Agric. Biol.*, 4, 454-458.

141. Fallah, A. (2006). "Abundance and distribution of phosphate solubilizing bacteria and fungi in some soil samples from north of Iran", 18th World Congress of Soil Science, Philadelphia, Pennsylvania, USA.

142. Khan, M.S., Zaidi, A. and Wani, P.A. (2007). "Role of phosphate-solubilizing microorganisms in sustainable agriculture - A review", *Agron. Sustain*. Dev., 27, 29-43.

143. Wani, P.A., Khan, M.S. and Zaidi, A. (2007a). "Effect of metal tolerant plant growth promoting *Bradyrhizobium* sp. (vigna) on growth, symbiosis, seed yield and metal uptake by green gram plants", *Chem.*, 70, 36–45.

144. Wani, P.A., Khan, M.S. and Zaidi, A. (2007b). "Impact of heavy metal toxicity on plant growth, symbiosis, seed yield and nitrogen and metal uptake in chickpea", *Aus. J. Exp. Agric.*, 47, 712–720.

145. Selvakumar, G., Mohan, M., Kundu, S., Gupta, A.D., Joshi, P., Nazim, S. and Gupta, H.S. (2008). "Cold tolerance and plant growth promotion potential of *Serratia marcescens* strain SRM (MTCC 8708) isolated from flowers of summer squash (*Cucurbita pepo*)", *Lett. Appl. Microbiol.*, 46, 171–175.

146. Saleem, M., Arshad, M., Hussain, S. and Bhatti, A.S. (2007). "Perspective of plant growth promoting rhizobacteria (PGPR) containing ACC deaminase in stress agriculture", *J. Ind. Microbiol. Biotechnol.*, 34, 635–648.

147. Ray, R.C., Ward, O.P., Singh, A. and Isobe, S. (2008). "Commercialization of microbial biotechnology in horticulture: summary, outlook of achievements, constraints and prospects.", In: Ray, R.C., Ward, O.P. (eds.)., *Microbial Biotechnology in horticulture, Science,* New Hampshire, 3, 341–365.

148. Babalola, O.O. (2010). "Beneficial bacteria of agricultural importance", *Biotechnol. Lett.*, 32, 1559–1570.

149. Berg, G. (2009). "Plant–microbe interactions promoting plant growth and health: perspectives for controlled use of microorganisms in agriculture", *Appl. Microbiol. Biotechnol.*, 84, 11–18.

150. Wang, Y. and Dai, C.C. (2011). "Endophytes: a potential resource for biosynthesis, biotransformation, and biodegradation", *Ann. Microbiol., 61*(2): 207–215.

151. Gil-Sotresa, F., Trasar-Cepedab, C., Leirosa, M.C. and Seoane, S. (2005). "Different approaches to evaluating soil quality using biochemical properties", *Soil Biol. Biochem.*, 37, 877–887.

152. Goldstein, A.H. (1986). "Bacterial solubilization of microbial phosphates: a historical perspective and future prospects", *Am. J. Altern. Agric.*, 1, 51–57.

153. Gaur, A.C. and Ostwal, K.P. (1972). "Influence of phosphate dissolving *Bacilli* on yield and phosphate uptake of wheat crop", *Indian J. Exp. Biol.*, 10, 393–394.

154. Chen, Y.P., Rehha, P.D., Arun, A.B. and Shen, F.T. (2005). "Phosphate solubilizing bacteria from subtropical soil and their tricalcium phosphate-solubilizing abilities", *Appl. Soil Ecol.*, 34, 33–41.

155. Vovk, N.I., Bazaeva, A.V. and Didenko, A.V. (2013). "Use of the phosphate-solubilizing bacterial preparation polymyxobacterin in pond aquaculture", *Turk. J. Fish Aquat. Sci.*, 13, 1–9.

156. Rodríguez, H.G.M., Rossolini, T., Gonzalez, J. Li, and Glick, B.R. (2006). "Isolation of a Gene from *Burkholderia cepacia* IS-16 encoding a protein that facilitates phosphatase activity", *Curr. Microbiol.*, 40, 362–366.

157. Goldstein, A.H. and Liu, S.T. (1987). "Molecular cloning and regulation of a mineral phosphate solubilizing gene from *Erwinia herbicola*", *Biotechnol.*, 5, 72–74.

158. Liu, T.S., Lee, L.Y., Tai, C.Y., Hung, C.H., Chang, Y.S., Wolfram, J.H., Rogers, R. and Goldstein, A.H. (1992). "Cloning of an *Erwinia carotovora* gene necessary for gluconic acid production and enhanced mineral phosphate solubilization in *Escherichia coli* HB101: Nucleotide sequence and probable involvement in biosynthesis of the coenzyme pyrroloquinoline quinone", *J. Bacteriol.*, 174, 5814–5819.

159. Babu-Khan, S., Yeo, T.C., Martin, W.L., Duron, M.R., Rogers R.D. and Goldstein, A.H. (1995). "Cloning of a mineral phosphate-solubilizing gene from *Pseudomonas cepacia*", *Appl. Environ. Microbiol.*, 61: 972–978.

160. Goosen, N., Horsman, H.P., Huinen, R.G. and van de Putte, P. (1989). "*Acinetobacter calcoaceticus* genes involved in biosynthesis of the coenzyme pyrrolo-quinolinequinone: nucleotide sequence and expression in *Escherichia coli* K-12", *J. Bacteriol.*, 171, 447–455.

161. Fraga, R., Rodriguez, H. and Gonzalez, T. (2001). "Transfer of the gene encoding the Nap A acid phosphatase from *Morganella morganii* to a *Burkholderia cepacia* strain", *Acta Biotechnol.*, 21, 59–369.

162. Goldstein, A.H. (1994). "Involvement of the quinoprotein glucose dehydrogenase in the solubilization of exogenous mineral phosphates by gram negative bacteria", In: Torriani-Gorni, A., Yagil, E. Silver, S. (eds.)., *Phosphate in Microorganisms: Cellular and Molecular Biology*. ASM Press, Washington, DC, 197–203.

163. Van Schie, B.J., De Mooy, O.H., Linton, J.D., Van Dijken J.P. and Kuenen, J.G. (1987). "PQQ-dependent production of gluconic acid by *Acinetobacter, Agrobacterium and Rhizobium* species", *J. Gen. Microbiol.*, 133: 867–875.

164. Kim, K.Y., Mc Donald, G.A. and Jordan, D. (1997). "Solubilization of hydroxypatite by *Enterobacter agglomerans* and cloned *Escherichia coli* in culture medium", *Biol. Fert. Soils.*, 24, 347–352.

165. Kim, K.Y., Jordan, D. and Krishnan, H.B. (1998b). "Expression of genes from *Rahnella aquatilis* that are necessary for mineral phosphate solubilization in *Escherichia coli*", *FEMS Microb. Lett.*, 159, 121–127.

166. Krishnaraj, P.U. and Goldstein, A.H. (2001). "Cloning of a *Serratia marcescens* DNA fragment that induces quinoprotein glucose dehydrogenase-mediated gluconic acid production in *Escherichia coli* in the presence of stationary phase *Serratia marcescens*", *FEMS Microbiol. Lett.*, 205, 215–220.

167. Paul, E.A. and Clark, F.E. (1989). "*Soil Microbiology and Biochemistry*". San Diego, CA: Academic Press, p. 273.

168. Dalal, R.C. and Diego, C.A. (1977). "Soil organic phosphorus", *Adv. Agron.*, 29, 83–117.

169. Anderson, G. (1980). "Assessing organic phosphorus in soils", In: *Role of Phosphorus in Agriculture*: (eds.)., Khasawneh, F.E., Sample, E.C. and Kamprath. E.J. ASA: Madison, WI, pp. 411–428.

170. Harley, J.L. and Smith, S.E. (1983). "*Mycorrhizal Symbiosis*", London, New York: Academic Press, p. 483.

171. Wanner, Y.B.L. (1987). "Cellular and molecular biology phosphate regulation of gene expression in *E. coli*", In: Niedhardt, F.C., Ingraham, J.L., Low, K.B., Magasanik, B., Schaechter, M., Umbarger, H.E. editors. *Escherichia coli* and *Salmonella typhimurium*. Washington, DC: ASM Press, pp. 1326–1333.

172. Rosenberg, H. (1987). "Phosphate transport in prokaryotes", In: Rosen, B., Silver, S. (Eds.)., Ion transport in prokaryotes. San Diego, CA: Academic Press, Inc,. pp. 205–48.

173. Thaller, M.C., Berlutti, F., Schippa, S., Iori, P., Passariello, C. and Rossolini, G.M. (1995b). "Heterogeneous patterns of acid phosphatases containing low-molecular-mass polypeptides in members of the family Enterobacteriaceae. *Int. J. Syst. Bacteriol.*, 4, 255–261.

174. Gugi, B., Orange, N., Hellio, F., Burini, J.F., Guillou, C., Leriche, F. and Guespin-Michel, J.F. (1991). "Effect of growth temperature on several exported enzyme activities in the psychrotropic bacterium *Pseudomonas fluorescens*", *J. Bacteriol.*, 173, 3814–3820.

175. Torriani-Gorini, A. (1994). "Regulation of phosphate metabolism and transport", In: Torriani-Gorini, A., Yagil, E. and Silver, S. (eds.)., *Phosphate in Microorganisms: Cellular and Molecular Biology*. Washington, DC: ASM Press,. pp. 1–4.

176. Thaller, M.C., Berlutti, F., Schippa, S., Lombardi, G. and Rossolini, G.M. (1994). "Characterization and sequence of PhoC, the principal phosphate-irrepressible acid phosphatase of *Morganella morganii*", *Microbiol.*, 140, 1341–1350.

177. Kertesz, M., Elgorriaga, A. and Amrhein, N. (1991). "Evidence for two distinct phosphonate degrading enzymes (C-P lyases) in *Arthrobacter* sp. GLP-1", *Biodegradation.*, 2, 53–59.

178. Kier, L.D., Weppelman, R. and Ames, B.N. (1977). "Resolution and purification of three phosphatases of *Salmonella typhimurium*", *J. Bacteriol.*, 130(1): 399–410.

179. Miller, S.I., Kukral, A.M. and Mekalanos, J.J. (1989). "A two-component regulatory system (phoP-phoQ) controls *Salmonella typhimurium* virulence", Proc Natl Acad Sci USA., 86(13): 5054–5058.

180. Pompei, R., Cornagli, G., Ingianni, A. and Satta, G. (1990). "Use of a novel phosphatase test for simplified identification of species of the tribe proteae", *J. Clin. Microbiol.*, 28, 1214–1218.

181. Pompei, R., Ingianni, A., Foddis, G., Di, Pietro G. and Satta G. (1993). "Patterns of phosphatase activity among enterobacterial species", Int. J. Syst. Bacteriol., 43, 174–178.

182. Reyes, L. and Scopes, R.K. (1991). "The use of multifunctional adsorbents to purify membrane-bound phosphatases from *Zymomonas mobilis*", *Bioseparation.*, 2(3):137–146.

183. Baoudene-Aassali, F., Baratti, J. and Michel, G.P.F. (1993). "Purification and properties of a phosphate irrepressible membrane-bound alkaline phosphatase from *Zymomonas mobilis*", *J. Gen. Microbiol.*, 139, 229–235.

4

Liquid Biofertilizers: Potential and Prospects

D.C. Bhattacharya

Department of Microbiology, Vidyasagar University, Midnapore – 721102, West Bengal, India

E-mail: *debu720@gmail.com*

ABSTRACT

The high population growth rate and huge demand for foodgrains have made it essential for the farmers to adopt intensive agricultural practices with high-yielding varieties. The demerits of long-term intensive agriculture have made it a technological imperative to depend on biofertilizers for sustainable soil health and fertility. In these contexts, the technical promises of liquid biofertilizers have been explored. The formulation, and its chemical additives have been mentioned with possible roles in the farmer's field. Some methods of applications, usual dosage, marker assisted monitoring of the performance of the introduced strains have been discussed.

Keywords: Biofertilizer, Inoculant, Liquid, Formulation, PGPR, PSM, N_2 Fixation, Sustainable Agriculture.

4.1 INTRODUCTION

In India, a population of 1.25 billion increasing at the rate of 17.69% (2001 -2011, decadal growth rate) will need an estimated amount of 257 million tonnes and 277 million tonnes of foodgrains at the end of the year 2016-17 and the year 2020-21 respectively. Also FAO estimates have projected an increase of 60% higher agricultural demand in the year 2030, and 85% of this additional requirement, will arise from the developing countries.

To arrange for this huge needs of the population, farmers will obviously opt for the high-yielding varieties of rice, wheat, etc. In short, the farming activities will definitely have to shift towards intensive agricultural practices, which in turn will exercise enormous pressure on the existing land and other natural resources for the availability of foodgrains and related agricultural commodities like pulses, oilseeds, etc.

As the intensive agricultural activities depend on the application of heavy doses of complex fertilizers, repeated applications of agro-chemicals, etc., in the long run an adverse impact can be expected on the highly fragile agro-ecological balance, already threatened by soil erosion, acidity and salinity of soils.

In intensive agriculture, the continuous removal of nutrients by crop plants leads to the depleted soil nutrient status along with irreversible damage to soil health by impairing soil biological activities. The situation will not only be detrimental to agriculture, but also to the sustainable environment in the rural areas.

In these contexts, biofertilizers can play an affordable technological solution to the challenges.

Biofertilizers are essential components of healthy and eco-friendly agriculture moulded in the principles of sustainable development goals (SDGs) to fulfil the targets emphasized in the Millennium development goals (MDGs).

4.2 BIOFERTILIZERS: DEFINITION AND SCOPE

Biofertilizers can broadly be defined as suitable carrier-based preparations containing live or latent microbial cells capable of N_2 fixation, phosphate solubilization/mobilization or cellulolytic activity, to be applied to the soil , seed or compost with the overall objective of enhancing plant nutrient availability.

Initial use of biofertilizers was reported by Dutch scientists in the year 1888 followed by the development of "Nitragin"- the first commercial biofertilizer developed by Nobbe & Hiltner in the year 1895.

Microorganisms used as biofertilizers include bacteria, algae and fungi. Based on the beneficial activity of microbes, biofertilizers can be further classified as N_2-fixing (e.g., *Rhizobium* sp., *Azospirillum* sp., *Azotobacter* sp., *Gluconacetobacter diazotrophicus,* etc.), phosphate solubilizing/mobilizing, commonly called PSM (*Bacillus megaterium* var *phosphaticum, B. subtilis, B. circulans, Aspergillus awamori,* etc). Recent studies have indicated potash mobilizer (*Frateuria aurentia*), sulphur oxidizer (*Thiobacillus thiooxidans*), zinc mobilisers (*Bacillus* sp, *Pseudomonas* sp) and manganese solubilizer (*Penicillium citrinum*) (Sharma et al., 2012).

PGPR (Plant growth promoting rhizobacteria), a new group of bacteria has also caught the attention of researchers in terms of its success in providing nutrients and growth promoting substances to the crop plants.

First definition of PGPR was given by Kloepper to describe soil bacteria that can colonize the roots of plants following inoculation onto seed and enhances the plant growth.

It has been reported that PGPRs have some selective advantage by virtue of their 1-aminocyclopropane-1-carboxylate (ACC) - deaminase activity, production of antioxidant enzymes, exopolysaccharides and organic solutes over related bacteria under stress environment.

In addition to N_2 fixation, phosphate mobilization, etc., PGPRs can effectively increase the rate of plant growth through the production of growth promoting hormones like IAA, GA_3, cytokinin like substances. They also reduce the deleterious effects of soil-borne fungal pathogens through distinct antibiosis mediated by amphisin, 2,4-diacetylphloroglucinol(DAPG), hydrogen cyanide, oomycin A, etc.

4.3 LIQUID BIOFERTILIZER: GENESIS OF THE CONCEPT

In the year 1982, Dr Teruo Higa at the University of Ryukyus Okinawa , Japan discovered a specific group of microorganisms with an unique ability to revive, restore and preserve themselves in an altered environment through fermentation. He named this group **EM** (Effective microorganisms).He further showed that EM helped to increase beneficial soil microbes and suppress harmful ones.

EM was originally formulated as dark brown liquid, pH approximately 3.5, with a yeasty vinegary smell, where microorganisms were alive but dormant. EM formulation can be activated by simple dilution with clean chlorine-free water. Activated EM was yellowish-brown in colour with a pleasant smell.

As EM formulations were found to be affordable and effective with high shelf-life, researchers in biofertilizer studies were encouraged with these developments. Soon attempts were made to prepare laboratory scale liquid biofertilizers in tune with the EM production process.

Though ingredients are different in case of liquid biofertilizer, the addition of suitable nutrients/medium, special cell protectants and amendments help the microorganisms to remain alive for a longer period with least damage from the fluctuation of temperature, storage or handling conditions.

4.4 NEED OF LIQUID BIOFERTILIZERS

In the carrier-based (solid) biofertilizer, the common carrier used is peat/lignite which has several implications on the survival of microorganisms for a longer storage period. At the time of production, a population of 10^8 cfu/mL is commonly found in the biofertilizers. But this population density decreases gradually as further storage is attempted at a temperature of 30°C or more.

Peat as a carrier material is also difficult to sterilize under steam sterilization, and also during sterilization toxic byproducts are produced that may cause decrease in microbial population.

It was observed that after four months of storage microbial count reduces to 10^6 cfu/mL, and at the end of six months , the effective microbial count is almost nil.

From a manufacturer's point, the biofertilizer plant has to be operated throughout the year in order to supply fresh biofertilizer to the farmers. As storage of biofertilizer is detrimental for sustaining microbial count (cfu/mL), unsold biofertilizer packets are either discarded or have to be re-inoculated with fresh microbial cultures.

Solid carrier-based biofertilizer production is a cost intensive process as it involves labour, energy intensive processing, drying, milling and neutralization before the product is commercialized.

One of the most important defects of carrier-based biofertilizers is their temperature and UV sensitivity and hence they cannot be maintained with 100% aseptic conditions. Thus carrier-based biofertilizers could not get much popularity even if there was proven benefit in terms of greater yield and nutrient availability for crop plants.

There are also the drawbacks of seed coat toxicity, seed coat damage, cell death due to desiccation and contamination from agrochemicals.

To nullify these bottlenecks, liquid biofertilizer technology came as a boon to the biofertilizer technologists.

4.5 LIQUID INOCULANTS: COMPOSITION AND CHARACTERISTICS

The performance of an inoculant is the most important criteria for the commercial success of the biofertilizer. This again depends on the type of formulation, ease of storage, applicability, etc. (Table 4.1).

Liquid biofertilizers are based on aqueous (broth cultures), mineral or organic oils, oil-in-water or polymer-based suspensions (Malusa et al., 2012; Xavier et al., 2004). In addition to usual nutrients, they contain special cell protectant and amendments that promote longer cell survival in package and after application to seed or soil (Brahmaprakash et al., 2012). Liquid biofertilizers have been promoted as being easier to handle and apply either on seed or in soil, and their popularity has increased in the recent years (Xavier et al., 2004; Stephens and Rask 2000).

Table 4.1 Some of the common formulations tried by the researchers in Liquid and related inoculants for PGPR

Formulation	*Additives*	*Microrganisms used*	*Plant species*	*Reference*
Liquid	CMC	*Trichoderma* sp. antagonists	wheat	Dal Bello et al., 2002
	CMC	*Pseudomonas aeruginosa*	ryegrass	Viji et al, 2003
	CMC	*Acinetobacter calcoaceticus, Bacillus licheniformis, Brevibacillus brevis, Micrococcus* sp.	*Jatropha curcas*	Jha and Sharaf., 2012
	Glycerol, PVP, Trehalose, FeEDTA	*Bradyrhizobium japonicum*	Soybean	Singleton et al., 2002
	PVP, FeEDTA	Several rhizobia; *B. japonicum, Bacillus megaterium*	Soybean	Albareda et al., 2008
	Unkown (commercial)	*Bradyrhizobium japonicum*	Soybean	Maurice et al., 2001
	Sucrose	Phosphate solubilizing bacteria: *Bacillus subtilis, B. amyloliquefaciens, Pseudomonas* sp.	Peanut, Rice	Taurian et al., 2010; Cong et al., 2009

Contd...

Contd...

	Gum Arabic	*Bradyrhizobium* sp., *Rhodobacter capsulatus, Rhizobium* sp.	*Acacia mangium;* Green gram; *Leucaena leucocephala;* Rice	Diouf et al., 2003; Gamal-Eldin and Elbanna, 2011; Wani et al., 2007
	Glycerol	*Pseudomonas fluorescens*	Tomato	Manikandan et al., 2010
Alginate	Humic acid	*Pseudomonas putida* and *Bacillus subtilis*	Lettuce	Rekha et al., 2007
Alginate	Skim milk	*Bacillus subtilis* and *Pseudomonas corrugata*	Maize	Trivedi et al., 2005

CMC-carboxymethyl cellulose, PVP-polyvinyl pyrrolidone, EDTA-Ethylenediamine tetraacetate

Liquid inoculants can be termed as an improved derivative of "no formulation" inoculants to nullify some of the limitations mentioned above. Essentially they are microbial cultures or suspensions amended with substances that may improve stickiness, stabilization, and surfactant and dispersal abilities. (Singleton et al., 2002).

The five most desirable characteristics of a good formulation are (Bashan, 1998):

(a) Chemical and physical characteristics: The carrier of contamination free-inoculants should be nearly sterile, chemically and physically uniform with high water holding capacity (for wet carriers). Consistency of raw material is an absolute requirement.

(b) Manufacturing qualities: The inoculants must be easy to manufacture, mix and package by microbial fermentation industry. It should allow addition of nutrients, have an easily adjustable pH, and be made of a reasonably low priced raw material in adequate supply and availability (Catroux et al., 2001).

(c) Farm-handling qualities: Good inoculants allow for ease of handling, provide rapid and controlled release of bacteria into the soil and can be applied with standard seeding machinery. This is very important because it is rare that farm practices change to accommodate a technology that delivers a high quality inoculant with specialized machinery, especially in conservative communities (Date, 2001).

(d) Environment–friendly: In accordance with contemporary environmental concerns over applications of substances that change soil characteristics, the inoculants should be non-toxic, biodegradable, leave no carbon footprint and be non-polluting. Applications should minimize environmental risks such as the dispersal of PGPR in the atmosphere or groundwater.

(e) Long storage quality: The inoculants should have sufficient shelf-life. One or two years at room temperature are often necessary for successful integration into agricultural distribution system in developed countries (Catroux et al., 2001; Deaker et al., 2004).

Practically, no single inoculant can have all these capacities for top-end quality for the manufacturer. However, a good inoculant must have as many as possible of these desired characteristics at reasonable cost with ease of availability. Unlike pharmaceutical, cosmetics and nanotechnology industries, efforts are not successful for agriculture and environmental purposes to create "super inoculants" most commonly due to their high costs.

4.6 COMMON ADVANTAGES OF LIQUID BIOFERTILIZER

The advantages of liquid biofertilizer over conventional carrier-based biofertilizer are listed below:

1. Production is easy and amenable.
2. It is possible to automate 100% of the production process.
3. 100% sterile conditions can be maintained from production to packaging.
4. Offers high cell count and higher shelf-life.
5. Higher temperature tolerance up to 45°C.
6. Higher cell count (10^9 cfu/mL) can be maintained up to 12 to 24 months depending on the storage temperature.
7. Lesser cost due to no requirement of pulverization, drying or milling. Sterilization is easier and cheap.
8. Better rate of survival on seeds and soil.
9. Multiple inoculants or microbial consortia can be accommodated within single formulation.
10. Effective even at low dosage (up to $1/10^{th}$ of the conventional carrier-based biofertilizer).
11. Easy to apply through mechanical means, fertigation, foliar spray, etc.

Disdavantages of Liquid Biofertilizers

1. Higher investment required for setting up a production unit.
2. Higher per unit cost.

4.7 EFFECT OF SOME ADDITIVES ON THE SURVIVAL OF MICROBIAL INOCULANTS IN LIQUID BIOFERTILIZER

Several compounds were tested for their ability to promote survival of *B. japonicum* after inoculation (Singleton et al., 2002). Cell protectants like arabinose, trehalose, glycerol, PVP were studied for their effect on biofertilizer organisms *Rhizobium* (strain RBG-314), *Azospirillum* (strain AZS-303), *B. megaterium* (strain AMT-1001) (Rao D.L.N. et al., 2008).

Polyvinylpyrrolidone (PVP): PVP may bind toxic compounds present in seed exudates that are mobilized during inoculation and seed germination. PVP has a high water-binding capacity and slows the drying of the inoculants after application. PVP coalesces into the ridges of seedcoat and provides a thicker layer of protection. Its sticky nature also enhances cell and inoculants adherence to seed (Fig. 4.1).

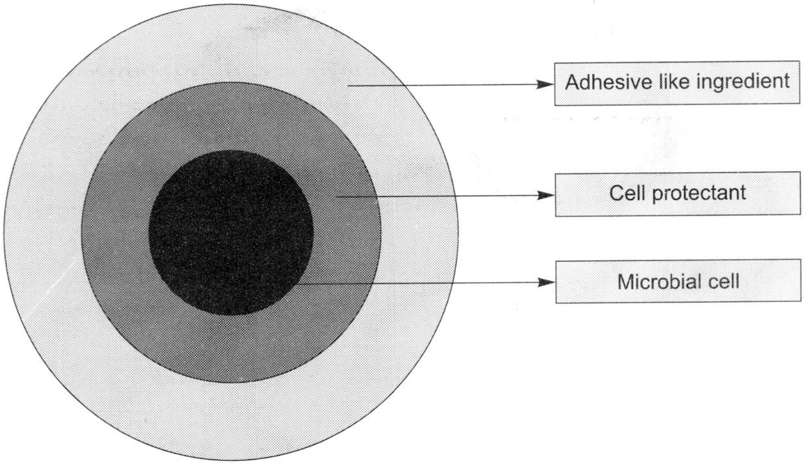

Fig. 4.1 Schematic diagram of a droplet of liquid inoculant.

FeEDTA: Reports have suggested that seed-released compounds bind iron in the yeast extract, making it unavailable to cells. Supplementary Fe may, therefore, replace Fe bound by seed exudates.

Glycerol: Glycerol is a carbon source for rhizobia. It also has high water-binding capacity and may protect cells from desiccation by slowing the rate of drying. Its flow characteristics are also rapid and helpful for an even coating of seeds.

Trehalose: This compound is widely described as an enhancer of cell tolerance to desiccation and osmotic and temperature stress. It acts by stabilising both enzymes and cell membranes, and is also a compatible osmoticum.

Glucose: Glucose promotes the production of the storage product glycogen. In the absence of glycogen, osmotically stressed cells are reported to release their supplies of trehalose, thought to be an undesirable outcome. Glucose also enhances the production of exopolysaccharides, which could protect the cells during the rapid drying at the time of inoculation. It had no apparent effect on cell survival.

Mannitol: Mannitol has many important roles in a liquid biofertilizer. It is a carbon source from all strains of *Bradyrhizobium*. Mannitol is reported to protect cells during rapid drying period, which they experience at inoculation both as a solute and in enhancing polysaccharide production. When the mannitol was removed from the media, the survival of *B. japonicum* was adversely affected. Its effect on survival appears to be maximum at approximately 1.0 g/lit (Singleton et al., 2002).

Polyethylene glycol (PEG): It is a low molecular weight, water-soluble compound with adhesive properties.

Polyvinyl alcohol (PVA): It is a high molecular weight compound, soluble in hot water. It has stabilizing properties for the cells. It is useful in reducing the extent of protein precipitation or the coagulation of cells.

Gum Arabic: It is a large molecular weight biopolymer with adhesive properties. It has properties of stabilization and emulsification, with a high water activity and limits heat transfer.

Sodium alginate: It is a large molecular weight, nontoxic compound that has adhesive properties, with high water activity and limits heat transfer. It is useful in long term survival of inoculants.

Tapioca flour: It is a large molecular weight biopolymer. It has adhesive properties and can be used as a binder and thickener. It also stabilizes, limits heat transfer with high water activity.

Carboxymethyl cellulose (CMC): It is a cellulose derivative containing carboxymethyl groups and is used for controlled release of different active principles. It acts as a thickener, stabilizing agent in suspension, and can support cells for a longer period due to retention of water.

4.8 CHARACTERISTICS OF SOME AVAILABLE LIQUID BIOFERTILIZERS

Rhizobium **sp:** It is commonly used in leguminous crops and pulses. It helps the crop to produce root nodules within which they reduce molecular nitrogen to ammonia

Physical characteristics of liquid (*Rhizobium*) biofertilizer:

- Colour is dull to off white
- Absence of bad odour.
- Absence of foamy substances.
- pH range 6.8 - 7.5

Azospirillum **sp:** Most commonly used in upland rice, millets, wheat and oilseeds. It helps to fix nitrogen through associative symbiosis. Common species are *A. lipoferum, A. brazilense* and *A. amazonense*. Soil characteristics of *Azospirillum* sp. is tested through their ability to reduce nitrate and denitrification. Some of the tropical grasses like *Digitaria* sp., *Panicum* sp. also contain *Azospirillum* in the roots.

Physical characteristics of liquid (*Azospirillum*) biofertilizer:

- Colour varies from blue to dull white.
- Presence of bad odour signifies contamination of liquid formulation due to handling.
- Presence of yellow gummy substance confirms appropriate quality.
- pH should not be acidic.

Azotobacter **sp:** Free living nitrogen fixing bacterium, mostly found in rice, wheat crop fields when there is a lack of organic matter in the soil. Mature culture of *Azotobacter* sp. is brownish due to the production of melanin pigment with the activity of enzyme tyrosinase, which oxidizes tyrosine.

A. chroococcum - brown black pigmentation in liquid formulation
A. beijerinckii - yellow to light brown pigmentation in liquid formulation.
A. paspali - green fluorescent pigmentation in liquid formulation.
A. vinelandii - green fluorescent pigmentation in liquid formulation.
A. macrocytogenes- pink pigmentation in liquid formulation.
A. insignis - greyish-blue pigmentation without gum formation.
A. agilies - green fluorescent pigmentation in liquid formulation.

Gluconacetobacter diazotrophicus: This bacterial isolate was reported from Brazil within sugarcane rhizosphere, stem and leaves. They are N_2-fixing endosymbiont that antagonize pathogens like *Xanthomonas albilineans.* They can fix up to 150 kg N/ha under microaerophilic conditions. Usually, they require a high sucrose medium (10%) for growth with a very low pH (3.0). They grow by secreting levansucrase, an extracellular enzyme enzyme, that can hydrolyze sucrose into fructose and glucose. It does not inhibit the effect of nitrate on N_2 fixation to some extent. Hence, an ideal choice for biofertilizer even in the environments applied with nitrate fertilizers. This phenomenon is due to the absence of nitrate reductase in *G. dizotrophicus.* They have recorded a capacity to fix up to 300 kg N/ha.

Possibilities of multiple inoculants have already been tried where (N_2-fixing + PSM) have given better response in terms of plant and yield. Also investigators are studying the combined effect of (PGPR + PSM) on crop growth and yield advantages over control.

4.9 APPLICATION METHODOLOGY FOR LIQUID BIOFERTILIZER

1. Seed Treatment
2. Root dipping
3. Soil application

Seed treatment: Seed treatment is the most common application tried for almost every type of inoculants. It was found to be effective and cheap. For handling large amounts of seeds, seeds can be coated within a bucket by mixing the inoculants directly with hand. Common inoculants are *Rhizobium, Azotobacter, Azospirillum* and PSM.

For small quantity of seeds, the inoculants can be mixed in a plastic bag, having a bigger size than the combined volume of seeds plus inoculants. This bag can trap some air above the mixture of seeds and inoculants. Thereafter, the bag is squeezed in such a way that all the seeds are wetted in an uniform way. Then the bag is opened and inflated with air with gentle shaking. After opening the seed packet, the seeds are dried under the shade for 20-30 minutes.

Root dipping: For paddy transplantation, vegetable crop seedling–root dipping is an usual practice.The appropriate amount of *Azospirillum*/PSM is mixed with 5-10 litres of water at a corner of the field. The roots of the seedlings are dipped within the mixture for at least half an hour before starting transplantation.

Soil Treatment: 200 mL of PSM is mixed with 400-600 kg of cowdung FYM , 1/2 bag of available rock phosphate. This mixture is kept overnight under shade with 50% moisture content. The preparation and application methodologies are summarized in Fig. 4.2.

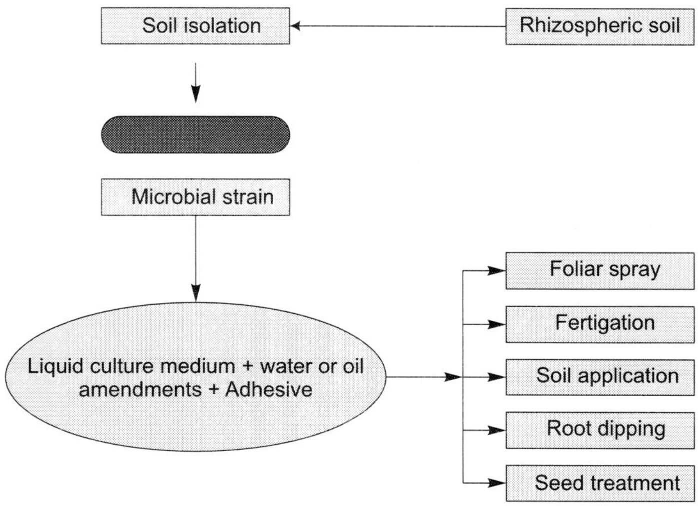

Fig. 4.2 Flow chart for preparation and application methods of liquid biofertilizers.

4.10 USUAL DOSAGE OF LIQUID BIOFERTILIZERS

For leguminous crops and pulses, seed treatment by *Rhizobium* at 200 mL/acre is sufficient. For leguminous plants and trees, however, soil treatment is needed, where *Rhizobium*@ 1-2 mL/plant is needed.

For cereal crops, wheat and barley, seed treatment is done by *Azotobacter/ Azospirillum* @ 200 mL/acre. The dose is same for mustard, sunflower and other oilseeds.

For common millets *Azotobacter* is applied for seed treatment at 200 mL/acre. *Azospirillum* is applied through seed treatment in sorghum and maize at 20 mL/acre. For plantation crops, seedling treatment is done by *Azotobacter* at 500 mL/acre. For fruitplants/agro-forestry, *Azotobacter* is applied as soil treatment at 2-3 mL/plant.

4.11 MOLECULAR MARKERS USED IN MONITORING OF MICROBIAL STRAINS

For measuring the suitability of a microbial inoculant to be used as a biofertilizer, one has to ascertain its adaptability and soil ecological performance before introducing the said strain. Here, molecular marker-based technologies are highly useful.

Table 4.2 summarises the work done in monitoring the microbial interactions.

Table 4.2 Molecular markers used for monitoring strains of microbial inoculants

Gene (s)	Product	Visual indicator	Used in monitoring of	Reference
gusA	β- glucuronidase	Blue/yellow colour	*Rhizobium*, *Pseudomonas*	Wilson et al., (1991)
lacZ	β- galactosidase	Blue colour	*Pseudomonas*, *Azospirillum*	Katupitiya et al., (1995)
xylE	Catechol 2, 3-dioxygenase	Yellow colour	*Pseudomonas*	Winstanley et al., (1991)
phoA	Alkaline phosphatase	Blue colour	*Rhizobium*	Reuber et al., (1991)
luxA, luc	Luciferase	Light	*Pseudomonas*, *Azospirillum*	De Weger et al., (1991)
celB	β- glucosidase	Blue colour	*Rhizobium*, *Azospirillum*	Sessitsch et al., (1998)
tfdA	2,4-dichlorophenoxy monooxygenase	Red colour	*Pseudomonas*	King et al., (1991)
gfp	Green Fluorescent protein	Green fluorescence	*Bacillus*, *Rhizobium*	Jaderlund (2008)

CONCLUSION

Since the introduction of high yielding varieties in India, agricultural production has increased but at the same time problems have come in the form of extreme soil pH, salinity, denuded top soil, desertification, etc. All of these things may not be solved with a single approach. But soil nutrient status may sufficiently be improved with judicious application of fertilizers augmented by suitable biofertilizers/inoculants.

As we discuss the formulation of biofertilizers, liquid biofertilizers have brought a whole new concept for the application and utility of biofertilizers. The liquid formulation has made the shelf life better if not best within the temperature range of up to 45°C.

In the advent of todays highly mechanized agriculture, one can easily accommodate the application methods of liquid inoculants. It can also be applied through fertigation or through foliar spray. But the most critical aspect of commercial success of liquid inoculants still happens to be the suitable strains of microbial species that are to be applied as biofertilizer. Some research progress has already been made, and we can expect better strains with higher nutrient supply, better survival rate and high multiplication capacity within soil.

More concerted endeavour is needed from microbiologists, agronomists, and soil chemists to manifest the real potential of liquid biofertilizers with special emphasis to keep it cost effective and suitable for sustainable agricultural developments.

REFERENCES

1. Albareda, M., Rodriguez-Navarro D.N., Camacho, M. and Temprano F.J. (2008). "Alternatives to peat as a carrier for rhizobia inoculants: solid and liquid formulations". *Soil. Biol. Biochem.,* 40, 2771–2779.
2. Bashan, Y. (1998). "Inoculants of plant growth-promoting bacteria for use in agriculture". *Biotechnology Advances* 16(4), 729–770.
3. Brahmaprakash, G.P. and Sahu, P.K. (2012). Review: Biofertilizers for Sustainability. *J. Indian Inst. of Sci.,* 92(1), 37–62.
4. Catroux, G., Hartmann, A. and Revellin, C. (2001). "Trends in rhizobial inoculant production and use". *Plant Soil,* 230, 21–30.
5. Cong, P.T., Dung, T.D., Hien, T.M., Hien, N.T., Choudhury, A.T.M.A, Keeskes, K.L. and Kennedy, I.R. (2009). "Inoculant plant growth-promoting microorganisms enhance utilization of urea-N and grain yield of paddy rice in southern Vietnam". *Eur. J. Soil. Biol.,* 45, 52–61.
6. Dal Bello G.M., Monaco, C.I. and Simon, M.R. (2002). "Biological control of seedling blight of wheat caused by *Fusarium graminearum* with beneficial rhizosphere microorganisms". *World. J. Microbiol. Biot.,* 18, 627–636.
7. Date, R.A. (2001). Advances in Inoculant Technology: a brief review. *Aust. J. Exp. Agr.,* 41, 321–325.
8. Deaker, R., Roughley, R.J. and Kennedy, J.R. (2004). "Legume seed inoculation technology- a review". *Soil. Biol. Biochem.* 36, 1275–1288.
9. De Weger, L.A., Dunbar, P., Mahafee, W.F., Lugtenberg, B.J.J. and Sayler, G.S. (1991). "Use of bioluminescence reporters to detect *Pseudomonas* spp. in the rhizosphere". *App. Environ. Microbiol.* 57, 3641–3644.
10. Diouf, D., Forestier, S., Neyra, M. and Lesueur, D. (2003). "Optimisation of inoculation of *Leucaena leucocephala* and *Acacia mangium* with rhizobium under greenhouse conditions". *Ann. For Sci.* 60, 379–384.
11. Gamal-Eldin, H. and Elbanna, K. (2011). "Field evidence for the potential of *Rhodobacter capsulatus* as biofertilizer for flooded rice". *Curr. Microbiol.,* 62, 391–395.
12. Jaderlund, L. (2008). "Fates and impacts of the genetically modified plant growth-promoting bacterium *Pseudomonas fluorescens* SB25". Doctoral thesis, Swedish Univ. of Agric Sci, Uppsala.
13. Jha, C.K. and Saraf, M. (2012). "Evaluation of multispecies plant-growth-promoting consortia for the growth promotion of *Jatropha curcas* L. J. Plant. Growth. Regul., 31, 588–598.
14. King, R.J. Short, K.A. and Seidler, R.J. (1991). "Assay for detection and enumeration of genetically engineered microorganisms which is based on the activity of a deregulated 2, 4-Dichlorophenoxyacetate monooxygenase". *App. Environ. Microbiol.,* 57(6), 1790–1792.
15. Katupitiya, S., New, P.B., Elmerich, C. and Kennedy, I.R. (1995). Improved N_2 fixation in 2, 4-D treated wheat roots associated with *A. lipoferum*: studies of colonization using reporter genes. *Soil. Biol. Biochem.,* 27, 447–452.

16. Kloepper, J.W. (1992). "Plant growth promoting rhizobacteria as biological control agents". In: *Soil Microbial Ecology, Applications in Agricultural and Environmental Management* (ed. F.B. metting Jr), Marcel Dekker, New York, USA, pp. 255–274.
17. Manikandan, R., Saravanakumar, D., Rajendran, L., Raguchander, T. and Samiyappan, R. (2010). "Standardization of liquid formulation of *Pseudomonas fluorescens* Pf1 for its efficacy against *Fusarium* wilt of tomato". *Biol. Control.,* 54, 83–89.
18. Malusa, E., Sas-Paszt, L. and Ciesielska, J. (2012). "Technologies for beneficial microorganisms inocula used as biofertilizers". *Sci. World. J.,* 2012:491206 doi:10.1100/2012/491206.
19. Maurice, S., Beauclair, P., Giraud, J.J., Sommer, G., Hartmann, A. and Catroux, G. (2001). "Survival and change in physiological state of *Bradyrhizobium japonicum* in Soybean (*Glycine max* L.Merril) liquid inoculants after long term storage". *World J. Microbiol. Biotech.,* 17, 635–643.
20. Rao, D.L.N., Trimurtulu, N. and Kumari L. (2008). "Improvement of Biofertilizer Quality". *Biofert, Newslett,* 16(2), 8–9.
21. Rekha, P.D., Lai, W.A., Arun, A.B. and Young, C.C. (2007). "Effect of free and encapsulated *Pseudomonas putida* CC-FR2-4 and *Bacillus subtilis* Cc-pg104 on plant growth under gnotobiotic conditions". *Bioresource Technol.,* 98, 447–451.
22. Reuber, T.L., Long, S.L. and Walker, G.C. (1991). Regulation of *Rhizobium meliloti exo* genes in free-living cells and in planta examined using Tn*phoA* fusions. *J. Bacteriol.,* 173, 426–434.
23. Sharma, P., Kaur, N. and Garg, K. (2012). Revitalizing soil health with biofertilizers. *Biofert, Newslett,* 20(2), 5–11.
24. Sessistsch, A., Hardarson, G., de Ves W.M. and Wilson, K.J. (1998). Use of marker genes in competition studies of *Rhizobium. Plant and Soil,* 204, 35–45.
25. Singleton, P., Keyser, H. and Sande, E. (2002). Development and evaluation of liquid inoculants. In: Herridge, D., (ed.) *Inoculants and Nitrogen Fixation of Legumes in Vietnam.* ACIAR Proceedings 109e.Canberra. pp. 52–66.
26. Stephens J.H.G. and Rask, H.M. (2000). "Inoculant production and formulation". *Field Crops Res.,* 65, 249–258.
27. Taurian, T., Anzuay, M.S., Angelini, J.G., Tonelli, M.L., Luduena, L., Pena, D., Ibanez and F., Fabra, A. (2010). "Phosphate-solubilizing peanut associated bacteria screening for plant growth-promoting activities". *Plant Soil,* 329, 421–431.
28. Trivedi, P., Pandey, A. and Palni, L.M.S. (2005). "Carrier-based preparations of plant growth-promoting bacterial inoculants suitable for use in cooler regions". *World J. Microb. Biot.* 21, 941–945.
29. Viji, G., Uddin, W. and Romaine, C.P. (2003). "Suppression of gray leaf spot (blast) of perennial ryegrass turf by *Pseudomonas aeruginosa* from spent mushroom substrate". *Biol. Control.,* 26, 233–243.

30. Wani, P.A., Khan, M.S. and Zaidi, A. (2007). "Effect of metal tolerant plant growth-promoting *Bradyrhizobium* sp(*vigna*) on growth, symbiosis, seed yield and metal uptake by greengram plants". *Chemosphere* 70, 36–45.

31. Wilson K.J., Giller K.E. and Jefferson, R.A. (1991). "β-glucuronidase (GUS) operon fusions as tool for studying plant-microbe interactions". In: *Advances in Molecular Genetics of Plant-Microbe Interactions.* Vol 1, eds. H. Hennecke & DPS Verma pp. 226–229 Kluwer Academic Publishers, Dordrecht.

32. Winstanley, C. Morgan, J.A.W., Pickup, R.W. and Saunders J.R. (1991). Use of *xylE* marker gene to monitor survival of recombinant *Pseudomonas* populations in lake water by culture on nonselective media. *App. Environ. Microbiol.* 57, 1905–1913.

33. Xavier, I.J., Holloway, G. and Leggett, M. (2004). Development of rhizobial inoculant formulations. Online Crop Manag Netw.doi:10.1094/CM-2004-0301-06-RV.

5

Understanding the Molecular Genetics of Microbial Phosphate Solubilization

Shilajit Barua* and Jishnu Adhikari

Department of Microbiology and Research and Development Centre, Vijaygarh Jyotish Ray College Bejoygarh, Jadavpur. Kolkata – 700032, West Bengal, India

**Corresponding Author: E-mail: shilajit_26@yahoo.co.in*

ABSTRACT

Phosphorus, next to nitrogen is the most critical essential element that influences plant growth and production throughout the world. Major reserves of phosphorus in the earth are marine sediments, terrestrial soils, dissolved inorganic phosphate in ocean, crustal rocks such as apatite and biota or biomass, which majorly remains unavailable to plants. Diverse groups of phosphate solubilizing microorganisms (PSMs) including bacteria, actinomycetes and fungi have been shown to solubilize insoluble inorganic phosphates to soluble forms through secretion of several organic acids and inorganic acids. Many of them also release phosphates from organic sources through by phosphatase, phytase and C-P lyase enzymes. solubilization is an important biological event in nature which is higly regulated at genetic level. Several classes of mineral phosphate solubilizing genes like, that are involved in mineral phosphate solubilization like *pqq, gab, gcd* has been characterized. On the other hand, cloning regulation and expression of several organic phosphatase and phytase genes has also been reported. Detailed knowledge of the genetic system and regulation of phosphate solubilization would help researchers to manipulate phosphate solubilization in microbes to be applied as bioinoculants for plant growth promotion.

Keywords: Phosphate Solubilization, PSM, Organic Acids, pqq, Phytase.

5.1 INTRODUCTION

Ensuring food sovereignty and food security to the ever increasing global population necessitates increased agricultural production, from a steadily decreasing and degrading land resource base, imposing considerable strain on agro ecosystems throughout the world [1]. Such degradation has resulted from extensive usage of chemical fertilizers. Many synthetic fertilizers contain acids, such as sulphuric acid and hydrochloric acid, which tend to increase the acidity of the soil, thereby reducing the soil's beneficial organism population and interferes with plant growth. Generally, healthy soils harbour sufficient number of microbes with diverse ecological niche, and contribute to increased soil fertility and plant growth promotion. They include nitrogen fixers, phosphate solubilizers, phytohormone producers as well as those that keep away plant pathogens by antagonistic action. Change in physico-chemical parameters of the soil due to continued application of chemical fertilizers detrimentally affects these microbial habitats. Chemical fertilizers have also been reported to detrimentally affect plant growth. For example, citrus trees tend to yield fruits that are lower in vitamin C when treated with synthetic fertilizer. Lack of trace elements in soil can also be attributed to the use of regular application of chemical fertilizers [2].

Environment-friendly biotechnological approaches offer alternatives to chemical fertilizers [3]. A substantial number of bacterial species, mostly those associated with the plant rhizosphere, may exert a beneficial effect upon plant growth [4]. This group of bacteria has been termed "plant growth promoting rhizobacteria" or PGPR [5]. Although plant growth promoting rhizobacteria occur in soil, usually their numbers are not high enough to compete with other bacteria commonly present in the rhizosphere. Therefore, for agronomic utility, inoculation of plants with target microorganisms at a much higher concentration than those normally found in soil is necessary to take advantages of their beneficial properties for plant yield enhancement. An increasing number of PGPR are successfully used as commercial biofertilizers for agricultural improvement [6]. Among PGPR, nitrogen fixers and phosphate-solubilizing bacteria (PSB) are the two major physiological group that are used as bioinoculants.

Phosphorus, next to nitrogen is the most critical essential element that influences plant growth and production throughout the world. It occurs as an important component of nucleic acids, high energy phosphate bonds, etc. It plays significant role during photosynthesis, root development, crop maturation and crop quality improvement. Plants fulfil their 'P' need from soil. Several plant growth-promoting rhizobacteria have the ability to solubilize insoluble inorganic phosphate compounds, such as tricalcium phosphate, dicalcium phosphate, hydroxyapatite and rock phosphate. Apart from P fertilization, microbial P-mobilization is probably the only possible way to increase available phosphate for upland crops [7]. Evidence of naturally occurring rhizospheric phosphorus solubilizing microorganism (PSM) dates back to 1903 [8]. Diverse bacterial genera have been reported to solubilize both organic and inorganic phosphates more efficiently than fungi [9]. Several mechanisms of microbial phosphate solubilization have been proposed by scientific workers throughout the world. This chapter aims at reviewing major genetic systems and molecular events that are involved in microbial phosphate solubilization, which would help scientific readers to carry out in-depth analysis of such microbiological event.

5.2 PHOSPHORUS IN SOIL AND SIGNIFICANCE OF PHOSPHATE SOLUBILIZATION

Phosphorus (P) is a major plant growth limiting element and unlike nitrogen it does not have any environmental reserve to make it biologically available [10]. Plant growth attributes like root development, stalk and stem strength, flower and seed formation, crop maturity and production, N-fixation in legumes, crop quality, and resistance to plant diseases are related to phosphorus nutrition. Dynamics of phosphorus in soil is characterized by physicochemical (sorption-desorption) and biological (immobilization-mineralization) processes. P is majorly applied to soil as fertilizers, which enters the immobile pools through precipitation reaction with highly reactive Al^{3+} and Fe^{3+} in acidic, and Ca^{2+} in calcareous or normal soils [11]. As per Isherword, 1998, efficiency of P fertilizer throughout the world is around 10 - 25%, and concentration of bioavailable P in soil is very low reaching the level of 1.0 mg kg^{-1} soil [12]. P, in soil exists as organic and inorganic phosphates. A large proportion is present in insoluble forms, and therefore, not available for plant nutrition. Chemical fertilizers are the major source of inorganic P in soil that majorly exists as insoluble mineral complexes and is inaccessible to plants. Organic matter, on the other hand, accounts for 20–80% of soil P is also an important reservoir of immobilized P. Soil microorganisms play a key role in soil P dynamics and subsequent availability of phosphate to plants [13].

Bacteria were found to be more effective in phosphorus solubilization than fungi [9]. Phosphate solubilizing microbial community in soil constitutes of 1 to 50% PSB, while phosphorus solubilizing fungi (PSF) are only 0.1 to 0.5% in P solubilization potential [14]. Among the soil bacterial communities, ectorhizospheric *Pseudomonas* and *Bacilli*, and endosymbiotic are effective phosphate solubilizers [14]. *Bacillus megaterium, B. circulans, B. subtilis, B. polymyxa, B. sircalmous, Pseudomonas striata,* and *Enterobacter* sp. are potent P solubilizing strains [16, 17].

PSB are ubiquitous with variation in forms and population in different soils. Population of PSB in soil depends on different soil properties (physical and chemical properties, organic matter, and P content) and cultural activities [18]. High proportion of PSM is found to be concentrated in the rhizosphere, and they are metabolically more active than other habitats [19].

Assimilation of soluble P by microbial biomass prevents it from adsorption or fixation [20]. Soil microbial communities enhance P availability to plants by mineralizing organic P in soil and by solubilizing precipitated phosphates [14; 21; 22]. In the presence of labile carbon, these bacteria serve as a sink for P by rapidly immobilizing it even in low P soils [23]. Subsequently, these PSB become a source of P to plants upon its release from their cells. The PSB and plant growth promoting rhizobacteria (PGPR) together could reduce P fertilizer application by 50% without any significant reduction of crop yield [24; 25].

5.3 MECHANISM OF PHOSPHATE SOLUBILIZATION

Phosphates in the environments exist in both organic and inorganic form. Microbes apply various strategies to solubilize both phosphate types, as summarized in Fig. 5.1.

5.3.1 Solubilization of Inorganic Phosphates

A major proposed mechanism of mineral phosphate solubilization is the action of organic acids synthesized by soil microorganisms [26; 27]. Culture filtrate of PSM exhibited

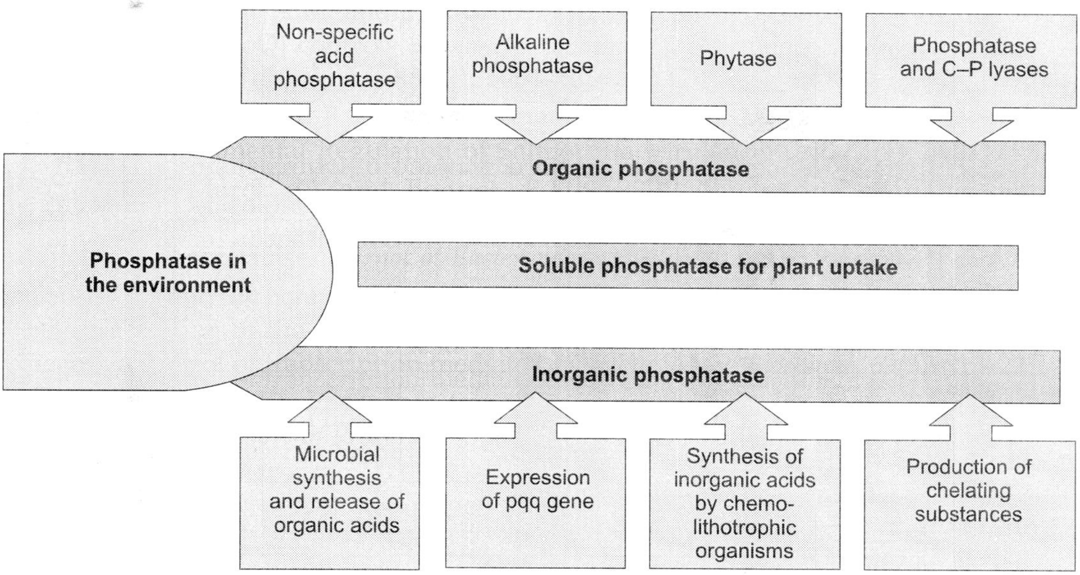

Fig. 5.1 Overview of the mechanism of phosphate solubilization in environment.

decrease in pH from an initial value, indicating production of acids [28]. The analysis of culture filtrates of PSMs has shown the presence of a number of organic acids such as malic, glyoxalic, succinic, fumaric, tartaric, alpha keto butyric, oxalic, citric, 2-ketogluconic and gluconic acid [2]. The production of different organic acids by different phosphate solubilizing bacteria has been well-documented (Table 5.1).

Table 5.1 Organic acid producing phosphate solubilizing microbes

Organic acid produced	*Producer phosphate solubilizing microbe*	*Reference*
Gluconic acid	*Pseudomonas* sp.	[28]
	Erwinia herbicola	[29]
	Pseudomonas cepacia	[30]
	Burkholderia cepacia	[31]
2-ketogluconic acid	*Rhizobium leguminosarum*	[32]
	Rhizobium meliloti	[33]
	Bacillus firmus	[34]
	Other soil bacteria	[35]
Mixtures of lactic, isovaleric, isobutyric, and acetic acids.	*Bacillus licheniformis*	[36]
	Bacillus amyloliquefaciens	[37]
Other organic acids, such as glycolic, oxalic, malonic, and succinic acid	Other Phosphate Solubilizing Bacteria	[28; 34]

Production of organic acids results in acidification of the microbial cell and its surroundings. Consequently, Pi may be released from mineral phosphate by proton substitution for Ca^{2+} [12]. Organic acids isolated from a culture of *Rhizobium leguminosarum* solubilized equivalent amount of P as that by whole culture [32]. In addition, treatment of culture filtrates from several *Rhizobium* strains with pepsin or removal of proteins by acetone precipitation did not affect phosphate release capacity, showing that this was not an enzymatic process. However, neutralization with NaOH destroyed the solubilization activity [33].

Direct periplasmic oxidation of glucose to gluconic acid (GA), and often 2-ketogluconic acid, forms the metabolic basis of the mineral phosphate solubilization phenotype in some Gram-negative bacteria [38]. No significant amounts of organic acid production could be detected from a phosphate solubilizer fungus, *Penicillium* sp. [28].

Besides organic acids, inorganic acids such as nitric and sulphuric acids are also produced by the nitrifying bacteria and *Thiobacillus* during the oxidation of nitrogenous or inorganic compounds of sulphur which react with calcium phosphate and convert them into soluble forms [38].

Change in pH and reduction potention of the surrounding due to extracellular oxidation of glucose via the quinoprotein glucose dehydrogenase to gluconic acid, as found in several Gram-negative bacteria, is the most efficient mechanism of mineral phosphate solubilization [39].

Gluconic acid biosynthesis is carried out by the glucose dehydrogenase (GDH) enzyme and the co-factor, pyrroloquinoline quinone (PQQ) (Fig. 5.2).

Azospirillum brasilense has been reported to produce gluconic acid in vitro when grown in media containing fructose as principal carbon source, which in turn can be induced by amendment of glucose. Under above conditions, acid production was accompanied by in vitro phosphate solubilization. Glucose is the precursor for synthesis of gluconic acid. This has suggested that phosphate solubilization in these strains is mediated by glucose or gluconic acid metabolism [41].

5.3.2 Genetics of Inorganic Phosphate Solubilization

In most bacteria, mineral phosphate-dissolving capacity has been shown to be related to the production of organic acid [42].

A gene involved in mineral phosphate solubilization (MPS) from the Gram-negative bacteria *Erwinia herbicola* was first cloned and expressed in *E. coli* HB101 by Goldstein and Liu [43]. Wild type *E. coli* can synthesize glucose dehydrogenase (GDH), but not PQQ, thus it does not produce GA. Expression of the gene from *E. herbicola* allowed production of GA and conferred the ability to solubilize hydroxyapatite. A protein similar to the gene III product of a *pqq* synthesis gene complex from *Acinetobacter calcoaceticus*, and *pqq*E of *Klebsiella pneumonie* [29] was encoded from the cloned 1.8 kb locus. The authors opined that the cloned DNA fragment functions as a PQQ synthase gene, and probably, some *E. coli* strains contain some cryptic PQQ synthase genes that could be complemented by this single open reading frame (ORF) isolated by them.

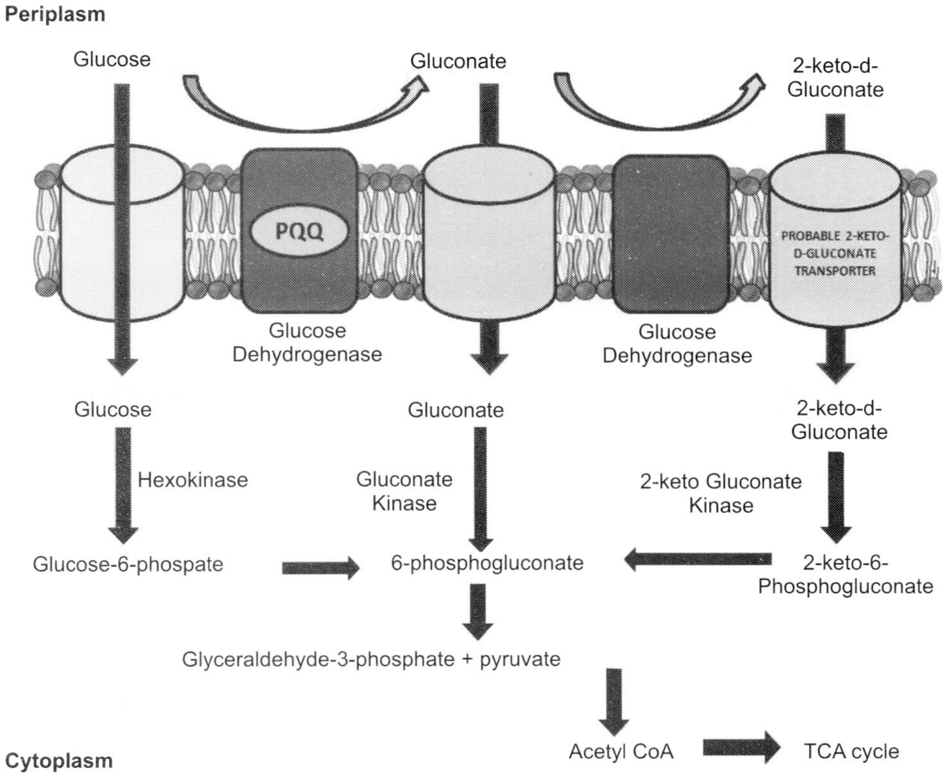

Fig. 5.2 Production of gluconic acid via the alternative extracellular oxidation pathway of glucose metabolism [2].

As per the review carried out by Mishra et al. [44], genetics of PQQ synthesis has been studied in many bacteria including *Acinetobacter calcoaceticus, Enterobacter intermedium* 60-2G, *Gluconobacter oxydans, Klebsiella pneumonia, Methylotrophs extorquens* AM1 and *Pseudomonas fluorescens* CHA0. The number of genes involved in the synthesis of PQQ ranges from 4 in *A. calcoaceticus* to 6–7 genes in *Methylotrophs* and *G. oxydans* [44]. The majority of the bacteria making PQQ contain 6 or 7 genes (pqqABCDEF/G) in an operon [45] (Fig. 5.3).

PQQ biosynthesis involves several enzymes encoded by these genes including *pqq*E that encodes a regulatory enzyme named as PQQ synthase [46]. The *pqq* A encodes a small peptide of generally 24 amino acids bearing tyrosine and glutamate, which get fused and form a precursor for subsequent transformations into PQQ [47]. This molecule remains attached with a precursor peptide and is cleaved off at a later step in the biosynthetic process by other enzymes of the pathway. Although the *pqq* A gene product is redundant for the synthesis of PQQ in some of the bacteria, its availability seems to be a rate-determining step in PQQ biosynthesis.

Escherichia coli is deficient in PQQ synthase and, therefore, has been used as a host for cloning the PQQ synthase gene from diverse group of bacteria. Transgenic *E. coli*

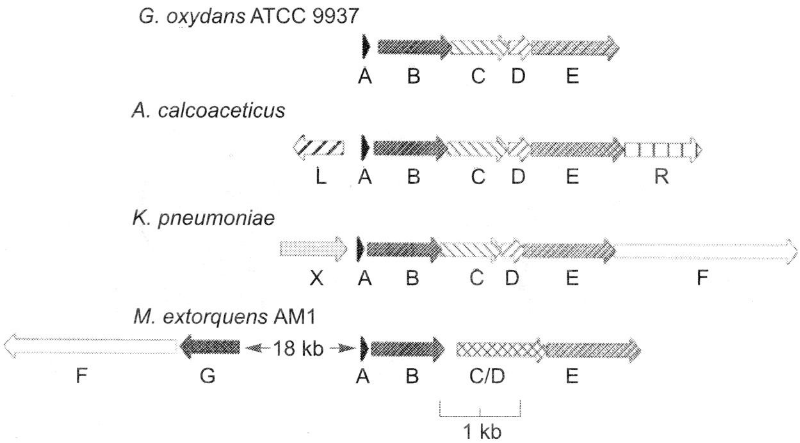

Fig. 5.3 PQQ biosynthetic operon in different bacteria [45].

producing PQQ, attributed to functional complementation of MPS activity, was found capable of solubilizing insoluble rock phosphates [44].

Another gene (*gab*Y) involved in GA production and MPS from *Pseudomonas cepacia* was cloned and deduced amino acid sequence showed no homology with previously cloned direct oxidation pathway (GA synthesis) genes, but was similar to histidine permease membrane-bound components [48]. The authors [48] also discovered that in the presence of *gab*Y, GA is produced only if the *E. coli* strain expresses a functional glucose dehydrogenase (gcd) gene. Based on the above observations it was speculated that this ORF could be related to the synthesis of PQQ by an alternative pathway, or the synthesis of a gcd co-factor is different from that of PQQ. The reported synergistic effect of exogenous PQQ and this gene might support this alternative. The gcd gene acted by regulating GA production under cell-signal effects. This was evident from the fact that a DNA fragment from *Serratia marcescens* was found to induce GA synthesis in *E. coli*, but showed no homology to *pqq* or gcd genes [49].

Other genes involved in the MPS phenotype studied so far seem not to be related to *pqq* DNA or gcd biosynthetic genes. A genomic DNA fragment from *Enterobacter agglomerans* showed MPS activity in *E. coli* JM109, although the pH of the medium was not altered [50]. These results indicate that acid production is an important way, but not the only mechanism of phosphate solubilization by bacteria [51]. A phosphoenol pyruvate carboxylase (*pcc*) gene from *Synechococcus elongatus* PCC 6301 was constitutively overexpressed in fluorescent pseudomonads to increase the supply of oxaloacetate – a crucial anabolic precursor and an intermediate in biosynthesis of organic acids implicated in phosphate (P) solubilization [52]

Different observations regarding the effect of soluble P on the expression of the phosphate-solubilizing activity has been put forward. Mineral phosphate solubilization trait in *E. herbicola* was induced by P starvation and repressed by elevated exogenous P levels (complete repression achieved at P concentrations 0.20 mM) [43]. A strain of

Burkholderia cepacia showed reduced expression of tricalcium phosphate solubilization at increasing phosphate concentration 0.2 mM, and finally failed to express any solubilization ability at P levels between 30 and 40 mM [53]. On contrary, Halder et al. [54] reported no effect of soluble P up to 6 mM on rock phosphate solubilization in cultures of *Rhizobium leguminosarum*. Repression of MPS activity in several phosphate solubilizing bacteria in presence of soluble P was observed while others exhibited no repression effect at concentrations up to 50 mM [55]. Hence, P availability could regulate mineral phosphate solubilization in some species and had no effect in others. All these findings demonstrate the complexity of MPS in different bacterial strains, but at the same time, offer a basis for better understanding of this process.

5.3.3 Solubilization of Organic Phosphates

Soil, plant and animal remains, which contain a large amount of organo-phosphorus compounds, are readily mineralized by microbial process. Decomposition of organic matter in soil is carried out by the action of numerous saprophytes, which produce the release of radical orthophosphate from the carbon structure of the molecule. Organo-phosphonates can also be mineralized by biodegradation [56]. The microbial mineralization of organic phosphorus is strongly influenced by environmental parameters, for instance moderate alkalinity favors the mineralization of organic phosphorus [57].

Degradation of organic phosphorus compounds depends mainly on the physicochemical and biochemical properties of their molecules. Nucleic acids, phospholipids, and sugar phosphates are broken down easily, while phytic acid, polyphosphates, and phosphonates are decomposed more slowly [52].

Phosphorus can be released from organic compounds in soil by three groups of enzymes:

- nonspecific phosphatases, which perform dephosphorylation of phospho-ester or phosphoanhydride bonds in organic matter;
- phytases, which specifically cause P release from phytic acid; and
- phosphonatases and C–P Lyases, enzymes that perform C–P cleavage in organophosphonates.

The phosphohydrolases are clustered in acid or alkaline. The acid phosphohydrolases, unlike alkaline phosphatases, show optimal catalytic activity at acidic to neutral pH values. Moreover, they can be further classified as specific or nonspecific acid phosphatases, in relation to their substrate specificity. Due to abundance of their specific substrate, activity of acid phosphatases and phytases are predominant in soil [58]. Activity of organic-P solubilizing enzyme results in mineralization of P from organic phosphates, which would otherwise be lost due to precipitation with soil particle ions [59].

Bacterial nonspecific acid phosphatases (phosphohydrolases) (NSAPs) belong to molecular class A, B, and C [60]. These enzymes seem to function as organic phosphoester scavengers (release inorganic phosphates from nucleotides and sugar phosphates, for example, while the organic by-products are incorporated into the cell), providing the cell with essential nutrients [61]. A comprehensive review of bacterial nonspecific acid phosphohydrolases has been carried out by Rossolini et al. [62].

Most phytases (myo-inositol hexakisphosphate phosphohydrolases) are high molecular weight acid phosphatases. Its specific substrate phytate is the primary source of inositol and the major stored form of phosphate in plant seeds and pollen. Nevertheless, monogastric animals are incapable of using the P bound in phytate because their gastrointestinal tracts have low levels of phytase activity. Thus, nearly all the dietary phytate phosphorus ingested by these species is excreted, resulting in phosphorus pollution in areas of intensive animal production [58]. Microbes secreting the enzyme phases are ecologically exploited to mineralize phosphate from phytate. Developing agricultural inoculants with high phytase production would be of great interest for improving plant nutrition and reducing P pollution in soil. [13].

The specific phosphohydrolases with different activities include 39-nucleotidases and 59-nucleotidases [63]; hexose phosphatases [64]; and phytases [65]. A specific group of P releasing enzymes are those able to cleave C-P bonds from organophosphonates [66].

Some phosphohydrolases are secreted outside the plasma membrane, where they are either released in a soluble form or retained as membrane-bound proteins, which allows them to act as scavenging enzymes on organic phosphoesters that are components of high molecular weight material (i.e., RNA and DNA) and cannot cross the cytoplasmic membrane. The macromolecules are first converted to low molecular weight components, and this process may occur sequentially, i.e., the transformation of RNA and DNA to nucleoside monophosphate via RNase and DNase respectively, followed by the release of P and organic by-products via phosphohydrolases, providing the cell with essential nutrients [12].

5.3.4 Genetics of Organic Phosphate Solubilization

Severals authors have identified cloned sequences and expressed different genes or even set of genes that are involved in organic phosphate solubilization. Rossolini et al. [62], isolated and characterized several acid phosphatase genes from Gram-negative bacteria. Sequence of class A phosphatase genes isolated from *M. morganii* and *Pseudomonas stuartii* exhibited high degree of similarity suggesting their evolution from a common ancestor. Various conserved domains and a signature sequence motif for each family (A, B, and C) of bacterial phosphatases have been confirmed. Genes encoding NSAP have been isolated from different bacteria and characterized by several authors (Table 5.2).

Table 5.2 List of acid phosphatase genes and their source

Name of the acid phosphatase gene	*Source organism*	*Type of NSAP*	*Reference*
olpA	*Chryseobacterium meningosepticum* CCUG 4310	Class C	[67]
aphA -Ec	*Escherichia coli* MG1655	Class B	[68]
*pho*N	*Salmonella enterica serovar typhimurium*	Class A	[69]

Contd...

Contd...

*pho*C	*Morganella morganii*	Major phosphate-irrepressible Class A	[70]
*nap*B		Minor phosphate-irrepressible class B	
*pho*D	*Zymomonas mobilis*	Pi irrepressible NSAP	[71]
apo	*Pseudomonas fluorescens* MF3	Temperature dependent NSAP with maximal expression at 17.58°C.	[72]
apy	*Shigella flexneri*	Thermoregulated ATP diphosphohydrolase.	[73]

Genes like *ush*A, *agp and cpd*B have been isolated from *E. coli,* each encoding specific but different organic phosphates [63; 74]. A 43-kDa acid-hexose phosphatase gene from *Providencia stuartii* and *Providencia rettgeri* [75], and also a gene cluster involved in the synthesis of specific P-releasing enzyme from organophosphonate substrates (C-P lyase) [76] has been cloned and characterized. *acpA* gene isolated from *Francisella tularensis* expresses an acid phosphatase with optimum activity at pH 6, with a wide range of substrate specificity justifying its agricultural importance in plant growth promotion [77]. Genes *pho*C (encodes for class I NSAP), *nap*A (encodes for class I NSAP) isolated from *Morganella morganii* were successfully cloned, expressed and characterized [67; 78]. Rodriguez et al. [79] characterized a phosphate solubilizing gene from genomic library of a phosphate solubilizing *Burkholderia cepacia* strain. The expressed product was characterized and were inferred to be probably a porin that is specifically involved in the uptake of phosphorylated compounds and provides the periplasmic phosphatase with increased access to its substrate, which is normally external to the bacterium. Two nonspecific periplasmic acid phosphatase genes (*nap*D and *nap*E) from *Rhizobium (Sinorhizobium) meliloti* was cloned and the expressed proteins were characterized in detail. Both NapD and NapE exhibited phosphatase activity against several organic phosphates like p-Nitrophenol-phosphate, Phenyl-phosphate, Ribose-5-phosphate, D,L-Myo-inosito1-1-phosphate, D- Glucose- 6-phosphate and α-Naphthy-1-phosphate [80; 81]. These genes could be potential targets for genetic manipulation in soil phosphate-solubilizing bacteria.

The cleavage of the C-P bond from organophosphonates by phosphonoacetaldehyde hydrolase and C-P lyases is also inducible only under conditions of phosphate limitation [82]. However, production of the PhoN enzyme was reported to be moderately induced by Pi starvation [69]. Later, it was opined that this gene is under the control of the *phoP-phoQ* two-component regulatory system [83], which promotes transcription of phoN and other PhoP activated genes under low environmental Mg^{2+} concentrations [84]. *Morganella morganii* produces a high-level phosphate-irrepressible production of acid phosphatase (HPAP phenotype), a major phosphate-irrepressible class A acid phosphatase (named PhoC), which is associated with a minor acid phosphatase (named NapA) [70]. In contrary, *pho*D gene in *Zymomonas mobilis* is expressed constitutively [71]. Acid phosphatase production due to expression of apo gene in *Pseudomonas fluorescens* was regulated by the growth temperature with maximal expression at 17.58°C [72], while production of apyrase, an ATP diphosphohydrolase in *Shigella flexneri,* other related virulent *Shigella* spp. and enteroinvasive *E. coli* strains was reported to be expressed in a thermoregulated manner [73]. In *E. coli* MG1655, production of the p27 enzyme (probably a *nap*A gene product), a class B acid phosphatase appears to be switched off when cells were grown on glucose, and turned on when growth was supported by alternative carbon sources [63]. In vivo expression of the *E. coli agp* gene, which encodes a periplasmic acid glucose-1-phosphate phosphatase, was proposed to be positively regulated by cyclic adenosine monophosphate (cAMP) and the cAMP receptor protein (CRP) [85]. Also, acid phosphatase gene (*app*A) in *E. coli* is reported to be negatively controlled by cAMP at pH 2.5 [86].

Pattern of alkaline phosphatase activity in bacteria varies from one genera to another, particularly among members of the family Enterobacteriaceae. Enzyme activity is detectable only under specific environmental conditions since production of these enzymes is often controlled by complex regulatory mechanisms. Comprehensive understanding of attributes, regulation and physiological as well as ecological role of these enzymes is still lacking, even in the most thoroughly investigated *Escherichia coli* and *Salmonella typhimurium.*

Alkaline phosphatase is generally phosphate-repressible phosphatases, whose production is regulated by inorganic phosphate (Pi) concentration. This was established by an experiment in which alkaline phosphatase (gene *pho*A) in *E. coli,* was suddenly and fully induced when the Pi concentration decreased from 100 mM to 0.16 mM. A Pi transport operon as a regulatory element, in addition to sensor activator operon was also reported [87]. In PHO regulon, the genes are controlled by Pi and activated by *Pho*B [88]. Pi repressible bacterial alkaline phosphatase has also been characterized in *Morganella morganii* [67], *Pseudomonas fluorescens* MF3, *Providencia stuartii,* and *Providencia rettgeri* [60; 89].

It was observed that the sequence in the –35 region of phoC (encoding for an acid phosphatase of *Zymomonas mobilis*) was remarkably similar to that of the 'pho box' in *E. coli.* 12 of 18 bases were conserved in *Zymomonas mobilis* phoC, and five conserved bases in the –10 region were identical, from which it can be inferred that regulation of the expression of phosphatase genes in most of the members of family Enterobacteriaceae is similar to that of *E. coli* [90].

These findings indicate that most of the alkaline phosphatases found in the family Enterobacteriaceae are Pi-repressible, while many of the acid phosphatases are Pi-irrepressible. Existing knowledge about phosphatase from members of Enterobacteriaceae constitutes a basis for better understanding and further exploration of the control systems for phosphatase expression in soil bacteria.

Studying the molecular genomics of phytases is of immense importance for the role in animal nutrition as well as improving crop productivity by mineralizing organic phosphates [91]. Several authors cloned and characterized genes encoding for different type of phytase (Table 5.3).

Table 5.3 List of few reported phytase gene

Name of the phytase gene	*Source organism*	*Characteristic feature of phytase*	*Reference*
phy	*Bacillus* sp. DS11	Thermostable	[92]
phy C	*Bacillus subtilis* strain VTT E-68013	Maximal activity at 55°C. Requires calcium for its activity and/or stability and was readily inhibited by EDTA.	[93]
*app*A	*Escherichia coli K-12* (substrain: MG1655)	Bifunctional enzyme having both acid phosphatase and phytase activity	[94]
*app*A2	*Escherichia coli*	Bifunctional enzyme having both acid phosphatase and phytase activity	[95]
nphy	*Bacillus subtilis*	Neutral phytase	[96]
*phy*L	*Bacillus licheniformis*	Neutral phytase	[97]
*phy*A	*Bacillus amyloliquefaciens* FZB45	Contributes to plant growth promotion in low P input environment	[98]

Experimentation on phytase gene and its expression revealed the importance of phytase activity in stimulation of plant growth under limited P in soil. Hence, phytase genes can be biotechnologically used to manipulate the P-solubilizing trait to PGPB strains used as agricultural inoculants.

All of the available evidence indicates that the regulation of phosphatase enzymes is a complex matter that requires considerable additional research.

5.4 BIOTECHNOLOGICAL MANIPULATION OF P SOLUBILIZING GENE

Kim et al. [92] showed that mps (mineral phosphate solubilizing) gene isolated from *Ranella aquatilis* when cloned and expressed in *E. coli* exhibited higher gibberellic acid production and hydroxyapatite dissolution in comparison with the donor strain. Expression of a mps gene in a different host could be influenced by the genetic background of the recipient strain, the copy number of plasmids present, and metabolic interactions. Interestingly, gene involved in MPS could be transferred to induce or improve phosphate-dissolving capacity in PGPB strains. Attempt was made to improve MPS in PGPB strains by biotechnologically incorporating PQQ synthetase gene from *Erwinia herbicola* [99]. This gene, isolated by Goldstein and Liu [43], was subcloned in a broadhost range vector (pKT230) expressed in *E. coli* and was subsequently transferred to PGPB strains of *Burkholderia cepacia* and *Pseudomonas aeruginosa*, using tri-parental conjugation.

Several of the exconjugants that were recovered in the selection medium showed a larger clearing halo in medium with tricalcium phosphate as the sole P source indicating that heterologous expression of this gene in the recombinant strains gave rise to improved MPS ability of these PGPBs.

A bacterial citrate synthase gene, when cloned and expressed in tobacco, resulted in citrate overproduction in tobacco roots. This lead to increase exudation of organic acids and P availability to the plant, thereby increasing leaf and fruit biomass in plant, when grown under P-limiting conditions. It also required less P-fertilizer to achieve optimal growth. This shows the putative role of organic acid synthesis genes in P uptake in plants [100]. More recently, genomic integration of the pcc gene of *Synechococcus* PCC in *Pseudomonas fluorescens* 7942 allowed phosphate solubilization in the recipient strain [58].

CONCLUSION

Microbial phosphate solubilization in terrestrial environments promotes P-uptake, thereby facilitating plant growth promotion. Hence, detailed investigation of this event is essential since phosphate solubilizing microbes can be used as potential bioinoculants for improving plant health and yield attributes.

Proper knowledge of physiological response of PSB would facilitate to adopt and apply improved bioinoculant development strategies. Detailed knowledge of the genetic system of phosphate-solubilizing bacteria is essential for carrying out genetic manipulation using modern biological tools. This would result in improvement of their phosphate-solubilizing capabilities and/or the introduction of such trait in strains with other plant growth promoting effects.

The first step towards the biotechnological improvement of PS-bioinoculant would involve cloning of genes involved in mineral phosphate solubilization, such as those influencing the synthesis of organic acids, as well as phosphatase encoding genes. Subsequent subcloning of these genes in appropriate vectors and their transfer and expression (or overexpression) in target host strains would result in improvement of phosphate solubilization capabilities by the selected strains.

Stability and performance of the transformed strains would be required to be further tested once they are inoculated under plantation system. Finally, experimentation by genetic manipulation should be carried out to ensure chromosomal integration of the gene for higher stability of the character and minimize the possibility of horizontal transfer of the inserted gene in soil.

REFERENCES

1. Tilak, K.V.B.R., Ranganayaki, N., Pal, K.K., De, R., Saxena, A.K., Nautiyal, C.S. and Johri, B.N. (2005). "Diversity of plant growth and soil health supporting bacteria". *Curr. Sci.*, 89(1), 136–150.
2. Ahmed, N. and Shahab, S. (2011). "Phosphate solubilization: their mechanism, genetics and application". *The Internet J. Microbiol*, 9(1).

3. Dobbelaere, S., Vanderleyden, J. and Okon, Y. (2003). "Plant growth-promoting effects of diazotrophs in the rhizosphere". *Cri. Rev. Plant Sci.*, 22(2), 107–149.
4. Glick, B.R. (1995). "The enhancement of plant growth by free-living bacteria".*Can. J. Microbiol.*, 41(2), 109–117.
5. Kloepper, J.W. and Schroth, M.N. (1978, August). Plant growth-promoting rhizobacteria on radishes. In *Proceedings of the 4th International Conference on Plant Pathogenic Bacteria* (Vol. 2, pp. 879–882).
6. Subba Rao, N.S. (1993). *Biofertilizers in Agriculture and Forestry*. Oxford and IBH Publishing Co. Pvt. Ltd. New Delhi. pp. 242.
7. Goldstein, A.H. (1986). Bacterial solubilization of mineral phosphates: historical perspective and future prospects. *American Journal of Alternative Agriculture*, 1(02), 51–57.
8. Khan, M.S., Zaidi, A. and Wani P.A. (2007). Role of phosphate-solubilizing microorganisms in sustainable agriculture: A review. *Agronomy and Sustainable Development*. 27, 29–43.
9. Alam, S., Khalil, S., Ayub, N. and Rashid, M. (2002). In vitro solubilization of inorganic phosphate by phosphate solubilizing microorganisms (PSM) from maize rhizosphere. *Int J Agric Biol*, 4, 454–458.
10. Ezawa, T., Smith, S.E. and Smith, F.A. (2002). P metabolism and transport in AM fungi. *Plant and Soil*, 244 (1-2), 221–230.
11. Gyaneshwar, P., Kumar, G.N., Parekh, L.J. and Poole, P.S. (2002). Role of soil microorganisms in improving P nutrition of plants. In: *Food Security in Nutrient: Stressed Environments: Exploiting Plants' Genetic Capabilities* (pp. 133-143). Springer Netherlands.
12. Goldstein, A.H. (1994). Involvement of the quinoprotein glucose dehydrogenase in the solubilization of exogenous phosphates by Gram-negative bacteria. *Phosphate in Microorganisms: Cellular and Molecular Biology*. ASM Press, Washington, DC, 197–203.
13. Richardson, A.E., Hadobas, P.A. and Hayes, J.E. (2001). Extracellular secretion of Aspergillus phytase from Arabidopsis roots enables plants to obtain phosphorus from phytate. *The Plant Journal*, 25(6), 641–649.
14. Chen, Y.P., Rekha, P.D., Arun, A.B., Shen, F.T., Lai, W.A. and Young, C.C. (2006). Phosphate solubilizing bacteria from subtropical soil and their tricalcium phosphate solubilizing abilities. *Applied Soil Ecology*, 34(1), 33–41.
15. Igual, J.M., Valverde, A., Cervantes, E. and Velázquez, E. (2001). Phosphate-solubilizing bacteria as inoculants for agriculture: use of updated molecular techniques in their study. *Agronomie*, 21(6-7), 561–568.
16. Subbarao, N.S. (1988). Phosphate solubilizing microorganism. *Biofertilizer in Agriculture and Forestry*. Regional Biofertilizer Development Centre, Hissar, India, 133–142.
17. Kucey, R.M.N., Janzen, H.H. and Leggett, M.E. (1989). Microbially mediated increases in plant-available phosphorus. *Advances in Agronomy*, 42, 199–228.

18. Kim, K.Y., Jordan, D. and McDonald, G.A. (1997). Effect of phosphate-solubilizing bacteria and vesicular-arbuscular mycorrhizae on tomato growth and soil microbial activity. *Biology and Fertility of Soils*, 26(2), 79–87.
19. Vazquez, P., Holguin, G., Puente, M.E., Lopez-Cortes, A. and Bashan, Y. (2000). Phosphate-solubilizing microorganisms associated with the rhizosphere of mangroves in a semiarid coastal lagoon. *Biology and Fertility of Soils*, 30(5-6), 460–468.
20. Khan, K.S. and Joergensen, R.G. (2009). Changes in microbial biomass and P fractions in biogenic household waste compost amended with inorganic P fertilizers. *Bioresource Technology*, 100(1), 303–309.
21. Kang, S.C., Hat, C.G., Lee, T.G. and Maheshwari, D.K. (2002). Solubilization of insoluble inorganic phosphates by a soil-inhabiting fungus Fomitopsis sp. PS 102. *Current Science*, 82(4), 439–442.
22. Pradhan, N. and Sukla, L.B. (2009). Solubilization of inorganic phosphates by fungi isolated from agriculture soil. *African Journal of Biotechnology*, 5(10).
23. Bünemann, E.K., Bossio, D.A., Smithson, P.C., Frossard, E. and Oberson, A. (2004). Microbial community composition and substrate use in a highly weathered soil as affected by crop rotation and P fertilization. *Soil Biology and Biochemistry*, 36(6), 889–901.
24. Jilani, G., Akram, A., Ali, R.M., Hafeez, F.Y., Shamsi, I.H., Chaudhry, A.N. and Chaudhry, A.G. (2007). Enhancing crop growth, nutrients availability, economics and beneficial rhizosphere microflora through organic and biofertilizers. *Annals of Microbiology*, 57(2), 177–184.
25. Yazdani, M., Bahmanyar, M.A., Pirdashti, H. and Esmaili, M.A. (2009). Effect of Phosphate solubilization microorganisms (PSM) and plant growth promoting rhizobacteria (PGPR) on yield and yield components of Corn (Zea mays L.). *Proc World Acad Sci Eng Technol*, 37, 90–92.
26. Halder, A.K., Mishra, A.K., Bhattacharyya, P. and Chakrabartty, P.K. (1990). Solubilization of rock phosphate by Rhizobium and Bradyrhizobium. *Journal of General and Applied Microbiology*, 36(2), 81–92.
27. Puente, M.E., Bashan, Y., Li, C.Y. and Lebsky, V.K. (2004). Microbial populations and activities in the rhizoplane of rock-weathering desert plants. I. Root colonization and weathering of igneous rocks. Plant Biology, 6(5), 629–642.
28. Illmer, P. and Schinner, F. (1992). Solubilization of inorganic phosphates by microorganisms isolated from forest soils. Soil Biology and Biochemistry, 24(4), 389–395.
29. Liu, S.T., Lee, L.Y., Tai, C.Y., Hung, C.H., Chang, Y.S., Wolfram, J.H., Rogers, R. and Goldstein, A.H. (1992). Cloning of an *Erwinia herbicola* gene necessary for gluconic acid production and enhanced mineral phosphate solubilization in *Escherichia coli* HB101: nucleotide sequence and probable involvement in biosynthesis of the coenzyme pyrroloquinoline quinone. *Journal of Bacteriology*, 174(18), 5814–5819.

30. Goldstein, A.H., Rogers, R.D. and Mead, G. (1993). Mining by microbe. *Nature Biotechnology*, 11(11), 1250–1254.

31. Song, O.R., Lee, S.J., Lee, Y.S., Lee, S.C., Kim, K.K. and Choi, Y.L. (2008). Solubilization of insoluble inorganic phosphate by Burkholderia cepacia DA23 isolated from cultivated soil. *Brazilian Journal of Microbiology*, 39(1), 151–156.

32. Halder, A.K., Mishra, A.K., Bhattacharyya, P. and Chakrabartty, P.K. (1990). Solubilization of rock phosphate by Rhizobium and Bradyrhizobium. *Journal of General and Applied Microbiology*, 36(2), 81–92.

33. Halder, A.K. and Chakrabartty, P.K. (1993). Solubilization of inorganic phosphate by Rhizobium. *Folia Microbiologica*, 38(4), 325–330.

34. Banik, S. and Dey, B.K. (1982). Available phosphate content of an alluvial soil as influenced by inoculation of some isolated phosphate-solubilizing micro-organisms. *Plant and Soil*, 69(3), 353–364.

35. Duff, R.B. and Webley, D.M. (1959). 2-Ketogluconic acid as a natural chelator produced by soil bacteria. *Chemistry and Industry*, 1376–1377.

36. Behera, B.C., Singdevsachan, S.K., Mishra, R.R., Dutta, S.K. and Thatoi, H.N. (2014). Diversity, mechanism and biotechnology of phosphate solubilising microorganism in mangrove—A review. *Biocatalysis and Agricultural Biotechnology*, 3(2), 97–110.

37. Zaidi, A., Khan, M.S., Ahemad, M. and Oves, M. (2009). Plant growth promotion by phosphate solubilizing bacteria. *Acta Microbiologica et Immunologica Hungarica*, 56(3), 263–284.

38. Goldstein, A.H. (1995). Recent progress in understanding the molecular genetics and biochemistry of calcium phosphate solubilization by gram negative bacteria. *Biological Agriculture and Horticulture*, 12(2), 185–193.

39. Gaur, A.C. and Gaind, S. (1999). Phosphate solubilizing microorganisms: An overview. *Current Trends in Life Sciences*, Today and Tomorrows Publishers, New Delhi.

40. Hilda, R., Gonzalez, T. and Selman, G. (2000). Expression of a mineral phosphate solubilizing gene fromErwina herbicolain two rhizobacterial strains. *J. Biotechnol.* 84, 155–161.

41. Rodriguez, H., Gonzalez, T., Goire, I. and Bashan, Y. (2004). Gluconic acid production and phosphate solubilization by the plant growth-promoting bacterium *Azospirillum* spp. *Naturwissenschaften*, *91*(11), 552–555.

42. Rodrìguez, H. and Fraga, R. (1999). Phosphate solubilizing bacteria and their role in plant growth promotion. *Biotechnology Advances*, 17(4), 319–339.

43. Goldstein, A.H. and Liu, S.T. (1987). Molecular cloning and regulation of a mineral phosphate solubilizing gene from *Erwinia herbicola. Nature Biotechnology*, 5(1), 72–74.

44. Misra, H.S., Rajpurohit, Y.S. and Khairnar, N.P. (2012). Pyrroloquinoline-quinone and its versatile roles in biological processes. *Journal of Biosciences*, 37(2), 313–325.

45. Felder, M., Gupta, A., Verma, V., Kumar, A., Qazi, G.N. and Cullum, J. (2000). The *pyrroloquinoline quinone* synthesis genes of *Gluconobacter oxydans*1. *FEMS Microbiology Letters*, 193(2), 231–236.

46. Goldstein, A., Lester, T. and Brown, J. (2003). Research on the metabolic engineering of the direct oxidation pathway for extraction of phosphate from ore has generated preliminary evidence for PQQ biosynthesis in *Escherichia coli* as well as a possible role for the highly conserved region of quinoprotein dehydrogenases. *Biochimica et Biophysica Acta (BBA)-Proteins and Proteomics,* 1647(1), 266–271.

47. Goosen, N., Huinen, R.G. and Van de Putte, P. (1992). A 24-amino-acid polypeptide is essential for the biosynthesis of the coenzyme pyrrolo-quinoline-quinone. *Journal of Bacteriology,* 174(4), 1426–1427.

48. Babu-Khan, S., Yeo, T.C., Martin, W.L., Duron, M.R., Rogers, R.D. and Goldstein, A.H. (1995). Cloning of a mineral phosphate-solubilizing gene from *Pseudomonas cepacia. Applied and Environmental Microbiology,* 61(3), 972–978.

49. Krishnaraj, P.U. and Goldstein, A.H. (2001). Cloning of a Serratia marcescens DNA fragment that induces quinoprotein glucose dehydrogenase-mediated gluconic acid production in *Escherichia coli* in the presence of stationary phase *Serratia marcescens. FEMS Microbiology Letters,* 205(2), 215–220.

50. Kim, K.Y., McDonald, G.A. and Jordan, D. (1997). Solubilization of hydroxyapatite by *Enterobacter agglomerans* and cloned *Escherichia coli* in culture medium. *Biology* and *Fertility of Soils,* 24(4), 347–352.

51. Illmer, P. and Schinner, F. (1995). Solubilization of inorganic calcium phosphates—solubilization mechanisms. *Soil Biology and Biochemistry,* 27(3), 257–263.

52. Buch, A., Archana, G. and Naresh Kumar, G. (2010). Heterologous expression of phosphoenolpyruvate carboxylase enhances the phosphate solubilizing ability of fluorescent pseudomonads by altering the glucose catabolism to improve biomass yield. *Bioresource Technology,* 101(2), 679–687.

53. Rodrìguez, H. and Fraga, R. (1999). Phosphate solubilizing bacteria and their role in plant growth promotion. *Biotechnology Advances,* 17(4), 319–339.

54. Halder, A.K., Mishra, A.K., Bhattacharyya, P. and Chakrabartty, P.K. (1990). Solubilization of rock phosphate by *Rhizobium* and *Bradyrhizobium. Journal of General and Applied Microbiology,* 36(2), 81–92.

55. Mikanova, O., Kubat, J., Simon, T., Vorisek, K. and Randova, D. (1997). Influence of soluble phosphate on P-solubilizing activity of bacteria. *Rostlinna Vyroba-UZPI (Czech Republic),* 43, 421–424.

56. McGrath, J.W., Wisdom, G.B., McMullan, G., Larkin, M.J. and Quinn, J.P. (1995). The Purification and Properties of Phosphonoacetate Hydrolase, a Novel Carbon-Phosphorus Bond-Cleavage Enzyme from Pseudomonas Fluorescens 23F. *European Journal of Biochemistry,* 234(1), 225–230.

57. Paul, E.A. (2006). *Soil Microbiology, Ecology and Biochemistry.* Academic Press.

58. Rodriguez, H., Fraga, R., Gonzalez, T. and Bashan, Y. (2006). Genetics of phosphate solubilization and its potential applications for improving plant growth-promoting bacteria. *Plant and Soil,* 287(1-2), 15–21.

59. Tarafdar, J.C. and Claassen, N. (1988). Organic phosphorus compounds as a phosphorus source for higher plants through the activity of phosphatases produced by plant roots and microorganisms. *Biology and Fertility of Soils*, 5(4), 308–312.

60. Thaller, M.C., Berlutti, F., Schippa, S., Iori, P., Passariello, C. and Rossolini, G.M. (1995). Heterogeneous patterns of acid phosphatases containing low-molecular-mass polypeptides in members of the family *Enterobacteriaceae*. *International journal of Systematic Bacteriology*, 45(2), 255–261.

61. Wanner, B.L. (1996). Phosphorus assimilation and control of the phosphate regulon. *Escherichia coli and Salmonella Cellular and Molecular Biology*,. Eds. FC Niedhardt, R Curtiss III, JL Ingraham, EC Lin, K.B. Low, B Magasanik, WS Reznikoff, M Riley, M Schaechter and HE Umbarger. 2nd edition, 1ASM Press, Washington, DC. pp. 1357–1381.

62. Rossolini, G.M., Schippa, S., Riccio, M.L., Berlutti, F., Macaskie, L.E. and Thaller, M.C. (1998). Bacterial nonspecific acid phosphohydrolases: physiology, evolution and use as tools in microbial biotechnology. *Cellular and Molecular Life Sciences CMLS*, 54(8), 833–850.

63. Burns, D.M. and Beacham, I.R. (1986). Nucleotide sequence and transcriptional analysis of the *E. coli* ushA gene, encoding periplasmic UDP-sugar hydrolase (5′-nucleotidase): regulation of the ushA gene, and the signal sequence of its encoded protein product. *Nucleic Acids Research*, 14(10), 4325–4342.

64. Pradel, E. and Boquet, P.L. (1988). Acid phosphatases of *Escherichia coli*: molecular cloning and analysis of agp, the structural gene for a periplasmic acid glucose phosphatase. *Journal of Bacteriology*, 170(10), 4916–4923.

65. Cosgrove, D.J., Irving, G.C.J. and Bromfield, S.M. (1970). Inositol phosphate phosphatases of microbiological origin. The isolation of soil bacteria having inositol phosphate phosphatase activity. *Australian Journal of Biological Sciences*, 23(2), 339–344.

66. Bujacz, B., Wieczorek, P., Krzysko-Lupicka, T., Golab, Z., Lejczak, B. and Kavfarski, P. (1995). Organophosphonate utilization by the wild-type strain of *Penicillium notatum*. *Applied and Environmental Microbiology*, 61(8), 2905–2910.

67. Thaller, M.C., Berlutti, F., Schippa, S., Lombardi, G. and Rossolini, G.M. (1994). Characterization and sequence of PhoC, the principal phosphate-irrepressible acid phosphatase of Morganella morganii. *Microbiology*, 140(6), 1341–1350.

68. Rossolini, G.M., Thaller, M.C., Pezzi, R. and Satta, G. (1994). Identification of an Escherichia coli periplasmic acid phosphatase containing of a 27 kDa-polypeptide component. *FEMS Microbiology Letters*, 118(1-2), 167–173.

69. Kier, L.D., Weppelman, R.O.G.E.R. and Ames, B.N. (1977). Resolution and purification of three periplasmic phosphatases of Salmonella typhimurium. *Journal of Bacteriology*, 130(1), 399–410.

70. Pompei, R., Ingianni, A., Foddis, G., Di Pietro, G. and Satta, G. (1993). Patterns of phosphatase activity among enterobacterial species. *International Journal of Systematic Bacteriology*, 43(1), 174–178.

71. Baoudene-Assali, F., Baratti, J. and Michel, G.P. (1993). Purification and properties of a phosphate-irrepressible membrane-bound alkaline phosphatase from *Zymomonas mobilis*. *Journal of General Microbiology*, 139(2), 229–235.
72. Burini, J.F., Gügi, B., Merieau, A. and Guespin-Michel, J.F. (1994). Lipase and acidic phosphatase from the psychrotrophic bacterium *Pseudomonas fluorescens*: two enzymes whose synthesis is regulated by the growth temperature. *FEMS Microbiology Letters*, 122(1-2), 13–18.
73. Bhargava, T., Datta, S., Ramachandran, V., Ramakrishnan, R., Roy, R.K., Sankaran, K. and Subrahmanyam, Y.V.B.K. (1995). Virulent Shigella codes for a soluble apyrase: identification, characterization and cloning of the gene. *Current Science*, 68(3), 293–300.
74. Beacham, I.R. and Garrett, S. (1980). Isolation of *Escherichia coli* mutants (cpdB) deficient in periplasmic 2′: 3′-cyclic phosphodiesterase and genetic mapping of the cpdB locus. *Journal of General Microbiology*, 119(1), 31–34.
75. Riccio, M.L., Rossolini, G.M., Lombardi, G., Chiesurin, A. and Satta, G. (1997). Expression cloning of different bacterial phosphataseencoding genes by histochemical screening of genomic libraries onto an indicator medium containing phenolphthalein diphosphate and methyl green. *Journal of Applied Microbiology*, 82(2), 177–185.
76. Ohtake, H., Wu, H., Imazu, K., Anbe, Y., Kato, J. and Kuroda, A. (1996). Bacterial phosphonate degradation, phosphite oxidation and polyphosphate accumulation. *Resources, Conservation and Recycling*, 18(1), 125–134.
77. Reilly, T.J., Baron, G.S., Nano, F.E. and Kuhlenschmidt, M.S. (1996). Characterization and sequencing of a respiratory burst-inhibiting acid phosphatase from *Francisella tularensis*. *Journal of Biological Chemistry*, 271(18), 10973–10983.
78. Maria, C.T., Lombardi, G., Berlutti, F., Schippa, S. and Gian, M.R. (1995). Cloning and characterization of the NapA acid phosphatase/phosphotransferase of *Morganella morganii*: identification of a new family of bacterial acid-phosphatase-encoding genes. *Microbiology*, 141(1), 147–154.
79. Rodríguez, H., Rossolini, G.M., Gonzalez, T., Li, J. and Glick, B.R. (2000). Isolation of a gene from *Burkholderia cepacia* IS-16 encoding a protein that facilitates phosphatase activity. *Current Microbiology*, 40(6), 362–366.
80. Deng, S., Summers, M.L., Khan, M.L. and McDermott, T.R. (1998). Cloning and characterization of a Rhizobium meliloti nonspecific acid phosphatase. *Archives of Microbiology*, 170(1), 18–26.
81. Deng, S., Elkins, J.G., Da, L.H., Botero, L.M. and McDermott, T.R. (2001). Cloning and characterization of a second acid phosphatase from Sinorhizobium meliloti strain 104A14. *Archives of microbiology*, 176(4), 255–263.
82. Kertesz, M., Elgorriaga, A. and Amrhein, N. (1991). Evidence for two distinct phosphonate-degrading enzymes (CP lyases) in *Arthrobacter* sp. GLP-1. *Biodegradation*, 2(1), 53–59.

83. Miller, S.I., Kukral, A.M. and Mekalanos, J.J. (1989). A two-component regulatory system (phoP phoQ) controls *Salmonella typhimurium* virulence. *Proceedings of the National Academy of Sciences*, 86(13), 5054–5058.

84. Véscovi, E.G., Soncini, F.C. and Groisman, E.A. (1996). Mg^{2+} as an Extracellular Signal: Environmental Regulation of Salmonella Virulence. *Cell*, 84(1), 165–174.

85. Pradel, E., Marck, C. and Boquet, P.L. (1990). Nucleotide sequence and transcriptional analysis of the *Escherichia coli* agp gene encoding periplasmic acid glucose-1-phosphatase. *Journal of Bacteriology*, 172(2), 802–807.

86. Touati, E. and Danchin, A. (1987). The structure of the promoter and amino terminal region of the pH 2.5 acid phosphatase structural gene *appA* of *E. coli*: a negative control of transcription mediated by cyclic AMP. *Biochimie*, 69(3), 215–221.

87. Rosenberg H. (1987). Phosphate transport in prokariotes. In: Rosen B, Silver S, editors. Ion *Transport in Prokaryotes*. San Diego, CA: Academic Press, London, pp. 205–248.

88. Torriani-Gorini A. (1994). Regulation of phosphate metabolism and transport. In: Torriani-Gorini A., Yagil E., Silver S., editors. *Phosphate in Microorganisms: Cellular and Molecular Biology*. Washington, DC: ASM Press. pp. 1–4.

89. Gügi, B., Orange, N., Hellio, F., Burini, J.F., Guillou, C., Leriche, F. and Guespin-Michel, J.F. (1991). Effect of growth temperature on several exported enzyme activities in the psychrotrophic bacterium *Pseudomonas fluorescens*. *Journal of Bacteriology*, 173(12), 3814–3820.

90. Pond, J.L., Eddy, C.K., Mackenzie, K.F., Conway, T., Borecky, D.J. and Ingram, L.O. (1989). Cloning, sequencing, and characterization of the principal acid phosphatase, the phoC+ product, from Zymomonas mobilis. *Journal of Bacteriology, 171*(2), 767–774.

91. Richardson, A.E., Pankhurst, C.E., Doube, B.M., Gupta, V.V.S.R. and Grace, P.R. (1994). Soil microorganisms and phosphorus availability. *Soil Biota: Management in Sustainable Farming Systems.*, 50–62.

92. Kim, Y.O., Lee, J.K., Kim, H.K., Yu, J.H. and Oh, T.K. (1998). Cloning of the thermostable phytase gene (phy) from Bacillus sp. DS11 and its overexpression in *Escherichia coli*. *FEMS Microbiology Letters*, 162(1), 185–191.

93. Kerovuo, J., Lauraeus, M., Nurminen, P., Kalkkinen, N. and Apajalahti, J. (1998). Isolation, characterization, molecular gene cloning, and sequencing of a novel phytase from *Bacillus subtilis*. *Applied and Environmental Microbiology*, 64(6), 2079–2085.

94. Golovan, S., Wang, G., Zhang, J. and Forsberg, C.W. (1999). Characterization and overproduction of the *Escherichia coli* appA encoded bifunctional enzyme that exhibits both phytase and acid phosphatase activities. *Canadian Journal of Microbiology*, 46(1), 59–71.

95. Rodriguez, E., Han, Y. and Lei, X.G. (1999). Cloning, Sequencing, and Expression of an *Escherichia coli* Acid Phosphatase/Phytase Gene *app*A2 Isolated from Pig Colon. *Biochemical and Biophysical Research Communications*, 257(1), 117–123.

96. Yao, B., Yuan, T.Z., Wang, Y.H., Cao, S.S., Wang, Y.R., Shi, X.Y. and Fan, Y.L. (2001). Cloning of neutral phytase gene *nphy* from *Bacillus subtilis* and its expression in *Escherichia coli*. *Chinese Journal of Biotechnology*, 17(1), 11–15.

97. Tye, A., Siu, F., Leung, T. and Lim, B. (2002). Molecular cloning and the biochemical characterization of two novel phytases from *B. subtilis* 168 and *B. licheniformis*. *Applied Microbiology and Biotechnology*, 59(2-3), 190–197.

98. Idriss, E.E., Makarewicz, O., Farouk, A., Rosner, K., Greiner, R., Bochow, H. and Borriss, R. (2002). Extracellular phytase activity of *Bacillus amyloliquefaciens* FZB45 contributes to its plant-growth-promoting effect. *Microbiology*, 148(7), 2097–2109.

99. Rodrìguez, H., Gonzalez, T. and Selman, G. (2000). Expression of a mineral phosphate solubilizing gene from *Erwinia herbicola* in two rhizobacterial strains. *Journal of Biotechnology*, 84(2), 155–161.

100. López-Bucio, J., de la Vega, O.M., Guevara-García, A. and Herrera-Estrella, L. (2000). Enhanced phosphorus uptake in transgenic tobacco plants that overproduce citrate. *Nature Biotechnology*, 18(4), 450–453.

6

Potential use of Rhizobacteria as Biofertilizer and Its Role in Increasing Tolerance to Drought Stress

Sonali Pandey, Anupama Verma and Dipjyoti Chakraborty*

Department of Bioscience and Biotechnology, Banasthali Vidyapith, Rajasthan – 304022, India

***Corresponding Author E-mail:** dciitkgp@gmail.com*

ABSTRACT

Drought and increasing salinity has significant and harmful impacts on land, water, vegetation quality, wildlife environments, agronomy and ecosystem especially for arid and semiarid areas where high evapo-transpiration rates expose plants to further adaptive pressure. Arid and semiarid areas occupy an increasing fraction of the earth's surface. Agriculture in semiarid regions requires the implementation of methods to balance nutrients and to conserve soil organic matter. Plant growth promoting rhizobacteria play a major role in increasing crop yield and they have additionally been implicated in providing enhanced stress tolerance both biotic and abiotic. There is an urgent need for research in microbial traits that are necessary and useful for proper utilization as biofertilizers in crop plants for specific microenvironments. Standardization of effective formulations, survivability and effectiveness under field conditions, compatibility with chemical treatment of seeds and mineral fertilizers and appropriate technology for applications need further research and development.

Keywords: Biofertilizer, Drought Stress, Plant-Microbe Interaction, Plant Growth Promoting Rhizobacteria, Sustainable Agriculture.

6.1 INTRODUCTION

6.1.1 Plant Stress

Plants are under constant biotic and abiotic stress that is expressed by a altered physiological state interrupting metabolic patterns resulting in aberrant physiology. In biological systems, stress is essentially an adverse force that prohibits the normal growth and developmental activities [1]. In plants, it is generally manifested by decrease in photosynthesis, growth and yield. Stress is broadly divided into two classes based on the causal factors, viz., those that are imposed by living organisms such as virus, bacteria, fungi, etc., are known as biotic stress; those that are caused due to non- living surroundings and environment such as drought, radiation, salt, ozone, heat, cold, oxidative stress, heavy metal toxicity, etc., are put under 'abiotic stress.

Plants are subjected at any given point of time to many of these stress factors generally in combination, adversely affecting their growth and productivity. Excessive production of ROS (reactive oxygen species) in plants is due to stress. These stress induced species (reduced or activated derivatives of O_2) contain free radicals such as OH^-, O^-_2, OH_2 and non-radical molecules such as H_2O_2 and cause different types of alterations in plants such as cellular damage, metabolic disorders and senescence processes.

6.1.1.1 Drought stress

Drought is a meteorological term and occurs due to water deficit condition. Although water stress also involves conditions of excess water, water deficit is of more common occurrence. Lower water availability than optimum occurs due to decreased rainfall, reduction in soil water availability and continuous loss of water by transpiration and evaporation, atmospheric conditions caused due to environmental factors. Decreased water content, loss of turgor, lower leaf water potential and closure of stomata, decreased cell enlargement and growth are the main characters of drought affected plants.

Metabolic pathways, physiological and biochemical processes especially photosynthesis, respiration, ion uptake, translocation, nutrient metabolism are adversely affected by drought stress which may also lead to plant death under severe conditions [2-7]. Cell elongation is more affected by drought than cell division. Drought disrupts the bilayer structure of membrane and creates porosity during desiccation. It reduces the activity of cytosolic and organelle proteins. Severe water deficiency may lead to total denaturation of these proteins. Dehydration of protoplasm produces cellular electrolytes and these electrolytes cause disruption of cellular metabolism.

6.1.2 Legume Plants, PGPRs

Plant species of the Leguminosae family form symbiotic relationship (symbiosis is defined as nodule formation by bacteria on roots and stems in legume crops) with N_2-fixing bacteria (nodule inducing bacteria), also known as *Rhizobium*. About 20,000 legume species form this type of relationship. Seeds of these species are highly nutritive, particularly rich in protein and oil content. Humans and animals depend on legume crops to fulfil their proteins and calorie requirements. These are an important component of traditional diet. Because of their importance, in many countries about 20-57% of their cultivable land has different varieties of legumes, i.e., Bambara groundnuts (*Vigna subterranea*), black gram (*Vigna mungo*), peas (*Pisum sativum*), green gram (*Vigna radiata*) and beans (*Phaseolus vulgaris*), etc. Production of legumes is also beneficial for the land because it increases soil organic matter, improves soil structure, porosity of the land, recycle nutrients, decreases soil pH, diversifies the rhizosphere microbes and breaks disease build up. Therefore, it is necessary to increase production of legumes. Application of chemical fertilizers is widespread to increase plant productivity, but these are pricey (available to farmers at hugely subsidized rates) and adversely affect the environment. Due to this, use of synthetic chemicals to increase the productivity of crop plants is slowly being phased out. To increase the yield, inoculation of different plant growth promoting microorganisms (PGPMs) either alone or in combination has been practiced.

Bacteria forming nitrogen fixation symbiosis with legumes are soil-borne bacteria. These bacteria reside in rhizosphere and form colony either on the root surface i.e. exo-root or penetrate into the root cortex, i.e., endo-root [8]. Process of nitrogen fixation occurs in the nodules where bacteria form bacteroids and fix atmospheric nitrogen.

6.1.2.1 Symbiosis

First pure bacterial culture from nodule suspension was obtained by Beijerinck (1888) and this culture facilitated the nodule formation in *Pisum* as well as in *Vicia;* this was later on called *Rhizobium leguminosarum* [9-10]. According to Bergey's Manual of Determinative bacteriology, all the bacteria that are capable to form nodules in legumes are placed in the family Rhizobiaceae [10]. *R. leguminosarum, R. meliloti, R. phaseoli, R. trifoli* are fast growing species and *R. japonicum, R. lupine* are slow growing species. There are many mechanisms to establish symbiotic relationship. Symbiotic relationship rely on various factors such as competitiveness, effectiveness, specificity and infectivity of the bacteria, and a complex signaling network that link the host and symbiont establishes symbiotic relationship. Quorum sensing is a key component of signalling network used to define the concentration of bacteria, for example, in and around the root and in plant root nodule. These microorganisms also support the plants to increase their tolerance against abiotic stress such as water deficit, low and high temperature, salt stress, injury, metal toxicity, etc. This view is supported by the concept that *Rhizobium, Bradyrhizobium, Azotobactor, Azospirillum, Bacillus, Pseudomonas,* etc. are found under different climatic conditions in saline and alkaline areas, desert ecosystems, highly eroded hill slopes and acid soils of India [11-12].

6.1.2.2 Rhizobacteria as potential biofertiliser

Rhizobacteria are rhizosphere-competent and form colonies on roots, generally found in three forms, i.e., free-living, parasitic and saprophytic. These relationships are beneficial for plant in different terms as growth, crop quality and productivity. Kloepper and Schroth (1978) first defined PGPRs (plant growth promoting rhizobacteria) which include different strains of various genera *Aeromonas, Acinetobacter, Alcaligenes, Azospirillium, Arthrobacter, Azotobacter, Bacillus, Beijerinckia, Azoarcus, Burkholderia, Clostridium, Erwinia, Enterobacter, Gluconacetobacter, Flavobacterium, Klebsiella, Pseudomonas, Rhizobium* and *Serratia,* etc. [13-19]. On the basis of plant-PGPR relationship, PGPRs are classified into symbiotic and free-living bacteria [20]. Many PGPRs are endophytic [21-22] and further differentiated into intra-cellular and extra-cellular endophytic PGPR (iPGPR and ePGPR). iPGPRs form nodules because they can easily penetrate in plant cells. On the other hand, ePGPRs are never found inside the cells; they reside in the rhizosphere and on the rhizoplane. There is a mutual relationship between plants and PGPRs. Plant exudates act as nutrients for PGPRs, i.e., fatty acids, amino acids, nucleotides, phenolic compounds, organic acids, plant growth regulators, sterols, putrescine, vitamins and sugars [23]. PGPR benefit plant through nitrogen assimilation, fixation and via solubilization of organic phosphates and inorganic phosphates. Direct and indirect interactions between PGPRs and plant root generally increase the nutrient absorption from the soil and also increase its availability

[24], limiting soil-borne pathogens, producing phytohormones [25] and removing pollutants that negatively affect plants [26]. The term biofertilizer refers to preparation containing live microbes which help in enhancing the soil fertility either by fixing atmospheric nitrogen, solubilization of phosphorus or decomposing organic wastes or by augmenting plant growth by producing growth hormones with their biological activities and includes several PGPRs.

6.2 REVIEW

Drought stress widely affects agricultural crops and reduces yield in terms of fresh and dry weight [27]. Reduced yield is observed in *Poncirus trifoliatae* seedlings [28], soya bean [29], *Petroselinum cripsum* [30], common bean and in green gram [31] against drought stress. Rhizobial inoculation is a preferred method to increase productivity in legumes. Different bacterial genera such as *Bacillus, Rhizobium, Pseudomonas, Paenibacillus, Pantoea, Burkholderia, Azospirillum, Achromobacter, Microbacterium, Variovorax, Enterobacter, Methylobacterium,* etc. are noticed under different abiotic stress conditions and support plant to resist the stresses. These bacteria can also be used as potential biofertilizer.

6.2.1 Plant Response to Drought

At the early stage of plant growth and development drought stress is very harmful to plants. Decreased water content adversely affects growth and development of plants. Some crops are very susceptible to drought in comparison to others such as rice (a submerged crop). Drought stress is expressed in several forms in the plant affecting all plant parts and has severe morphological, physiological and biochemical implications [29, 32, 33].

6.2.1.1 Leaves and roots in drought

In comparison to root growth, leaf growth is more sensitive to drought. Optimal leaf area is a key factor that influences the productivity of plants. The effect of drought is reflected by the reduced leaf size in many plants such as soya bean [34] and *Populus* [35]. Reduction in leaf area supports the plant to tolerate drought stress because of decreased transpiration. Generally, prolific root system accelerates growth at the initial phase by extracting the water from shallow soil layers which may be lost via evaporation process [36]. In maize and wheat there is no significant reduction observed in root growth against drought stress condition [37]. Increased root growth is reported in sunflower and *Catharanthus roseus* against water deficit [37,38]. Higher productivity is the main goal of growing crops. Water deficit during bud initiation in comparison to grain filling stage is more harmful for plants, and finally reduces yield [39]. In accordance, drought reduces seed yield in soya bean [29] because of reduction in pod number and seeds per unit area.

6.2.1.2 Stomatal response to drought

Closure of stomata is the first response of plants against drought stress which prevents further water loss due to decrease in transpiration [40]. There are two processes, metabolic dependent and metabolic independent, for the closure of stomata. Hydro passive closure of stomata occurs when water evaporates directly from guard cells without any metabolic involvement. On the other hand, hydroactive closure of stomata is metabolically dependent [41] and requires ions and metabolites. ABA is an important regulator of this process. ABA, ethylene and cytokinins are involved in root-shoot signalling. Basically two pathways. ABA dependent and ABA independent, are engaged to control osmotic stress caused by drought and high salt. Genetic studies suggest that both of these pathways crosstalk with each other [42].

6.2.1.3 Photosynthetic machinery in drought

Drought and salinity influence primary processes of plant such as photosynthesis and cell growth [43]. Generally, decreased content of Chl a, Chl b and carotenoids are noticed against water deficit conditions in plants [27] such as in cotton and *Catharanthus roseus* [44]. Oxidative stresses that occur due to several stress conditions [45] harm the photosynthesis process [46]. Drought and salinity limit mesophyll transport of CO_2 and it accumulates into the chloroplast as shoot and root produce hormones that are responsible for closure of stomata and finally change photochemical processes and carbon metabolism of leaf. The adverse effect of the stress depends on the strength and period of strain, species and age of the plant [47-49]. Decrease in stomatal conductance during mild stress condition allows plant water saving and improve water use efficiency to protect against stress. It is observed that salinity [50] and drought stress [49,51] decrease reduced mesophyll conductance to CO_2. To prevent irreversible damage of photosynthetic machinery, plants respond quickly to drought stress. Closure of stomata is the first response of plants to prevent this damage. In severe drought stress, decreased Rubisco activity is observed that ultimately results in limited photosynthesis [52]. Studies suggest that decrease in photosynthesis activity primarily occurs due to CO_2 deficiency.

6.2.1.4 Sugars and other osmolytes in drought stress

Osmotic adjustment is one of the processes by which plants respond to drought. Glutamate, proline, glycine betaine, mannitol, carnitine, sorbitol, polyols, fructans, trehalose, oligosaccharides and inorganic ions like K^+, sucrose, etc., are used as osmolytes to decrease cellular osmotic potential of plants and protect plants under water deficit conditions [53]. Accumulation of osmolytes protects macromolecules stabilizing protein tertiary structure and scavenging reactive oxygen species [54]. Some sugars such as raffinose and galactinol function as osmo-protectants, synthesized in response to drought [55]. Mannitol (a sugar alcohol) scavenges the reactive oxygen species, hydroxyl radicals and stabilizes enzyme structure [56]. Osmolytes avoid the intramolecular hydrogen bond formation in macromolecules by the formation of hydrogen bond with them. Trehalose (non-reducing disaccharide) stabilizes membrane structure and macromolecules during drought stress. Accumulation of various osmolytes like betains, proline, ectoine, etc. has been studied during various environmental stresses [57].

6.2.1.5 Reactive oxygen species (ROS) in drought

Generation of ROS in plants is observed during the normal metabolic processes which are expedited under stress conditions. In most plant species ROS production increases due to disruption of metabolic processes by different abiotic stresses like water deficit, salt stress, flooding, cold and heat [58]. ROS is mainly divided in two classes, free radical forms (O_2^-, $OH^\cdot$, OH_2) and non-radical forms (H_2O_2, etc). During the hypersensitive response of plants, H_2O_2 can also induce programmed cell death [59]. DNA, protein, lipid, chlorophyll, and membrane are damaged by higher accumulation of ROS [60]. Therefore, ROS are detrimental to plants. Chloroplast, microbodies and mitochondria are the main location of ROS production [61]. Stress conditions like water deficit, heat, salinity and cold etc., limit the CO_2 fixation. Sometimes this combination among themselves or with high light increase the production of ROS in chloroplast, enhance the glycolate oxidase reaction in peroxisomes and ETC in mitochondria [61]. Antioxidants work as redox buffer and also regulate ROS homeostasis. They also affect gene expression associated with different environmental stresses [61].

6.2.1.6 Antioxidant enzymes in drought

Different types of physiological and biochemical adjustments are exhibited by plants at cellular and organism level to tolerate drought stress. Production of ROS during drought stress is very harmful for biological macromolecules. Amino acids, polyols, glycine betaine, organic acids, proline, etc., are different types of osmolytes necessary for plants to maintain drought stress. Auxins, salicylic acid, gibberrellins, abscisic acid and cytokinin also act as antioxidant substances. In accordance, polyamines, citrulline and several enzymes are used to decrease the effect of drought. Basically, plant antioxidant defence system consists of enzymatic and non-enzymatic molecules. It controls the oxidative damage caused due to an excess amount of reactive oxygen species and removes active oxygen species [61]. As generation of reactive oxygen species increases during drought stress, response of enzymatic and non-enzymatic defence system is also enhanced simultaneously [62]. Ascorbate peroxidase (APX), superoxide dismutase (SOD), glutathione reductase (GR), peroxidase (POX) and catalase (CAT) are enzymes of antioxidative systems. These scavenge H_2O_2, neutralize free radicals and oxy- intermediates [63]. Most effective enzymes that convert superoxide radical (O^-_2) and H_2O_2 to water and O_2 and prevent cellular damage in plants are CAT and SOD [64]. Ascorbate and α-tocopherol occur in chloroplast to scavenge reactive oxygen radicals [65].

6.2.1.7 Non-enzymatic defence system

Various phytochemicals, secondary metabolites and vitamins provide protection to oxidative stress [66-67]. The non-enzymatic system consists of glutathione, carotenoids, ascorbic acid, α-tocopherol, phenolic compounds and ubiquinol, etc. These are antioxidant molecules that protect plants from oxidative stress [68]. Lipophilic molecules, i.e., flavonoids, lipoic acid, α-tocopherol, ubiquinol, etc. occur in plant membrane and protect it against lipid peroxidation. Mitochondrial membrane contains ubiquinol and lipoic acid. They serve as potent inhibitors of lipid peroxidation.

6.2.1.7.1 α-tocopherol

Commonly known as vitamin E, α- tocopherol is a lipophilic antioxidant and increases in plants during abiotic stresses [69]. Thylakoid membrane of chloroplast and protoplastid are the main sources of α-tocopherol. Because of its lipophilic nature, it occurs in the lipid portion of the cell membrane by interacting with the polyunsaturated acyl group of lipid [70]. These molecules have chromanol ring structure [71]. α β γ δ is relative antioxidant activity of the four tocopherol isomers. Plants and algae synthesize tocopherol, and generally it is found in all plant parts. α-tocopherols protect the plant cell from oxidative stress and also stop lipid peroxidation as reported in soya bean [72].

6.2.1.7.2 Carotenoids

Carotenes are antioxidant molecules and susceptible to oxidative destruction. Chloroplast is the main source of β-carotene, linked with PSI and PSII. β-carotene as an accessory pigment in chloroplast protects phytochemical processes from oxidative damage. It is therefore essential for chloroplast functioning. Generation of singlet oxygen is also prevented by β-carotene in photosynthetic cells. β-carotene provides this protection via triplet chlorophyll quenching and avoiding the production of $^1O_2$.

6.2.1.7.3 Ascorbic acid

Ascorbate is a water-soluble reducing substrate necessary for the cell cycle and elongation of cells. It is an important antioxidant molecule generally found in the apoplast, organelles and cells of plants [73] and also in plant nodules at concentration of 1-2 mM. ROS produced by photosynthetic and respiratory processes are scavenged by ascorbic acid [74]. Ascorbic acid scavenges superoxide, hydroxyl radicals and singlet oxygen and reduces hydrogen peroxide to water. In combination with other antioxidant molecules, it enhances the enzyme activity of many enzymes and decreases the injuries that occur due to oxidative damage [75]. Ascorbic acid is extensively used as reductant for the removal of H_2O_2 such as in apoplast [60]. Ascorbic acid removes free radicals, regenerates tocopherol from tocopheroxyl radical and protects cellular membrane from oxidative damage [76].

6.2.1.7.4 Reduced glutathione

Glutathine is a water-soluble tripeptide molecule found in cytosol, mitochondria, ER and vacuole. It is a major source of non-protein thiols [77]. High reductive potential of GSH occurs due to its central nucleophilic cysteine residue. GSH act as antioxidant which removes cytotoxic H_2O_2. During the scavenging of H_2O_2, the ratio of reduced (GSH) to oxidized (GSSG) is changed. It is a key component in certain pathways of redox signalling [78]. Content of GSH in various cellular compartments may vary and depends upon the intensity of stress [79]. It also reacts with hydroxyl radical, singlet oxygen and superoxide radical. It has the capacity to produce ascorbic acid via the ascorbate-glutathione cycle.

6.2.1.7.5 Polyphenolic compounds

Polyphenolic compounds are metal chelators [80] and are used to eliminate radical species or to neutralize lipid radicals [81]. Apoplast and vacuole are the main sites for phenolics. An important phenolic hormone is salicylic acid, which provides resistance to plants during stress conditions [82]. The concentration of phenolics in nodules is about 0.4- 4 mM. To eliminate H_2O_2 in apoplast and vacuole, phenolics are oxidized to phenoxyl radicals through peroxidase [83]. Flavonoids are stress protecting agents and also possess pharmacological properties [84, 85]. Flavonoids are of two types: preformed and induced. Induced compound is formed in plants during physical injury, infection and stress. Synthesis of these compounds is increased under various stress conditions.

6.2.1.8 Antioxidant enzymes

Since plants have sessile nature, they need certain mechanisms that protect them from unfavourable environment and help them to maintain fitness. Antioxidant enzymes are ascorbate peroxidase (APX), DHA reductase (DHAR), peroxidase (POX), glutathione reductase (GR), catalase (CAT) and superoxide dismutase (SOD), etc [86]. APX, CAT and SOD are required to eliminate superoxide radicals [87]. Many of these are found in cell membranes, i.e., monodehydroascorbate reductase, MnSOD of peroxisomes and the APX of mitochondria [88, 89] and provide protection against ROS. SOD, CAT and APX occur in apoplast leaves of plants like barley and are necessary for scavenging of superoxide and hydrogen peroxide [90].

6.2.1.8.1 Superoxide dismutase

SODs (metalloenzymes) dismutate the superoxide radical into O_2 and hydrogen peroxide. The first line of defence against ROS inside a cell is built by SOD [77]. There are different types of SODs such as CuZnSOD, FeSOD and MnSOD that depend on the metal at active site. These are nuclear encoded. SOD genes are sensitive to environmental stresses when ROS production is increased. Apoplast, infection threads and cytosol adjacent to cell wall extensively contain CuZnSOD. MnSOD is predominated in infected cells. These SODs have distinct location in plants to perform their specific functions. Growth of cell wall in the meristem, infection threads, apoplast and response of plant to bacterial infection are the basic function of CuZnSOD. In accordance, functioning of symbiotic tissue in nodule is maintained by MnSOD. SOD prevents oxidation of GSH and in this way GSH act as a free radical scavenger. Combined action of SOD and GSH form an intracellular defence system against abiotic stress [75]. ROS scavenging is a complex process and requires SOD along with other peroxidases [64]. Therefore, SOD is essential for plant stress tolerance [91].

6.2.1.8.2 Ascorbate peroxidase

APX is an intracellular enzyme which detoxifies H_2O_2 to water by ascorbate. Cytosol and mitochondria of nodules are the main sites of APX [89]. In photosynthetic organisms, combined actions of APX and ascorbate glutathione cycle restrict the accumulation of H_2O_2 at toxic level [75,92]. Higher level of APX is required to protect plants against oxidative stress [92]. Protection against salt stress to the membrane is also provided by APX. APX along with phenolics and reduced ascorbate scavenge H_2O_2 in plant vacuole.

6.2.1.8.3 Catalase

Catalase has an energy efficient mechanism to scavenge H_2O_2 because it does not consume reductant [64]. CAT activity is higher in subcellular sites like peroxisome where H_2O_2 level is very high. Catalase has lower affinity for H_2O_2 as compared to other enzymes. Modulation of catalase activity is observed during drought stress.

6.2.1.8.4 Peroxidases

Activity of POX increases during drought stress [93]. Joint action of CAT, POX and APX is necessary to prevent oxidative damage. SOD converts superoxide radicals into H_2O_2 and this H_2O_2 is scavenged by these enzymes [94].

6.2.1.9 Stress signaling pathway

Under hyperosmotic stress activation of several MAPKs is observed. Induction of SIMK (salt stress inducible MAPK) in alfalfa and SIPK (salicylic inducible MAPK) in tobacco is observed under hyperosmotic stress [95]. The stress signaling pathway functions to increase the accumulation of osmolytes that protect from stress damage by maintaining the osmotic balance. Under drought and salinity, sugars (soluble) such as fructose, glucose, sucrose, etc. act as signaling molecules and by interacting with hormones form signaling network in plants [45,96]. Generally, these stresses increase soluble sugar content in plants and decrease content of starch [97], but reduction in soluble sugars is also observed under severe drought [98].

6.2.2 Drought Tolerance Through PGPR

The mechanism by which plants respond to drought stress needs to be elucidated to grow stress tolerant plants [87, 99]. It is a very complex process because of interaction between factors affecting and being affected by the specific stress, various physiological, biochemical and molecular phenomena resulting in altered plant growth [100]. Along with water stress, nutrient availability to the plant is severely affected under drought conditions and can be improved by the use of suitable microorganisms [101]. In the past few decades, several PGPRs have been isolated and characterized for use in sustainable agriculture (Table 6.1). Growth promoting bacteria, through complex interactions with plant which are mostly species specific, show significant increase in yield of agronomically important crops [133,134]. PGPRs have largely been reported from the genera *Azotobacter, Arthrobacter, Azospirillum, Enterobacter, Bacillus, Pseudomonas, Streptomyces* and *Serratia* [135].

Table 6.1 PGPRs conferring drought stress tolerance in legumes with potential of use as biofertilizer

S. No.	Organism	Crop	Mechanism	References
1.	*Variovorax paradoxus*	Pea (*Pisum sativum*)	Synthesis of ACC-deaminase	[102, 103]
2.	*Paenibacillus polymyxa* and *Rhizobium tropici*	Common bean (*Phaseolus vulgaris*)	Change in hormone balance stomatal conductance	[104]
3.	*Pseudomonas* sp.	Pea (*Pisum sativum*)	Decreased ethylene production	[105]
4.	*Azospirillum brasilense*	Common bean (*P. vulgaris*)	Antioxidant activities	[106]
5.	Rhizobacteria	Common bean (*Phaseolus vulgaris*)	Biochemistry of nodules	[107]
6.	*Bradyrhizobium japonicum*	Chickpea (*Cicer arietinum*)	Root-nodule symbiosis	[108]
7.	*Rhizobium tropici, Ensifer meliloti, Rhizobium etli*	Common bean (*Phaseolus vulgaris*)	Antioxidant activities	[109]

Contd...

Contd...

8.	Legume species	Cowpea (*Vigna unguiculata*)	SOD	[110]
9.	*Mesorhizobium ciceri*	Chickpea (*Cicer arietinum*)	Antioxidant activities	[111, 112]
10.	*Bradyrhizobium* sp.	Soybean (*Glycine max*)	-	[113]
11.	*Sinorhizobium* sp.	Common bean (*Phaseolus vulgaris*)	Symbiosis	[114]
12.	*Bradyrhizobium japonicum*	Soybean (*Glycine max*)	Symbiotic partnership	[115]
13.	*Rhizobium* sp.	Faba bean (*Vicia faba*)	Symbiotic partnership	[116-118]
14.	*Rhizobium* sp.	Pea (*Pisum sativum*)	Nitrogen fixation	[119]
15.	*Sinorhizobium meliloti* and *S. medicae*	Alfalfa (*Medicago sativa*)	Symbiotic nitrogen fixation	[120]
16.	*Rhizobium etli*	Common bean (*Phaseolus vulgaris*)	Over expressing Trehalose-6-Phosphate synthase in Rhizobia	[121]
17.	*Rhizobium*	Common bean (*Phaseolus vulgaris*)	Ureide accumulation	[122]
18.	Nitrogen-fixing bacteria	Soybean (*Glycine max*)	γ-aminobutyric acid (GABA) synthesis	[123]
19.	Nitrogen-fixing bacteria	Chickpea (*Cicer arietinum*)	-	[124]
20.	Nitrogen-fixing bacteria	Alfalfa (*Medicago sativa*)	Overexpression of the Arabidopsis H^+-PPase	[125]
21.	*Bradyrhizobium*	Grain legumes	Microsymbiont	[126]
22.	Nitrogen-fixing bacteria	Common bean (*Phaseolus vulgaris*)	Activity of nitrate reductase	[127]
23.	Nitrogen-fixing bacteria	Cowpea (*Vigna unguiculata*)	-	[128]
24.	*Bradyrhizobium japonicum*	Soybean (*Glycine max*)	Acetylene reduction	[129]
25.	Nitrogen-fixing bacteria	Common bean (*Phaseolus vulgaris*)	-	[130]
26.	*Bradyrhizobium* sp.	Mungbean (*Vigna radiata*)	-	[131]
27.	Nitrogen-fixing bacteria	Comman bean (*Phaseolus vulgaris*)	Antioxidant enzyme activities	[132]

Cytokinins, auxins, gibberellins, ethylene and abscisic acid (ABA) are phytohormones that are produced by PGPRs. These hormones promote plant growth either directly or with other secondary metabolites of bacteria. Volatile organic compounds are also produced by PGPRs that are important for plant immunity and development. *Herbaspirillum, Acetobacter, Rhizobium* and *Bacillus* species produce gibberellins ($GA_{1,}$ $GA_{3,}$ $GA_{4,}$ GA_{20}) [136]. Some species of *Staphylococcus, Pseudomonas* and *Micrococcus* generate auxins (IAA) [137]. *Azotobacter chroococcum* produces cytokinins (zeatin riboside or isopentenyl adenine) [138, 139]. Some organic compounds (volatile) like aldehydes, ketones and hydrocarbons play important role during the communication between plants and microorganisms, and also promote defence mechanisms and growth of plant [140-142]. In fields those are affected by high salt, drought and heavy metals, decreased productivity is observed. To increase productivity PGPRs are applied in field. PGPRs support growth of plant by two mechanisms: (1) direct and (2) indirect. In direct mechanism, growth of plant increases when plant pathogen and other rhizosphere microorganisms are absent. In indirect mechanisms, plant growth is enhanced by the removal of deleterious effects of plant pathogen.

Different types of adaptation mechanisms are used by plant to tolerate drought. Phytohormones are very important in this process [143]. Osmolytes, enzymes, siderophores, antibiotics, nitric oxide, organic acids, etc. are also produced by PGPRs. Nodulation of legume plant is affected by ethylene (a negative regulator). 1-aminocyclopropane- 1- carboxylic acid (ACC) is a precursor of ethylene. So it is necessary to maintain the ACC level in plants. PGPR has ACC deaminase enzyme which degrades ACC and promotes root growth [22, 102, 151-153]. It can use ACC as a nitrogen source. In this way PGPRs decrease plant ethylene level and support plant growth [19].

PGPRs are highly valuable for agriculture because they provide cross protection against multiple stresses by systemic identification of bacterial strains. The basis of cross protection is the non specificity of plant perception by PGPRs. The various mechanisms by which PGPRs enhance plant growth are summarized in Table 6.2. Glycine betaine produced by PGPRs provide cross protection against water deficiency as well as frost and salinity. In accordance, increased level of SOD is a general response to drought stress. K^{+}, glutamate, trehalose, proline, glycine betaine and ectoine are produced by rhizobacteria to maintain cytoploasmic osmolarity during drought stress. Membrane derived oligosaccharides situated in periplasmic space of bacteria maintain the turgor pressure of the periplasm. These highly anionic polysaccharides are too large to diffuse through the porin proteins.

Table 6.2 Mechanisms by which PGPR enhance plant growth

S. No.	*Mechanism*	*Reference*
1.	Suppression of pathogen	[144]
2.	Enhance bioavailability of minerals in soils	[135, 145, 146]
3.	Nitrogen fixation	[147]
4.	Enhance abiotic stress tolerance	[135, 148, 149]
5.	Stimulate the production of phytohormone by plants	[150]

Different physiological factors that determine the survival of rhizobia during desiccation, viz., survival rate of *Sinorhizobium meliloti* USDA 1021 is increased when they were dried in stationary phase with an increasing drying temperature and availability of chloride and sulphate [154].

6.2.2.1 PGPR as biofertilizer

There are several interpretations of the term 'biofertilizer' in different texts [155]. Here, the term biofertilizer is referred to as a product that contains living organisms and by adaptation of coordinated mechanisms either directly or indirectly facilitates plant growth. In sustainable agriculture systems, biofertilizers offer an alternative to mineral fertilizers and increase plant growth and yield. Members of Rhizobiaceae have been used as inoculum in non-leguminous plants like wheat, rice and maize also [156].

Different rhizobial strains respond differentially to drought. It is observed that common bean (*Phaseolus vulgaris*) inoculated with three rhizobial strains *Rhizobium etli* A32 (sensitive), *Rhizobium tropici* CIAT899 (tolerant) and *Ensifer meliloti* 4H41 (highly tolerant) exhibited differential response under drought stress [109]. Among the above three strains, *E. meliloti* 4H41 was the best strain as it was less affected by drought in comparison to other strains. It maintained the highest shoot biomass production and large size nodules characterized by stability of the level of lipid peroxidation (MDA), amounts of total soluble proteins and leghaemoglobin contents, increased activity of APX and POX.

Similarly, in chickpea, different strains of *Bradyrhizobium japonicum* (Thal- 8, Tal- 620, Dulawala) that respond differently to drought rendered varied degrees of stress tolerance to the inoculated plants [108]. Plants inoculated with strain Thal-8 were most effective in the root-nodule symbiosis and also partially alleviated decreased growth and yield imposed by drought stress. It increased the root biomass, number of nodules/plant, weight of seeds, GA and IAA contents of leaves of chickpea plants. Rhizobial strains like Thal-8 can protect chickpeas by enhancing the accumulation of osmoprotectants like free proline within leaves. Proline accumulation provides a mechanism for better plant survival under drought stress.

Nodulation is adversely affected under water deficient conditions, as root hair colonization and infection is hampered [157,158]. Some species included under the Rhizobiaceae can survive severe salinity and drought due to adaptive strategies [157-159]. Stability of MDA and leghaemoglobin content is essential as these are highly sensitive and are most affected by ROS overproduction and oxidative stress [159-162]. Tolerant strains of Rhizobia can also stabilize SOD and APX activities to prevent osmotic stress mediated nodule damage [163-164]. Incidentally, peroxidases are considered to be the first enzymes that respond to drought stress due to their localization on the periphery of the nodules [165]. In common bean, certain strains of *Rhizobium tropici* are able to tolerate upto 300 mM of NaCl [166] and sometimes even as high as 750 mM NaCl concentration in synthetic medium [167]. Selection of stress tolerant plant varieties inoculated with stress tolerant Rhizobia is rational for improving legume production under stress.

More recently, the beneficial effects of PGPR inoculation to particular crop plants as biofertilizers have been attributed to a consortium of microorganisms present in the soil and their synergistic effect rather than the outcome of a single microbial strain. Bacterium of the genus *Azospirillum* has been observed to be associated with rhizosphere microorganisms to fix atmospheric nitrogen and contribute to nitrogen nutrition of the plants [168]. Similarly, *Pseudomonas, Bacillus, Rhizobium* in association have also been considered to be the most effect phosphate solubilizers [169-171]. In soils poor in phosphorus, supplements of phosphate fertilizer are costly and soil microorganisms according to their ability may be utilized in complementation to each other to solubilize and mobilize it to the plant [172]. Indeed, most of the failures of Rhizobacteria have more to do with how the inoculants interact with the existing soil biota in addition to its compatibility to a particular crop plant [173]. Metabolites exuded from crop plants play an important role in the microbe-plant interaction and are essential for particular group of microorganisms to be bound in a beneficial association with the plant [174]. *Burkholderia* sp. are known to differentially colonize roots of maize to varying degree in accordance with the particular type of root exude [175]. The major problem in commercialization of PGPRs as biofertilizer has been in increasing production of microorganisms with management of the quality, stability, and effectiveness of the product. Standardization of effective formulations, survivability and effectiveness under field conditions, compatibility with chemical treatment of seeds and mineral fertilizers and appropriate technology for applications need further research and development.

CONCLUSION

Drought and increasing salinity have significant and harmful impacts on land, water, vegetation quality, wildlife environments, agronomy and ecosystem especially for arid and semiarid areas where high evapo-transpiration rates expose plants to further adaptive pressure. Arid and semiarid areas occupy an increasing fraction of the earth's surface. Legume plants are preferred for cultivation in these areas. They fix nitrogen with *Rhizobium*. With climate change and an increasing world population, there is an urgent need to develop the diverse range of nodulated legumes native to dry environments. With modern methods this goal is readily achievable ensuring sustainable of nitrogen fixation and in the endosymbionts (rhizobia) that induce nodulation in them. Agriculture in semiarid regions requires the implementation of methods to balance nutrients and to conserve soil organic matter. Several leguminous plants that uniquely combine optimal nutritional profiles, high tolerance to environmental stresses, high biomass productivity and most nutrient and moisture contributions to the soil can be utilized in several diverse ways while the high genetic variability that exists within the cultivated and wild relatives remains to be explored for further use. There is an urgent need for research in microbial traits that are necessary and useful for proper utilization as biofertilizers in crop plants for specific microenvironments.

ACKNOWLEDGEMENTS

The authors thank Professor Aditya Shastri, Vice Chancellor, Banasthali Vidyapith, India for providing the necessary facilities. The work is funded through a UGC, Govt. of India,

Major Research Project to DC. SP thankfully acknowledge UGC for fellowship. The authors declare no conflict of interest.

REFERENCES

1. Jones, H.G. and Jones, M.B. (1989). "Introduction: some terminology and common mechanisms", In: H.G., Jones, T.J., Flowers, M.B., Jones (Eds.), *Plants Under Stress,* Cambridge University Press, Cambridge. pp. 1–10.
2. Jaleel, C.A., Gopi, R., Sankar, B., Gomathinayagam, M. and Panneerselvam, R. (2008a). "Differential responses in water use efficiency in two varieties of *Catharanthus roseus* under drought stress", *Comp. Rend. Biol.,* 331, 42–47.
3. Jaleel, C.A., Gopi, R. and Panneerselvam, R. (2008b). "Growth and photosynthetic pigments responses of two varieties of *Catharanthus roseus* to triadimefon treatment", *Comp. Rend. Biol.,* 331, 272–277.
4. Jaleel, C.A., Manivannan, P., Lakshmanan, G.M.A., Gomathinayagam, M. and Panneerselvam, R. (2008c). "Alterations in morphological parameters and photosynthetic pigment responses of *Catharanthus roseus* under soil water deficits", *Colloids Surf. B: Biointerfaces,* 61, 298–303.
5. Jaleel, C.A., Manivannan, P., Murali, P.V., Gomathinayagam, M. and Panneerselvam, R. (2008d). "Antioxidant potential and indole alkaloid profile variations with water deficits along different parts of two varieties of *Catharanthus roseus*", *Colloids Surf. B: Biointerfaces,* 62, 312–318.
6. Jaleel, C.A., Sankar, B., Murali, P.V., Gomathinayagam, M., Lakshmanan, G.M.A. and Panneerselvam, R. (2008e). "Water deficit stress effects on reactive oxygen metabolism in *Catharanthus roseus;* impacts on ajmalicine accumulation", *Colloids Surf. B: Biointerfaces,* 62, 105–111.
7. Farooq, M., Basra, S.M.A., Wahid, A., Cheema, Z.A., Cheema, M.A. and Khaliq, A. (2008). "Physiological role of exogenously applied glycine betaine in improving drought tolerance of fine grain aromatic rice (*Oryza sativa* L.)", *J. Agron. Crop Sci.,* 194, 325–333.
8. Nehl, D.B., Allen, S.J. and Brown, J.F. (1996). "Deleterious rhizosphere bacteria: an integrating perspectice", *Appl. Soil Ecol.,* 5, 1–20.
9. Beijerinck, M.W. (1888). "Cultivation of *Bacillus radicicola* from the Nodules", *Bot. Ztg.,* 46, 740–750.
10. Frank, B. (1889). "About the fungal symbiosis of legumes", *Ber. Dtsch. Ges. Bot.,* 7, 332–346.
11. Tilak, K.V.B.R., Ranganayaki, N., Pal, K.K., De, R., Saxena, A.K., Nautiyalm, C.S., Mittalm, S., Tripathi, A.K. and Johri, B.N. (2005). "Diversity of plant growth and soil health supporting bacteria", *Curr. Sci.,* 89(1), 135–150.
12. Upadhyay, S.K., Singh, D.P. and Saikia, R. (2009). "Genetic diversity of plant growth promoting rhizobacteria from rhizospheric soil of wheat under saline conditions", *Curr. Microbiol.,* 59(5), 489–496.

13. Kloepper, J.W. and Schroth, M.N. (1978). "Plant growth-promoting rhizobacteria on radishes", In: Proc. of the 4th Internat. Conf. on Plant Pathogenic Bacter. Station de Pathologie Vegetale et Phytobacteriologie, INRA, Angers, France. 2, 879–882.
14. Sudhakar, P., Chattopadhyay, G.N., Gangwar, S.K. and Ghosh, J.K. (2000). "Effect of foliar application of *Azotobacter, Azospirillum* and *Beijerinckia* on leaf yield and quality of mulberry (*Morus alba*)", *J. Agr. Sci.,* 134, 227–234.
15. Bertrand, H., Nalin, R., Bally, R. and Cleyet-Marel, J.C. (2001). "Isolation and identification of the most efficient plant growth-promoting bacteria associated with canola (*Brassica napa*)", *Biol. Fertil. Soils,* 33, 152–156.
16. Bonaterra, A., Ruz, L., Badosa, E., Pinochet, J. and Montesinos, E. (2003). "Growth promotion of Prunus rootstocks by root treatment with specific bacterial strain", *Plant Soil,* 255, 555–569.
17. Murphy, J.F., Reddy, M.S., Ryu, C.M., Kloepper, J.W. and Li, R.H. (2003). "Rhizobacteria-mediated growth promotion of tomato leads to protection against Cucumber mosaic virus", *Phytopathol.,* 93, 1301–1307.
18. Joo, G.J., Kim, Y.M., Kim, J.T., Rhee, I.K., Kim, J.H. and Lee, I.J. (2005). "Gibberellins-producing rhizobacteria increase endogenous gibberellins content and promote growth of red peppers", *J. Microbiol.,* 43, 510–515.
19. Saleem, M., Arshad, M., Hussain, S. and Bhatti, A.S. (2007). "Perspective of plant growth promoting rhizobacteria (PGPR) containing ACC deaminase in stress agriculture", *J. Ind. Microbiol. Biot.,* 34, 635–648.
20. Khan, A.G. (2005). "Role of soil microbes in the rhizospheres of plants growing on trace metal contaminated soils in phytoremediation", *J. Trace Elem. Med. Biol.,* 18, 355–364.
21. Kloepper, J.W. and Beauchamp, C.J. (1992). "A review of issues related to measuring colonization of plant roots by bacteria", *Can. J. Microbiol.,* 38, 1219–1232.
22. Long, H.H., Schmidt, D.D. and Baldwin, I.T. (2008). "Native bacterial endophytes promote host growth in a species-specific manner; phytohormone manipulations do not result in common growth responses", *Plos One,* 3(7), e2702.
23. Vilchez, S., Manzanera, M. and Ramos, J.L. (2000). "Control of expression of divergent *Pseudomonas putida* put promoters for proline catabolism", *Appl. Environ. Microbiol.,* 66, 5221–5225.
24. Elmerich, C.N. and Newton, W.E. (2007). "*Associative and Endophytic Nitrogen-fixing Bacteria and Cyanobacterial Associations*", Springer, The Netherlands.
25. Glick, B.R. (1995). "The enhancement of plant growth by free-living bacteria", *Can. J. Microbiol.,* 41, 109–117.
26. Kuiper, I., Bloemberg, G.V. and Lugtenberg, B.J.J. (2001). "Selection of a plant bacterium pair as a novel tool for rhizostimulation of polycyclic aromatic hydrocarbon-degrading bacteria", *Mol. Plant Microbe. Interact.,* 14, 1197–1205.
27. Farooq, M., Wahid, A., Kobayashi, N., Fujita, D. and Basra, S.M.A. (2009). "Plant drought stress: effects, mechanisms and management", *Agron. Sustain. Dev.,* 29, 185–212.

28. Wu, Q.S., Xia, R.X. and Zou, Y.N. (2008). "Improved soil structure and citrus growth after inoculation with three arbuscular mycorrhizal fungi under drought stress", *Eur. J. Soil Biol.*, 44, 122–128.
29. Specht, J.E., Chase, K., Macrander, M., Graef, G.L., Chung, J., Markwell, J.P., Germann, M., Orf, J.H. and Lark, K.G. (2001). "Soybean response to water. A QTL analysis of drought tolerance", *Crop Sci.*, 41, 493–509.
30. Petropoulos, S.A., Daferera, D., Polissiou, M.G. and Passam, H.C. (2008). "The effect of water deficit stress on the growth, yield and composition of essential oils of parsley", *Sci. Hort.*, 115, 393–397.
31. Webber, M., Barnett, J., Finlayson, B. and Wang, M. (2006). "Pricing China's Irrigation Water", Working Paper, School of Anthropology, Geography and Environmental Studies, The University of Melbourne, Victoria, Australia.
32. Bhatt, R.M. and Srinivasa Rao, N.K. (2005). "Influence of pod load response of okra to water stress", *Indian J. Plant Physiol.*, 10, 54–59.
33. Shao, H.B., Chu, L.Y., Shao, M.A., Abdul Jaleel, C. and Hong-Mei, M. (2008). "Higher plant antioxidants and redox signaling under environmental stresses", *Comp. Rend. Biol.*, 331, 433–441.
34. Zhang, M., Duan, L., Zhai, Z., Li, J., Tian, X., Wang, B., He, Z. and Li, Z. (2004). "Effects of plant growth regulators on water deficit-induced yield loss in soybean", *Proceedings of the 4th International Crop Science Congress,* Brisbane, Australia.
35. Wullschleger, S.D., Yin, T.M., DiFazio, S.P., Tschaplinski, T.J., Gunter, L.E., Davis, M.F. and Tuskan, G.A. (2005). "Phenotypic variation in growth and biomass distribution for two advanced-generation pedigrees of hybrid poplar" *Can. J. Res.*, 35, 1779–1789.
36. Johansen, C., Baldev, B., Brouwer, J.B., Erskine, W., Jermyn, W.A., Li- Juan, L., Malik, B.A., Ahad Miah, A. and Silim, S.N. (1992). "Biotic and abiotic stresses constraining productivity of cool season food legumes in Asia, Africa and Oceania.", In: Muehlbauer, F.J., W.J., Kaiser (eds.), *Expanding the Production and Use of Cool Season Food Legumes*, Kluwer Academic Publishers, Dordrecht, The Netherlands. pp. 75–194.
37. Tahir, M.H.N., Imran, M. and Hussain, M.K. (2002). "Evaluation of sunflower (*Helianthus annuus* L.) inbred lines for drought tolerance", *Int. J. Agric. Biol.*, 3, 398–400.
38. Sacks, M.M., Silk, W.K. and Burman, P. (1997). "Effect of water stress on cortical cell division rates within the apical meristem of primary roots of maize", *Plant Physiol.*, 114, 519–527.
39. Prabhudeva, T.V., Chalapathi, M.M., Thimmegowda, S., Devakhumar, N., Rao, G.G. and Mallikarjuna, K. (1998). "Soil moisture stress and drought susceptibility index in sunflower", *Indian Agric.*, 42, 287–289.
40. MansWeld, T.J. and Atkinson, C.J. (1990). "Stomatal behaviour in water stressed plants", In: R.G., Alscher, J.R., Cumming (Eds.), *Stress Responses in Plants: Adaptation and Acclimation Mechanisms,* Wiley-Liss, New York, pp. 241–264.

41. Lake, J.A., Woodward, F.I. and Quick, W.P. (2002). "Long-distance CO_2 signalling in plants", *J. Exp. Bot.*, 53, 183–193.
42. Xiong, L. and Zhu, J.K. (2001). "Abiotic stress signal transduction in plants: molecular and genetic perspectives", *Physiol. Plant*, 112, 152–166.
43. Munns, R., James, R.A. and La¨uchli, A. (2006). "Approaches to increasing the salt tolerance of wheat and other cereals", *J. Exp. Bot.*, 57, 1025–1043.
44. Massacci, A., Nabiev, S.M., Pietrosanti, L., Nematov, S.K., Chernikova, T.N., Thor, K. and Leipner, J. (2008). "Response of the photosynthetic apparatus of cotton (*Gossypium hirsutum*) to the onset of drought stress under field conditions studied by gas-exchange analysis and chlorophyll fluorescence imaging", *Plant Physiol. Biochem.*, 46, 189–195.
45. Chaves, M.M. and Oliveira, M.M. (2004). "Mechanisms underlying plant resilience to water deficits: prospects for water-saving agriculture", *J. Exp. Bot.*, 55, 2365–2384.
46. Ort, D.R., (2001). "When there is too much light", *Plant Physiol.*, 125, 29–32.
47. Lawlor, D.W. and Cornic, G. (2002). "Photosynthetic carbon assimilation and associated metabolism in relation to water deficits in higher plants", *Plant Cell Environ.*, 25, 275–294.
48. Chaves, M.M., Maroco, J.P. and Pereira, J.S. (2003). "Understanding plant responses to drought – from genes to the whole plant", *Funct. Plant Biol.*, 30, 239–264.
49. Galme´s, J., Medrano, H. and Flexas, J. (2007a). "Photosynthetic limitations in response to water stress and recovery in Mediterranean plants with different growth forms", *New Phytol.*, 175, 81–93.
50. Centritto, M., Loreto, F. and Chartzoulakis, K. (2003). "The use of low [CO_2] to estimate diffusional and non-diffusional limitations of photosynthetic capacity of salt-stressed olive saplings", *Plant Cell Environ.*, 26, 585–594.
51. Scartazza, A., Lauteri, M., Guido, M.C. and Brugnoli, E. (1998). "Carbon isotope discrimination in leaf and stem sugars, water-use efficiency and mesophyll conductance during different developmental stages in rice subjected to drought", *Aust. J. Plant Physiol.*, 25, 489–498.
52. Bota, J., Flexas, J. and Medrano, H. (2004). "Is photosynthesis limited by decreased Rubisco activity and RuBP content under progressive water stress?", *New Phytol.*, 162, 671–681.
53. Ramanjulu, S. and Bartels, D. (2002). "Drought- and desiccation-induced modulation of gene expression in plants", *Plant Cell Environ., 25,* 141–151.
54. Zhu, J.K. (2001). "Plant salt tolerance", *Trends Plant Sci.*, 6, 66–71.
55. Taji, T., Ohsumi, C., Iuchi, S., Seki, M., Kasuga, M., Kobayashi, M., Yamaguchi-Shinozaki, K. and Shinozaki, K. (2002). "Important roles of drought and cold-inducible genes for galactinol synthase in stress tolerance in *Arabidopsis thaliana*", *Plant J.*, 29, 417–426.
56. Shen, B., Jensen, R.G. and Bohnert, H.J. (1997). "Mannitol protects against oxidation by hydroxyl radicals", *Plant Physiol.*, 115, 527–532.

57. Rontein, D., Basset, G. and Hanson, A.D. (2002). "Metabolic engineering of osmoprotectant accumulation in plants", *Metab. Eng.*, 4, 49–56.
58. Mittler, R. (2002). "Oxidative stress, antioxidants and stress tolerance", *Trends Plant Sci.*, 7, 405–410.
59. Neill, S.J., Desikan, R., Clarke, A., Hurst, R.D. and Hancock, J.T. (2002). "Hydrogen peroxide and nitric oxide as signalling molecules in plants", *J. Exp. Bot.*, 53, 1237–1247.
60. Blokhina, O., Virolainen, E. and Gagerstedt, K.V. (2003). "Antioxidants, oxidative damage and oxygen deprivation stress: a review", *Ann. Bot. (Lond.)*, 91, 179–194.
61. Foyer, C.H. and Noctor, G. (2005). "Oxidant and antioxidant signaling in plants: a re-evaluation of the concept of oxidative stress in a physiological context", *Plant Cell Environ.*, 29, 1056–1071.
62. Gressel, J. and Salun, E. (1994). "In Causes of Photooxidative Stress and Amelioration of Defence Systems in Plants", (eds. Foyer, C. H. and Mullineaux, P. M.), CRC Press, Boca Raton, FL. 237–274.
63. Sang, H.L., Ahsan, N., Lee, K.W., Kim, D.H., Lee, D.G., Kwak, S.S., Kwon, S.Y., Kim, T.H. and Lee, B.H. (2007). "Simultaneous over expression of both CuZn superoxide dismutase and ascorbate peroxidase in transgenic tall fescue plants confers increased tolerance to a wide range of abiotic stresses", *J. Plant Physiol.*, 64, 1626–1638.
64. Scandalios, J.G. (1997). "Molecular genetics of SOD in plants", In: Scandalios JG (ed) Oxidative stress and the molecular biology of antioxidant defense. Cold Spring Harbor Laboratory Press, Cold Spring Harbor, 527–568.
65. Smirnoff, N. and Wheeler, G.L. (2000). "AA in plants: biosynthesis and function", *Crit. Rev. Plant Sci.*, 19, 267–290.
66. Eraslan, F., Inal, A., Gunes, A. and Alpaslan, M. (2007). "Impact of exogenous salicylic acid on the growth, antioxidant activity and physiology of carrot plants subjected to combined salinity and boron toxicity", *Sci. Hortic.*, 113, 120–128.
67. Hong-Bo, S., Chu, L.Y., Zhao-Hua, L. and Cong-Min, K. (2008d). "Primary antioxidant free radical scavenging and redox signaling pathways in higher plant cells", *Int. J. Biol. Sci.*, 4, 8–14.
68. Seong, E.S., Cho, H.S., Choi, D., Joung, Y.H., Lim, C.K., Hur, J.H. and Wang, M.H. (2007). "Tomato plants overexpressing CaKR1 enhanced tolerance to salt and oxidative stress", *Biochem. Biophys. Res. Commun.*, 363, 983–988.
69. Munne-Bosch, S. and Algere, L. (2003). "Drought induced changes in the redox state of α-tocopherol, ascorbate and the diterpene cornosic acid in the chloroplasts of labiatae species differing in carnosic acid contents", *Plant Physiol.*, 131, 1816–1825.
70. Wang, X. and Quinn, P.J. (2000). "The location and function of vitamin E in membranes", *Mol. Membr. Biol.*, 17, 143–156.
71. Yamaguchi, J., Iwamoto, T., Kidam, S., Masushige, S., Yamada, K. and Esashi, T. (2001). "Tocopherol associated protein is a ligand dependent transcriptional activator", *Biochem. Biophys. Res. Commun.*, 285, 295–299.

72. Clement, S.A., Tan, C.C., Guo, J.L., Kitta, K. and Suzuki, Y.J. (2002). "Roles of protein kinase and alpha-tocopherol in regulation of signal transduction for GATA-4 phosphorylation in HL-1 cardiac muscle cells", *Free Radic. Biol. Med.*, 32, 341–349.

73. Giovannoni, J.J. (2007). "Completing a pathway to plant vitamin C synthesis", *Proc. Natl. Acad. Sci.*, USA 104. 22, 9109–9110.

74. Guo, Z., Tan, H., Zhu, Z., Lu, S. and Zhou, B. (2005). "Effect of intermediates on ascorbic acid and oxalate biosynthesis of rice and in relation to its stress resistance", *Plant Physiol. Biochem.*, 43, 955–962.

75. Asada, K. (1999). "The water–water cycle in chloroplasts: scavenging of active oxygens and dissipation of excess photons", *Annu. Rev. Plant Physiol. Plant Mol. Biol.*, 50, 601–639.

76. Laing, W.A., Michele, A.W., Janine, C. and Sean, M.B. (2007). "The missing step of the L-galactose pathway of ascorbate biosynthesis in plants, an L-galactose guanyltransferase, increases leaf ascorbate content", *Proc. Natl. Acad. Sci.*, 104(22), 9534–9539.

77. Alscher, R.G., Erturk, N. and Heath, L.S. (2002). "Role of superoxide dismutases (SODs) in controlling oxidative stress in plants" *J. Exp. Biol.*, 53, 1331–1341.

78. Pastori, G., Mullineaux, P. and Foyer, CH. (2000). "Post transcriptional regulation prevents accumulation of glutathione reductase protein and activity in the bundle sheath cells of maize. Implication on the sensitivity of maize to temperatures", *Plant Physiol.*, 122, 667–675.

79. Pei, Z.M., Murata, Y., Benning, G., Thomine, S., Klusener, B., Allen, G.J., Grill, E. and Schroeder, J.I. (2000). "Calcium channels activated by hydrogen peroxide mediated abscissic acid signaling in guard cells", *Nature*, 406, 731–734.

80. Ksouri, R., Megdiche, W., Falleh, H., Trabelsi, N., Boulaaba, M., Smaoui, A. and Abdelly, C. (2008). "Influence of biological, environmental and technical factors on phenolic content and antioxidant activities of Tunisian halophytes", *C.R. Biol.*, 331(11), 865–873.

81. Moran, J.F., Klucas, R.V., Grayer, R.J., Abian, J. and Becana, M. (1997). "Complexes of iron with phenolic compounds from soybean nodules and other legume tissues: prooxidant and antioxidant properties", *Free Radic. Biol. Med.*, 22, 861–870.

82. Alvarez, M.E. (2000). "Salicylic acid in the machinery of hypersensitive cell death and disease resistance", *Plant Mol. Biol.*, 44, 429–442.

83. Santiago, L.J.M., Louro, R.P. and De Oliveira, D.E. (2000). "Compartmentation of phenolic compounds and phenylalanine ammonia lyase in leaves of *Phyllanthus tenellus*. Roxb and their induction by copper sulphate", *Ann. Bot. (Lond.)*, 86, 1023–1032.

84. Kong, J.M., Chia, L.S., Goh, N.K., Chia, T.F. and Brouillard, R. (2003). "Analysis and biological activities of anthocyanins", *Phytochem.*, 64, 923–933.

85. Yilmaz, Y. and Toledo, R.T. (2004). "Health aspects of functional grape seed constituents", *Trends Food Sci. Technol.*, 15422–15433.

86. Vranova, E., Inze, D. and Van Breusegem, F. (2002). "Signal transduction during oxidative stress", *J. Exp. Bot.*, 53, 1227–1236.
87. Zhao, C.X., Guo, L.Y., Jaleel, C.A., Shao, H.B. and Yang, H.B. (2008). "Prospects for dissecting plant-adaptive molecular mechanisms to improve wheat cultivars in drought environments", *Comp. Rend. Biol.*, 331, 579–586.
88. Corpas, F.J., Barroso, J.B. and del Río, L.A. (2001). "Peroxisomes as a source of reactive oxygen species and nitric oxide signal molecules in plant cells", *Trends Plant Sci.*, 6, 145–150.
89. Iturbe-Ormaetxe, I., Matamoros, M.A., Rubio, M.C., Dalton, D.A. and Becana, M. (2001). "The antioxidants of legume nodule mitochondria", *Mol. Plant-Microbe Interact.*, 14, 1189–1196.
90. Vanacker, H., Carver, T.L.W. and Foyer, C.H. (1998). "Pathogen-induced changes in the antioxidant status of the apoplast in barley leaves", *Plant Physiol.*, 117, 1103–1114.
91. Chew, O., Whelan, J. and Miller, A.H. (2003). "Molecular definition of the ascorbate–glutathione cycle in *Arabidopsis* mitochondria reveals dual targeting of antioxidant defences in plants", *J. Biol. Chem.*, 278, 46869–46877.
92. Shigeoka, S., Ishikawa, T., Tamoi, M., Miyagawa, Y., Takeda, T., Yabuta, Y. and Yoshimura, K. (2002). "Regulation and function of ascorbate peroxidase isoenzymes", *J. Exp. Bot.*, 53(372), 1305–1319.
93. Zhang, J., Jia, W., Yang, J. and Ismail, A.M. (2006). "Role of ABA in integrating plant responses to drought and salt stresses", *Field Crop. Res.*, 97, 111–119.
94. Chaitanya, K.V., Sundar, D., Masilamani, S. and Ramachandra Reddy, A. (2002). "Variation in heat stress-induced antioxidant enzyme activities among three mulberry cultivars", *Plant Growth Regul.*, 36, 175–180.
95. Munnik, T., Ligterink, W., Meskiene, I., Calderini, O. and Beyerly, J., et al. (1999). "Distinct osmosensing protein kinase pathways are involved in signaling moderate and severe hyperosmotic stress", *Plant J.*, 20, 381–388.
96. Rolland, F., Baena-Gonzalez, E. and Sheen, J. (2006). "Sugar sensing and signaling in plants: conserved and novel mechanisms", *Annu. Rev. Plant Biol.*, 57, 675–709.
97. Chaves, M.M. (1991). "Effects of water deficits on carbon assimilation", *J. Exp. Bot.*, 42, 1–16.
98. Pinheiro, C., Chaves, M.M. and Ricardo, C.P. (2001). "Alterations in carbon and nitrogen metabolism induced by water deficit in the stems and leaves of *Lupinus albus* L.", *J. Exp. Bot.*, 52, 1063–1070.
99. Reddy, A.R., Chaitanya, K.V. and Vivekanandan, M. (2004). "Drought induced responses of photosynthesis and antioxidant metabolism in higher plants", *J. Plant Physiol.*, 161, 1189–1202.
100. Razmjoo, K., Heydarizadeh, P. and Sabzalian, M.R. (2008). "Effect of salinity and drought stresses on growth parameters and essential oil content of *Matricaria chamomile*", *Int. J. Agric. Biol.*, 10, 451–454.

101. De Freitas, J.R., Banerjee, M.R. and Germida, J.J. (1997). "Phosphate-solubilizing rhizobacteria enhance the growth and yield but not phosphorus uptake of canola (*Brassica napus* L.) ", *Biol. Fertil. Soils*, 24, 358-364.

102. Belimov, A.A., Dodd, I.C., Hontzeas, N., Theobald, J.C., Safronova, V.I. and Davies, W.J. (2009). "Rhizosphere bacteria containing 1-aminocyclopropane-1-carboxylate deaminase increase yield of plants grown in drying soil via both local and systemic hormone signaling", *New Phytol.*, 181, 413–423.

103. Chen, L., Dodd, I.C., Theobald, J.C., Belimov, A.A. and Davies, W.J. (2013). "The rhizobacterium *Variovorax paradoxus* 5C-2, containing ACC deaminase, promotes growth and development of *Arabidopsis thaliana* via an ethylene-dependent pathway", *J. Exp. Bot.*, 3–9.

104. Figueiredo, M.V.B., Burity, H.A., Martínez, C.R., and Chanway, C.P., (2008). "Alleviation of drought stress in the common bean (*Phaseolus vulgaris* L.) by co-inoculation with *Paenibacillus polymyxa* and *Rhizobium tropici*", *Appl. Soil Ecol.*, 40(1), 182–188.

105. Arshad, M., Shaharoona, B. and Mahmood, T. (2008). "Inoculation with *Pseudomonas* spp. containing ACC-deaminase partially eliminates the effects of drought stress on growth, yield, and ripening of pea (*Pisum sativum* L.)", *Pedosphere*, 18(5), 611–620.

106. German, M.A., Burdman, S., Okon, Y., and Kigel, J., (2000). "Effects of *Azospirillum brasilense* on root morphology of common bean (*Phaseolus vulgaris* L.) under different water regimes", *Biol. Fertil. Soils*, 32, 259–264.

107. García, N.A.T., Iribarne, C., López, C., Cervera, J.A. and Lluch, C. (2005). "Physiological implications of trehalase from *Phaseolus vulgaris* root nodules: partial purification and characterization", *Plant Physiol. Biochem.*, 43(4), 355-61.

108. Bano, A., and Rand, B. and Dazzo, F. (2010). "Adaptation of chickpea to desiccation stress is enhanced by symbiotic rhizobia", *Symbiosis*, 50, 129–133.

109. Mhadhbi, H., Chihaoui, S., Mhamdi, R., Mnasri, B., Jebara, M. and Mhamdi, R. (2011). "A highly osmotolerant rhizobial strain confers a better tolerance of nitrogen fixation and enhances protective activities to nodules of *Phaseolus vulgaris* under drought stress", *Afr. J. Biotechnol.*, 10(22), 4555–4563.

110. Hernandez, J, S., Delrio, L.S., and Sevilla, F., (1994). "Salt stress-induced changes in superoxide dismutase isozymes in leaves and mesophyll protoplasts from *Vigna unguiculata* (L.) Walp", *New Phytol.*, 126(1), 37–44.

111. Esfahani, M.N. and Mostajeran, A. (2011). "Rhizobial strain involvement in symbiosis efficiency of chickpea–rhizobia under drought stress: plant growth, nitrogen fixation and antioxidant enzyme activities", *Acta Physiol. Plant*, 33, 1075-1083.

112. Esfahani, M.N., Sulieman, S., Schulze, J., Yamaguchi-Shinozaki, K., Shinozaki, K. and Tran, L.P. (2014). "Mechanisms of physiological adjustment of N_2 fixation in *Cicer arietinum* L. (chickpea) during early stages of water deficit: single or multi-factor controls", *The Plant J.*, 79, 964–980.

113. Bethlenfalvay, G.J., Brown, M.S. and Franson, R.L. (1990). "Glycine-Glomus-*Bradyrhizobium* symbiosis", *Plant Physiol.*, 94, 723–728.

114. Cervera, J.A.H., Mellado, J.C., Laguerre, G. and Tichy, H.V. (1999). "At least five rhizobial species nodulate *Phaseolus vulgaris* in a Spanish soil", *FEMS Microbiol. Ecol.*, 30(1), 87–97.

115. Clement, M., Lambert, A., Herouart, D. and Boncompagni, E. (2008). "Identification of new up-regulated genes under drought stress in soybean nodules", *Gene*, 426(1-2), 15–22.

116. Katerji, N., Mastrorilli, M., Lahmer, F.Z., Maalouf, F. and Oweis, T. (2011). "Faba bean productivity in saline-drought conditions", *Eur. J. Agron.*, 35(1), 2–12.

117. Kenasa, G., Jida, M. and Assefa, F. (2014). "Characterization of Phosphate Solubilizing Faba Bean (*Vicia faba* L.) Nodulating Rhizobia Isolated from Acidic Soils of Wollega, Ethiopia", *Sci. Technol. Arts Res. J.*, 3(3), 11–17.

118. Hewedy, O.A., Eissa, R.A., Elzanaty, A.M., Nagaty, H.H. and Abd Elbary, M.I. (2014). "Phenotypic and Genotypic Diversity of Rhizobia Nodulating Faba Bean from Various Egyptian Locations", *J. Bioproces. Biotechniq.*, 4–5.

119. Marino, D., Frendo, P., Ladrera, R., Zabalza, A., Puppo, A., Arrese-Igor, C. and González, E.M. (2007). "Nitrogen fixation Control under Drought Stress. Localized or systemic?" *Plant Physiol.*, 143(4), 1968–1974.

120. Elboutahiri, N., Alami, I.T. and Udupa, M.S. (2010). "Phenotypic and genetic diversity in *Sinorhizobium meliloti* and *S. medicae* from drought and salt affected regions of Morocco", *BMC Microbiol.*, 10, 15.

121. Suárez, R., Wong, A., Ramírez, M. and Iturriaga, G. (2008). "Improvement of Drought Tolerance and Grain Yield in Common Bean by over expressing Trehalose-6-Phosphate Synthase in Rhizobia", *M.P.M.I.*, 21(7), 958–966.

122. Alamillo, M.J., Díaz-Leal, J.L., Sánchez-Moran, M.V. and Pineda, M. (2010). "Molecular analysis of ureide accumulation under drought stress in *Phaseolus vulgaris* L.", *Plant Cell Environ.*, 33, 1828–1837.

123. Serraj, R., Shelp, B.J. and Sinclair, T.R. (1998). "Accumulation of γ-aminobutyric acid in nodulated soybean in response to drought stress", *J. Physiol. Plant.*, 102(1), 79–86.

124. Allah, M.Z., Jenkinson, D.M. and Vadez, V. (2011). "A conservative pattern of water use, rather than deep or profuse rooting, is critical for the terminal drought tolerance of chickpea", *J. Exp. Bot.*, 62(12), 4239–4252.

125. Bao, A.K., Wang, S.M., Wu, G.Q., Xi, J.J., Zhang, J.L. and Wang, C.M. (2009). "Overexpression of the Arabidopsis H+-PPase enhanced resistance to salt and drought stress in transgenic alfalfa (*Medicago sativa* L.)", *Plant Sci.*, 176(2), 232–240.

126. Hungria, M. and Vargasb, M.A.T. (2000). "Environmental factors affecting N_2 fixation in grain legumes in the tropics, with an emphasis on Brazil", *Field Crop Res.*, 65(2-3), 151–164.

127. Ashraf M. and Iram A. (2005). "Drought stress induced changes in some organic substances in nodules and other plant parts of two potential legumes differing in salt tolerance", *Flora*, 200, 535–546.

128. Pandey, P.K. (1983). "Influence of defoliation on seed yield in cowpea [*Vigna unguiculata* (L.) Walp] in a sub-tropical environment", *Field Crop. Res.*, 7, 249–256.

129. Subramanian, P., Kim, K., Krishnamoorthy, R., Sundaram, S. and Sa, T. (2014). "Endophytic bacteria improve nodule function and plant nitrogen in soybean on co-inoculation with *Bradyrhizobium japonicum* MN110", *Plant Growth Regul.*, DOI 10.1007/s10725-014-9993-x.

130. Henne, C.E.R., Langenegger, A., Mani, J., Schenk, N., Zumsteg, A. and Feller, U. (2010). "Interactions between temperature, drought and stomatal opening in legumes", *Environ. Exp. Bot.*, 68, 37–43.

131. Moradia, A., Ahmadia, A. and Zadeh, H. (2008). "The effects of different timings and severity of drought stress on gas exchange parameters of mungbean", *Desert*, 13, 59–66.

132. Terzi, R., Sağlam, A., Kutlu, N., Nar, H. and Kadioğlu, A. (2009). "Impact of soil drought stress on photochemical efficiency of photosystem II and antioxidant. Enzyme activities of *Phaseolus vulgaris* Cultivars", *Turk. J. Bot.*, 34, 1–10.

133. Silva, V.N., Silva, L.E.S.F. and Figueiredo, M.V.B. (2006). "Rhizobia of action with rhizobacteria growth promoter in plants in cowpea crop (*Vigna unguiculata* L.Walp)", *Acta. Agron. Sci.*, 28, 407–412.

134. Araujo, F.F. (2008). "Seed inoculation with *Bacillus subtilis* formulated with oyster meal and growth of corn, soybean and cotton", *Cienc. Agrotec.*, 32, 456–462.

135. Dimkpa, C.O., Merten, D., Svatoš, A., Büchel, G. and Kothe, E. (2009a). "Metal-induced oxidative stress impacting plant growth in contaminated soil is alleviated by microbial siderophores", *Soil Biol. Biochem.*, 41, 154–162.

136. MacMillan, J. (2002). "Occurrence of gibberellins in vascular plants, fungi, and bacteria", *J. Plant Growth Regul.*, 21, 242–243.

137. Ali, B., Sabri, A.N. and Hasnain, S. (2010). "Rhizobacterial potential to alter auxin content and growth of *Vigna radiata* (L.)", *World J. Microbiol. Biotechnol.*, 26, 1379–1384.

138. Nieto, K.F. and Frankenberger, W.T., Jr. (1990). "Influence of adenine, isopentyl alcohol and *Azotobacter chroococcum* on the growth of *Raphanus sativus*", *Plant Soil*, 127, 147–156.

139. Ortiz-Castro, R., Valencia-Cantero, E. and Lopez-Bucio, J. (2008). "Plant growth promotion by *Bacillus megaterium* involves cytokinin signaling", *Plant Signal Behav.*, 3, 263-265.

140. Kai, M., Haustein, M., Molina, F., Petri, A., Scholz, B. and Piechulla, B. (2009). "Bacterial volatiles and their action potential", *Appl. Microbiol. Biotechnol.*, 81, 1001–1012.

141. Schulz, S. and Dickschat, J.S. (2007). "Bacterial volatiles: the smell of small organisms", *Nat. Prod. Rep.*, 24, 814–842.

142. Schulz, S., Dickschat, J.S., Kunze, B., Wagner-Dobler, I., Diestel, R. and Sasse, F. (2010). "Biological activity of volatiles from marine and terrestrial bacteria", *Mar. Drugs*, 8, 2976–2987.

143. Potters, G., Pasternak, T.P., Guisez, Y., Palme, K.J. and Jansen, M.A.K. (2007). "Stress-induced morphogenic responses: growing out of trouble?", *Trends Plant Sci.*, 12, 98–105.

144. Sikora, R.A., Schafer, K. and Dababat, A.A. (2007). "Modes of action associated with microbially induced in planta suppression of plant-parasitic nematodes", *Australas. Plant Path.*, 36, 124–134.

145. Carrillo-Castaneda, G., Munoz, J.J., Peralta-Videa, J.R., Gomez, E. and Gardea-Torresdey, J.L. (2003). "Plant growth-promoting bacteria promote copper and iron translocation from root to shoot in alfalfa seedlings", *J. Plant Nutr.*, 26, 1801–1814.

146. Carrillo-Castaneda, G., Munoz, J.J., Peralta-Videa, J.R., Gomez, E. and Gardea-Torresdey, J.L. (2005). "Modulation of uptake and translocation of iron and copper from root to shoot in common bean by siderophore-producing microorganisms", *J. Plant Nutr.*, 28, 1853–1865.

147. Dobbelaere, S., Vanderleyden, J. and Okon, Y. (2003). "Plant growth promoting effects of diazotrophs in the rhizosphere", *Crit. Rev. Plant Sci.*, 22, 107–149.

148. Yuwono, T., Handayani, D. and Soedarsono, J. (2005). "The role of osmotolerant rhizobacteria in rice growth under different drought conditions", *Austr. J. Agric. Res.*, 56, 715–721.

149. Sziderics, A.H., Rasche, F., Trognitz, F., Sessitsch, A. and Wilhelm, E. (2007). "Bacterial endophytes contribute to abiotic stress adaptation in pepper plants (*Capsicum annuum* L.)", *Can. J. Microbiol.*, 53, 1195–1202.

150. Lazarovits, G. and Nowak, J. (1997). "Rhizobacteria for improvement of plant growth and establishment", *Hortsci.*, 32, 188–192.

151. Glick, B.R., Penrose, D.M. and Li, J.P. (1998). "A model for the lowering of plant ethylene concentrations by plant growth-promoting bacteria", *J. Theor. Biol.*, 190, 63–68.

152. Burd, G.I., Dixon, D.G. and Glick, B.R. (2000). "Plant growth-promoting bacteria that decrease heavy metal toxicity in plants", *Can. J. Microbiol.*, 46, 237–245.

153. Belimov, A.A., Dodd, I.C., Safronova, V.I., Hontzeas, N. and Davies, W.J. (2007). "*Pseudomonas brassicacearum* strain Am3 containing 1-aminocyclopropane-1-carboxylate deaminase can show both pathogenic and growth-promoting properties in its interaction with tomato", *J. Exp. Bot.*, 58, 1485–1495.

154. Vriezen, J.A.C., de Bruijn, F.J. and Nusslein, K. (2006). "Desiccation responses and survival of *Sinorhizobium meliloti* USDA 1021 in relation to growth phase, temperature, chloride and sulphate availability", *Lett. Appl. Microbiol.*, 42, 172–178.

155. Vessey, J.K. (2003). "Plant growth-promoting rhizobacteria as biofertilizers", *Plant Soil*, 255, 571–586.

156. Dobereiner, J. (1997). "Biolgical nitrogen fixation in the tropics: social and economic contributions", *Soil Biol. Biochem.*, 29, 771–774.

157. Zahran, H.H. (1999). "Rhizobium legume symbiosis and nitrogen fixation under severe conditions and in an arid climate", *Microbiol. Mol. Biol. Rev.*, 63, 968-989.

158. Vriezen, J.A.C., de Bruijn, F.J. and Nusslein, K. (2007). "Responses of rhizobia to desiccation in relation to osmotic stress, oxygen, and temperature", *Appl. Environ. Microbiol.*, 73, 3451–3459.
159. Mhadhbi, H., Mhamdi, R., Jebaram, M., Limamm, F. and Aouani, M.E. (2009b). "Legume-rhizobia symbiotic interaction under salt and drought constraints: Generation of reactive oxygen species and protective role of antioxidant enzymes within nodules, *Advances in Plant Physiology*", In: Hemantaranjan, A. (ed.), Scientific Publishers (India), Jodhpur, 2009, 1-21.
160. Redondo, F.J., de la Peña, T.C., Morcillo, C.N., Lucas, M.M. and Pueyo, J.J. (2009). "Overexpression of flavodoxin in bacteroids induces changes in antioxidant metabolism leading to delayed senescence and starch accumulation in alfalfa root nodules", *Plant Physiol.*, 149, 1166–1178.
161. Aydi, S.S., Aydi, S., Gonzalez, E. and Abdelly, C. (2008). "Osmotic stress affects water relations, growth, and nitrogen fixation in *Phaseolus vulgaris* plants", *Acta Physiol. Plant,* 30, 441–449.
162. Hafsi, C., Romero-Puertas, M.C., del Rı´o, L.A., Abdelly, C. and Sandalio, L.M. (2010). "Antioxidative response of *Hordeum maritimum* L. to potassium deficiency", *Acta Physiol. Plant,* DOI 10.1007/s11738-010- 0537-3.
163. Mhadhbi, H., Fotopoulos, V., Djebali, N., Polidoros, A.N. and Aouani, M.E. (2009a). "Behaviours of *Medicago truncatula-Sinorhizobium meliloti* symbioses under osmotic stress in relation with symbiotic partner input. Effects on nodule functioning and protection", *J. Agron. Crop. Sci.*, 195, 225–231.
164. Jebara, S., Drevon, J.J. and Jebara, M. (2010). "Modulation of symbiotic efficiency and nodular antioxidant enzyme activities in two *Phaseolus vulgaris* genotypes under salinity", *Acta Physiol. Plant,* 32, 925–932.
165. Matamoros, M.A., Dalton, D.A., Ramos, J., Clemente, M.R., Rubio, M.C. and Becana, M. (2003). "Biochemistry and molecular biology of antioxidants in the rhizobia-legume symbiosis", *Plant Physiol.*, 133, 449–509.
166. Mnasri, B., Mrabet, M., Laguerre, G., Aouani, M.E. and Mhamdi, R. (2007b). "Salt tolerant rhizobia isolated from a Tunisian oasis that are highly effective for N2-Fixation with *Phaseolus vulgaris* constitute a novel biovar (bv. mediterranense) of *Sinorhizobium meliloti*", *Arch. Microbiol,* 187, 79–85.
167. Mnasri, B., Aouani, M.E. and Mhamdi, R. (2007a). "Nodulation and growth of common bean (*Phaseolus vulgaris*) under water deficiency", *Soil Biol. Biochem.*, 39, 1744–1750.
168. Boddey, R.M. and Dobereiner, J. (1995). "Nitrogen fixation associated with grasses and cereals: recent progress and perspectives for the future", *Fert. Res.*, 42, 241–250.
169. Rodriguez, H. and Fraga, R. (1999). "Phosphate solubilizing bacteria and their role in plant growth promotion", *Biotechnol. Adv.*, 17, 319–339.
170. Adesemoye, A.O., Torbert, H.A. and Kloepper, J.W. (2008). "Enhanced plant nutrient use efficiency with PGPR and AMF in an integrated nutrient management system", *Can. J. Microbiol.*, 54, 876–886.

171. Rodrı´guez-Dı´az, M., Rodelas, B., Pozo, C., Martı´nez-Toledo, M.V. and Gonza´lez-Lo´pez, J. (2008). "A review on the taxonomy and possible screening traits of plant growth promoting rhizobacteria", In: Ahmad, I., Pichtel, J., Hayat, S., (eds.), *Plant-Bacteria Interactions: Strategies and Techniques to Promote Plant Growth,* Wiley, UK.

172. Richardson, A.E. (2000). "Prospects for using soil microorganisms to improve the acquisition of phosphorus by plants", *Aust. J. Plant Physiol.,* 28, 897–906.

173. Remans, S., Blair, M.W., Manrique, G., Tovar, L.E., Rao, I.M., Croomenborghs, A., Torres, G.R., El-Howeity, M., Michiels, J. and Vanderleyden, J. (2008). "Physiological and genetic analysis of root responsiveness to auxin-producing plant growth-promoting bacteria in common bean (*Phaseolus vulgaris* L.)", *Plant Soil,* 302, 149–161.

174. Khalid, A., Arshad, M. and Kahir, Z.A. (2004). "Screening plant growth-promoting rhizobacteria for improving growth and yield of wheat", *Appl. Soil. Ecol.,* 96, 473–480.

175. Dalmastri, C., Chiarini, L., Cantale, C., Bevinino, A. and Tabacchioni, S. (1999). "Soil type and maize cultivar affect the genetic diversity of maize root-associated *Burkholderia cepacia* populations", *Microb. Ecol.,* 38, 273–284.

7

Biofertilizer: An Alternative Fertilizer and Constraints in Its Technology

Rashmi Srivastava, Dinesh Chandra and A.K. Sharma

Department of Biological Sciences, College of Basic Sciences and Humanities
Govind Ballabh Pant University of Agriculture and Technology
Pantnagar – 263 145, U.S. Nagar, Uttarakhand
E-mail: *rsri_10@yahoo.co.in*

ABSTRACT

Looking at the tremendous use of chemical fertilizers to enhance yield of various crops which, in turn, causes degradation of soil health, the farming community may shift to integrated nutrient management system (combination of chemical fertilizer + biofertilizer) so as to maintain soil health. Although the use of biofertilizer may help in maintaining status of health but its production and usage should be done critically. There are many problems which are being faced in selection of organism, production, storage, marketing and distribution of various biofertilizers. This chapter underlines the details of constraints in biofertilizer technology.

Keywords: Biofertilizer, Constraints.

7.1 INTRODUCTION

The contamination of the soil due to excessive use of chemical fertilizers and pesticides is one of the major problems being faced in agriculture [7]. As a result of continuous use of agrochemicals, soil becomes degraded year after year, making it difficult to sustain the soil fertility. Increasing use of chemical fertilizers in agriculture makes country self dependent in food production but it deteriorates environment and causes harmful impact on living beings. Due to insufficient uptake of these fertilizers by plants, the unused remaining fertilizers reach water bodies through rainwater, which in turn, causes eutrophication in water bodies and affects living beings including growth inhabiting microorganisms. These chemicals not only cause immense damage to soil health but also create a chain of ecological and economic problems. The excessive use of chemical fertilizers in agriculture is not only costly but also has adverse effects on soils. It depletes water holding capacity, soil fertility and disparity in soil nutrients. Then arose the need to develop some low cost effective and eco-friendly fertilizers which work without disturbing nature. This thought gave rise to the advent of biofertilizers. Now, certain species of microorganisms are widely used which have unique properties to provide natural products, and serve as a good substitute of chemical fertilizers. Biofertilizer comprises a number of microorganisms (bacteria, fungi and algae) which are beneficial for agriculture. Biofertilizers are cost-effective, eco-friendly and can increase crop yield. They improve soil efficiency to sustain crop. Biofertilizers enhance the nutrient availability to crop plants (by processes like fixing atmospheric nitrogen or dissolving phosphorus present in the soil) and also impart better health to plants and soil thereby enhancing crop yields in a moderate way. It is a natural method without any problems like salinity, alkalinity, soil erosion, etc. In the vast areas of low input agriculture and oil seeds production, as also in crops like sugarcane, etc., these products will be of much use to give sustainability to production. Microbes are effective in inducing plant growth as they secrete plant growth promoters (auxins, abscisic acid, gibberellic acid, cytokinins, ethylene) and enhance seed germination and root growth. They also play a considerable role in decomposition of organic materials and enrichment of compost. In view of the priority for the promotion of organic farming and reduction of chemical residues in the environment, special focus has to be given for the production of biofertilisers.

7.2 LIST OF SOME ORGANISMS COMMONLY USED AS BIOFERTILIZERS

7.2.1 Rhizobia

Legume plants have root nodules, where atmospheric nitrogen fixation is done by bacteria belonging to genera *Rhizobium, Bradyrhizobium, Sinorhizobium, Azorhizobium* and *Mesorhizobium,* collectively called rhizobia. When rhizobial culture is inoculated in field, pulse crop yield can be increased due to rhizobial symbiosis [9]. *Rhizobium* can fix 15-20 KgN/ha and increase crop yield upto 20%. *Azorhizobium* is a stem nodule forming bacteria and fixes nitrogen symbionts of the stem nodule which also produces large amount of IAA that promotes plant growth. *Bradyrhizobium* is reported as a good nitrogen fixer. *Bradyrhizobium* strain can enhance total organic carbon, nitrogen, phosphorus and potassium in the soil, increases plant growth and consequently plant biomass, causes reduction in the weed population and increases soil microbial population.

7.2.2 Diazotrophs

These are aerobic chemolithotrophs and anaerobic photoautotrophs. These are non nodule forming bacteria. They include a number of microorganisms, viz., *Acetobacter diazotrophicus, Azotobacter, Azospirillum* and *Herbaspirillum,* collectively known as diazotrophs. They are the free living aerobic, photoautotrophic, non-symbiotic bacteria. They secrete vitamin-B complex, gibberellins, napthalene, acetic acid and other substances that inhibit certain root pathogens and improve root growth and help in uptake of plant nutrients. *Azotobacter indicum* occurs in acidic soil in sugar cane plant roots. It can be applied in cereals, millets, vegetables and flowers through seed, seedling and soil treatment. *Azospirillum* is found to be very effective in increasing 10-15% yield of cereal crops and fixes nitrogen upto 20-40%. Different *A. brasiliense* strains, upon inoculation in wheat cause an increase in seed germination, plant growth, plumule and radicle length. *Herbaspirillum* species occur in roots, stems and leaves of sugar cane and rice. They produce growth promoters (IAA, Gibberillins, Cytokinins) and enhance root development and uptake of plant nutrients.

7.2.3 Cyanobacteria (Blue-Green Algae)

Nostoc, Anabaena, Oscillatoria, Aulosira, Lyngbya, etc. are the prokaryotic organisms and phototropic in nature. They play an important role in enriching paddy field soil by fixing atmospheric nitrogen and supply vitamin B complex and growth promoting substance which makes the plant grow vigorously. Cyanobacteria fix 20-30 kg N/ha and increase 10-15% crop yield when applied at 10 kg/ha *Azolla* – *Anabaena* symbiosis is a free floating aquatic fern found on water surface having a cyanobacterial symbiont *Anabaena azollae* in their leaves. It fixes atmospheric nitrogen in paddy field and excretes organic nitrogen in water during its growth and also immediately upon trampling. *Azolla* contributes nitrogen, phosphorus (15-20 kg/ha/month), potassium (20-25 kg/ha/month) and organic carbon, etc. and increases 10-20% yield of paddy crops and also suppresses weed growth. *Azolla* also absorbs traces of potassium from irrigation water and can be used as green manure before rice planting. *Azolla* sp. are metal tolerant and hence, can be applied near heavy metal polluted areas.

7.2.4 Phosphate Solubilizing Microbes

Pseudomonas fluorescens, Bacillus megaterium var. *phosphaticum, Acrobacter acrogens, Nitrobacter* sp., *Escherichia freundii, Serratia* sp., *Pseudomonas striata, Bacillus polymyxa* are the bacteria having phosphate solubilising ability. Some fungi also have phosphate solubilizing ability, e.g., *Aspergillus niger, Aspergillus awamori, Penicillium digitatum,* etc. Phosphobacterin are the bacterial fertilizers containing cells of *Bacillus megaterium* var. *phosphaticum,* prepared first by USSR scientists. They increased about 10 to 20% crop yield [8] and also produced plant growth promoting hormones which helped in phosphate solubilizing activity of soil. Al- Rehiayani et al. [1] found that *B. megaterium* reduced penetration of *M. chitwodi* and *Pratylenchus penetrans* into potato roots by 50%. Padgham and Sikora [18] stated that treatment with *Bacillus megaterium* resulted in 40% reduction in nematode penetration and gal formation compared with non treated rice plants. Khan et al. [15] stated that biofertilizers, based on plant growth promoting microorganisms, particularly phosphate-solubilizing microorganisms in place of inorganic fertilizers, could also be used in nematode disease management.

7.2.5 Growth Promoting Rhizobacteria (PGPR)

Some other microbes, e.g., *Bacillus* sp., *Pseudomonas fluorescence, Serratia* sp. and *Ochrobactrum* sp. are able to promote growth of plants. *Pseudomonas fluorescence* application to the black pepper enhanced uptake of nutrients which increased plant biomass. Bevivino et al. [4] found that rhizobacteria could stimulate plant growth directly by producing growth hormones and improving nutrient uptake and indirectly by changing microbial balance in the rhizosphere in favour of beneficial microorganisms.

7.2.6 Mycorrhiza

Mycorrhizas are developed due to the symbiosis between some specific root inhabiting fungi and plant roots and are used as biofertilizers. They absorb nutrients such as manganese, phosphorus, iron, sulphur, zinc, etc. from the soil and pass it to the plant. Mycorrhizal fungus increases the yield of crops by 30-40% and also produces plant growth promoting substances. They occur commonly in the roots of crop plants. AM fungal hyphae enhance the uptake of phosphorus and other nutrients that are responsible for plant growth stimulation including root and shoot length. AM also enhances the growth of black pepper and protects from *Phytophthora capsici*, *Radopholus similis* and *Meloidogyne incognita* [2]. AM fungi enhance water uptake in plants and also provide heavy metals tolerance to plants. Bagyaraj et. al. [3] reported that inoculation of tomato roots with root knot nematodes enhanced infection and spore production by AM fungus, *Glomus fasciculatum*. Inoculation of tomato plants with this fungus significantly reduced the number and size of the root knot gals produced by root knot nematode, *Meloidogyne incognita* and improved plant growth. Hajra et. al. [11] reported that leaf area and plant height were increased in mycorrhizal than non-mycorrhizal plants, while they showed a sharp decrease in nematode infected plants. The same plants showed less water content due to xylem vessel damage. In mycorrhizal plants, roots had large amount of carbohydrates indicating transfer of photosynthates to fungal partner. Nematode-infected roots have least amount of carbohydrates, showing a great sink of carbon to rhizosphere.

The uses of biofertilizers were advocated by various Krishi Vigyan Kendras, which are in direct relation with farmers' and are the pathway for lab to land programme. They do try to communicate between the investigators of biofertilizers from lab to the end users of biofertilizers. If the usage and cost effectivity could be assessed carefully, there can be assured benefits in production and usage of biofertilizers. The benefits usually obtained by the use of biofertilizers will not be as visible as that of chemical fertilizers. As biofertilizers are not dramatic and the results cannot be obtained in short time period because of which farmers' are not aware of its significance as enhancer of soil health of farmland. In the context of increasing awareness about the use of natural products and organic agriculture, these products will have good scope. Further, the organically grown produce fetches higher prices both in domestic and export markets. It is estimated that the production of biofertilisers in the country by the existing units is about 7500 to 9000 TPA. This is far below the potential requirement of 7.6 lakh TPA by the year 2000-2001 as estimated by the National Biofertiliser Development Centre (NBDC), Ghaziabad. So far, the Ministry of Agriculture has supported establishment of 67 biofertiliser units in different parts of the country. This estimated demand of NBDC is based on the cultivated area of the country and treatment of the total seed sown at the rate of 200g biofertilizer per 10 kg of seed. Although this assumption reflects only the macro level requirement, even if 50% of the cultivated area is to be brought under biofertilizer application, there will be a wide gap between the actual production and the requirements. The current trends indicate that there is a steady increase in the demand.

The technology used is indigenous and the scientific aspects of production are standardized by Agricultural Universities and Research Laboratories of GOI. Machineries and laboratory equipment are available from various manufacturers and should be of BIS standards. At present biofertilizers are produced using these microorganisms in large-scale industrially and are available for field application [5]. Biofertilizer production technology comprises three important steps: (1) development of strains, (2) upscale of biomass, and (3) preparation of inoculants. Since the beginning of biofertilizer production in India, pure bacterial broth with high cell count is blended aseptically with sterilized carrier such as peat, charcoal and/or lignite so as to obtain a moist powdered formulation having high population of desired microbes [10, 13, 14, 17, 21]. It is generally recommended that product should be free from contaminants, and having a microbial load of approximately 10^7 cells per gram carrier can give optimum results of plant growth promotion in designated crop following recommended method of application [12]. Further, in order to use efficient strains, focus is required to be paid to obtain region, soil and crop-specific strains and make them easily available to production units as the up-scaling of biomass is done in industries by entrepreneurs [6, 16]. As biofertilizers are live microbial preparation of very high cell count, the desired organisms are carefully monitored during the production process so as to obtain contamination free microorganisms. These problems in biofertilizer production technology have been addressed by several workers [19, 20, 23], however, proper procedure with cost effective calculation of biofertilizer production and marketing, especially at small scale level in rural areas has not been worked out to attract entrepreneurs to adopt biofertilizer technology as agri-business. Though the biofertilizer technology is a low cost, ecofriendly technology, several constraints limit the application or implementation of the technology. These constraints may be environmental, technological, infrastructural, financial, human resources, unawareness, quality, marketing, etc. Despite significant improvement/refinement in biofertilizer technology over the years, the progress in the field of biofertilizer production technology is below satisfaction due to below-mentioned constraints at different levels.

7.3 CONSTRAINTS OF STRAIN AND ITS GROWING CONDITIONS

The success of any biofertilizer in the field primarily depends upon the strain of microorganisms used in the product. Unavailability of appropriate and location specific strains which are efficient, competitive with indigenous microflora and cooperative with wide range of host under diverse environmental conditions, survival and persistence in soil, mobility in soil, ability to survive in moisture deficient, high temperature and adverse soil conditions, etc. can be used efficiently as biofertilizer. The strain, besides possessing specific attributes should also have the ability to colonize the soil and rhizosphere, be able to successfully compete with the native soil microorganisms and should have enough capacity to survive in the soil for long time in association with other soil microorganisms. Identification of strains as suited to the agro-eco system, particularly the soil pH and moisture conditions should also be one of the parameters to be used as biofertilizer. Unavailability of potential regional strains is one of the major crisis. The specificity and competitive ability of the strain is the key point on which the efficacies of the organism relay with respect to hosting soil and plant variety. Through research, specific strains as suited to a particular soil and environmental conditions should be identified and processed for use as biofertilizers. Biofertilizers of specific culture should be used for specific crop. In India large number of research institutes belonging to Agriculture Universities, Conventional Universities, ICAR Institutes and other organizations are involved in isolation and identification of these microbial strains.

In the next stage the culture has to be mass multiplied in two levels namely (i) at primary level using shakers in flasks; and (ii) secondary stage multiplication in fermentors. The manpower that has to be used in this step should be highly qualified, and hands on training for this purpose is highly essential to meet out the specialized demands for mass multiplication of bacterial cultures. The important factor in this is the preparation of growing medium in which the culture is mass multiplied. Composition of growth media for growing the desired microbe, which is to be used as biofertilizer, should be standardized with intricacies. One should assess the C : N ratio of media, ingredient that limits the growth at certain stage, pH, aeration of media, its growth stages and above all the formulation of media should not alter the plant growth promoting properties of microbes viz., enhancement of agronomic parameters, nutrient absorption efficiency, production of various metabolites, etc. After the media is formulated and sterilized in fermentor, it is inoculated using the "starter" cultures multiplied in the flasks at definite ratios, usually 5%. The bacterium growing medium is called broth and it is continuously aerated by passing sterile air from compressors. To deliver a very high population of microorganisms in biofertilizers, it is very much essential to cultivate these microorganisms under appropriate conditions to achieve very high population per unit of growing medium. Usually a final cell count of $>10^9$ cells per mL of broth should be achieved. This is being done in laboratories under controlled conditions in small glass containers (for small-scale production) or large-scale fermentors (for large-scale production units). In this process, in some cases the fermentation is taken up with specialized formulations with cell immobilizers added at different stages of growth. Finally, prepared broth with immobilized cells having a cell count of $>10^{10}$ is harvested and packed in bottles. In some other cases, the vegetative cells after being cultivated to desired level are converted into cysts or spores. These spores are further treated to keep them in dormant position. Final preparation is packed in bottles. If sterile conditions can be arranged, then instead of bottles plastic pouches can also be prepared.

7.3.1 Carrier Constraints

Once the optimum growth of microbial cells is achieved in the multiplication process, it has to be mixed with the suitable carrier material, which can provide ideal home for these microorganisms for about 6 months to 12 months time. The first step in this process involves the selection of suitable carrier materials. Unavailability of good quality carrier material or use of different carrier materials by different producers without knowing the quality of the materials is one of the constraints. As a result of intensive researches in this field, many materials have been identified as suitable carrier materials. Among them, peat, charcoal, lignite, charcoal-soil mixture, charcoal FYM mixture, vermiculite and kaolin (mixed with silica or hydrogel) have been identified as good carrier materials. Depending upon the availability and cost, different production units are using different carrier materials. Peat of good quality (more than 75% carbon) is a rare commodity in India. Nilgiri peat is of poor quality (below 50% carbon). According to the availability and cost at the production site, choice is only with lignite and charcoal in India. As per the suitability, the order is peat>lignite>charcoal>FYM>soil>rice husk. Good quality carrier must have good moisture holding capacity, it should be free from toxic substances, sterilisable and must have a readily adjustable pH from 6.4-7.0. Under Indian conditions, where extremes of soil and weather conditions prevail, there is yet no suitable carrier material identified capable of supporting the growth of biofertilizers. Better growth of bacteria is obtained in sterile carrier, and the best method of sterilization is gamma irradiation (while using autoclave, lime mixed lignite is filled upto two-thirds capacity of steel trays and sterilized at 121°C for 1-2 hours for three days) for carrier material. For preparation of finished goods the pure bacterial liquid containing very high population of required microorganisms is mixed with the carrier material to obtain moist powdered formulation, which is packed in polythene bags and supplied to the farmers. Depending upon the facilities, mixing of bacterial liquid with carrier material is being done either manually or under complete sterile conditions. Packets prepared by manual mixing method have shorter shelf life (2-3 months), while packets prepared under complete sterile conditions have longer shelf life (6-12 months).

7.3.2 Infrastructural Constraints

The development of bacterial inoculants is claimed to be cheaper, although the large-scale screening of strains with biological activity is still a tedious job. To develop a bio-fertilizer unit almost all the necessary equipment of high quality needs to be installed. One R&D laboratory should also be attached with biofertilizer unit. Equipment like fermentor, laminar airflow station, autoclave, boiler, air compressor, coolers, vacuum pump, blending machines, centrifuge, reactor, mass flow meter, cylinders of oxygen and nitrogen is required. One should have enough storage space in the godown which should be maintained at 20°C temperature. Sophisticated weighing and packing machines are also required. The whole production unit should be connected to uninterrupted power supply which is a major constraint. All the above-mentioned machines and conditions are essential for proper running of the biofertilizer unit. The unit should have its own transport vehicles for bringing the materials as well as for distributing the finished products. The plant must have technical collaboration with Agricultural Universities and research stations.

Poor inoculant quality is not due to poor production facilities but can be caused by poor standards, transport, inadequate distribution and storage facilities. Poor storage and distribution facilities at manufacturer end, distributor and dealer levels are the major constraints in the availability of good quality product in market. The inoculants after production and mixing with the potential selected carrier should be stored at 20°C for obtaining better shelf life. However, if the packets containing the bioinoculants get exposed to temperature above 40°C, the efficiency of the organisms gets diminished and even vanishes. Thus, a major drawback in India is the natural temperature as it is a tropical country and the temperature of most of the places and in most of the season goes beyond the appropriate temperature. It should be given a thought on how to maintain the temperature of the packets at 20°C. If a cool chain for delivering the packets containing biofertilizers at the farmers' field could be managed, the quality product could be delivered to end users.

7.3.3 Financial Constraints

Broadly, the capital cost includes the cost of land, development of land, fencing, civil works (plant building, office, godown, etc.), plant and machinery, preliminary and preoperative expenses etc. Capital cost of a model biofertilizer unit with a capacity of 150 TPA will be Rs. 75 lakh. The project cost comprises of land and land development, civil structures, plant and machinery, account of misc. fixed assets, preliminary and preoperative expenses, vehicles, contingency and margin money for working capital. The items of income include sale of biofertilizers, while the expenditure includes the cost of raw material, transportation and commission, power, fuel, packing distribution, wages and salary, repairs and maintenance, insurance, advertisement and other overheads. Yet, one should bear in mind that this type of future agriculture will probably increase the cost of development of bacterial inoculants, since many types of inoculants should be developed, and although fermentation is costlier method of production, the fermentation plant is more versatile. Ministry of Agriculture, Department of Agriculture and Cooperation, Government of India is implementing a central sector scheme, viz., "National Project on Development and use of Biofertilizers". Under this project, a subsidy up to Rs. 20 lakh is provided for setting up a biofertilizer production unit of 150 TPA. It is proposed to release the subsidy through back ended subsidy procedure. If the units are getting the subsidy, the amount will be adjusted to the last few instalments of bank loan. The farmer should be aware of the facilities to be provided by the government for biofertilizer production unit so as to avail the facility and get the benefit.

7.3.4 Environmental Constraints

Successful introduction of biofertilizers on farmers' field is restricted to certain crops and locations, either because of micro-environmental and biological reasons or because of the human factor. With regard to the last category, poor inoculant quality is not necessarily the result of small-scale production activity that can be caused by poor standards of transport, inadequate distribution and storage facilities. Inoculants can be stored for six months or longer at 20°C but lose their effectiveness within few hrs at 40°C or higher. No established sale network for biofertilizers exists except in a few cases. In practice, State Department of Agriculture places supply order mostly with their own production units, from where packets are transported to district headquarters before they reach the field. During delivery from production units to sowing field high temperature to which these tiny microbes are exposed result in their death, which in turn results in poor quality biofertilizer. The situation occurs frequently in most states where packets are transported during summer (+ 40°C) for use in kharif crops. Such high temperatures do not prevail in South Indian States, especially in Tamil Nadu, and consumption of biofertilizer has increased manifold there during the last two years. Encouraged with this, the North Indian States are also trying to use biofertilizers for their farmlands. But lack of cool chain and hot climate hinders the efficient usage of biofertilizers. There is a need for awareness of proper inoculant storage conditions at each point in the distribution chain. Unfortunately, this is not available at the distribution outlets; It needs immediate improvement.

7.3.5 Field Level Constraints

The benefits of biofertilizer level application depend on several soil and crop management aspects in field conditions. There is a low level of acceptance of biofertilizers at farmers' level. Because of slow release of nitrogen, the response in not immediate as is visible in case of chemical fertilizers. The constraints for incomplete expression of biofertilizers can be a combination of several factors such as soil and climate problems, unfavourable pH, high temperature, drought, salinity, high nitrate level, presence of certain elements at toxic level or deficiency of P, Co, Cu, Mo, etc., competition from native strains and compatibility with the host plant and genotypes, supply of energy (carbohydrate) to the plant roots, prevalence of plant diseases, etc., lack of knowledge and availability of quality products, use of substandard inoculants, lack of proper handling or faulty inoculation techniques, any adverse effect of agrochemicals and unfavourable conditions such as waterlogging.

Regarding BGA, despite years of research, actual use of algal inoculation is still minimal. Present production of 200 ton BGA is sufficient to treat only 20000 hectare at 10 kg/hectare. Its potentiality shown at Nagpur, Maharashtra is not reflected in the fields of UP and MP and Bihar. The reason for this may be the imbalance between testing and field studies; the utilization of grain yields is the only criterion in inoculation experiments (without collecting any information on the environmental characteristics, indigenous N fixing BGA, fate of inoculated BGA, etc.). Underestimation of the potentiality of indigenous strains and possibly sub-optimal dose of inoculant is also one of the possible hindrance in not harnessing the efficiency of the biofertilizers.

As the production and distribution of biofertilizers is seasonal, the commercial production units may suffer from lack of demand. The biofertilizers are required only at the sowing time during kharif and rabi season, so the production units become operative before sowing. Rest of the year vegetables, legumes, millets, etc. are planted as short duration crops and demand also diminishes. Thus, the units have the load only at start of sowing of rice and wheat. Some seasonal crops may also require biofertilizers, but when their ratio is compared with staple crops, its nominal.

7.3.6 Marketing Constraints

Lack of storage and marketing facilities and the supply chain currently suffers lack of infrastructure and high costs linked to handling small quantities for growing niche markets. Marketing of biofertilizers is not easy, as the product contains living organisms. It involves the quality of product, its usefulness, packaging, consumer demand, faith and also promotional aggressiveness of the producer. The use of biofertilizer in India started in early 1970s and since then even today consumption has not picked up basically due to unawareness of farmers' and sub-standard material in the market, since farmers' have not widely accepted the biofertilizers. Its off-take depends on the indent given by different State Government. Production share of State Government is 54% followed by institutional agencies 28% and 16% by private companies. This clearly indicates that the State Government have their own production units and the placing of indent depends on them, it is very difficult for other producers to market their product in those states. Because of these, most of the private companies who distributed biofertilizers during 1970s have closed their units because of not having indent orders from State Government. But in near past a lot of private companies are now getting involved in the business of production of biofertilizers. Considering the constraints in marketing of biofertilizers, a financial support is being extended from National Project on Biofertilizer. It is expected that the consumption of biofertilizer will increase to a desirable level. Intensive extension activities through compact demonstration programme, wide publicity through mass media can help in creating awareness among farmers on biofertilizers which may also help in marketing. Lack of retail outlets or the market network for the producers is also a constraint. Retail fertilizer dealers do not keep biofertilizers mainly because of the short shelf life, limited demand and lack of storage facilities is one of the possible cause for non-availability of right inoculant at the right place in right time.

7.4 QUALITY CONTROL

Quality of inoculants in the biofertilizer pack is one of the most important factors resulting in their success or failure and acceptance or rejection by the farmers. Basically, the quality means the presence of right type of microorganism in active form and in desired numbers. The stages requiring quality control are during mother culture stage, during carrier selection, during broth culture stage, while mixing of broth with carrier, during packing and during storage. Testing of the culture is usually done by taking a sample from the finished product for comparison with standard specification at the time of mixing of broth with carrier. In India, the Indian Standard Institution has developed standard for *Rhizobium* and *Azotobacter* [12]. The standard prescribed is that the inoculants shall contain a minimum 10^7 cells per gram of carrier on dry mass basis within 15 days before the expiry date marked on the packet when the inoculants is stored at 25-30°C [13, 23]. If hazards imposed by harsh storage and transport conditions under tropical climatic regime are minimized, either by decentralization of manufacturing units or by rapid transportation to farmers at the time of showing/planting, the quality of biofertilizers reaching the end users can be improved to a greater extent. No regulation or act on the quality of the products is also one of the causes of sub-standard inoculants.

Biofertilizers tend to mutate during fermentation and thereby raising production and quality control mechanisms is need of hour. Extensive research work on this aspect is urgently needed to eliminate such undesirable changes. Different microbiological and biotechnological tools should be used to come up with some possible answers. The tool should be so efficient that it differentiates the desired microbe from others residing in the niche. Some of molecular marker or lipid content could be used to mark the strain and this technique should be used to identify the presence and quality of the strain used.

7.5 AWARENESS ON THE TECHNOLOGY

The farmers are backbone of the country, which largely depends on the farming system for feeding its population. To maintain the supply and demand ratio of food and feed farmers are bound to use chemical fertilizers. Chemical fertilizers and pesticides do give immediate and visible response but in turn they degrade the soil and change the diversity of useful plant growth promoting microbial population. The usage of chemical fertilizers also causes damage to the ecosystem by continuous application of inorganic fertilizer. The land on continuous use of inorganic fertilizers loses its viability. The farmers' are unaware of the fact that excessive application and long-term usage of chemical fertilizer will make the land infertile in due course of time. However, a combinatorial usage of chemical and microbial fertilizer does maintain the microbes in niche of the rhizosphere of plants which helps in improving the soil health. In spite of considerable efforts in recent years, majority of farmers' in India are not aware of biofertilizers—their usefulness in increasing crop yields sustainably. Proper attention of farmers' towards the technology is still needed in order to manifest the results at field level. Communication gap and miscommunication to the farmers by some commercial producers also is a considerable constraint. Moreover, the establishment of microorganisms resisting the antagonistic activity of the native microbes is always challenging. Soil nature, temperature, humidity, pH, local insect population, etc., are some other acclimatization problems of which farmers' should be aware so as to select the proper biofertilizer. Unawareness of the farmers' about the benefits of the biofertilizer technology is also one of the major factors that contribute to its less usage. Consistent awareness of the farmers by conducting community programmes is still welcome. Inadequate staff and that too not technically qualified who can attend to technical problems of the farmers', who are not given proper instructions about the usage of biofertilizers, is also a major problem. Most important constraints for adoption of biofertilizers in India have been identified as lack of knowledge of the extension personnel and farmers' about inoculation technology, effective inoculant delivery/supply system and lack of sincere policy to exploit biofertilizers successfully [24]. Seed treatment for direct sown crop and seedling root dip for the transplanted crops are methods of inoculation of biofertilizers that give better results [22].

CONCLUSION

Regular use of chemical fertilizers and pesticides has deteriorated the soil health and led to its degradation. The way which could help in its restoration was the usage of biofertilizers. Although it's a time taking process but has proven to be effective in many ways and at many places. Criterion for selection of strain that could be used as biofertilizer should be strictly followed and judged before its use as per climate, soil physicochemical properties and the functionality that a particular strain to be used as biofertilizer should carry. One important aspect that one should consider is that its functionality should not be lost due to its multiplication in the fermentor and if possible it should be enhanced using factors that could impact its growth. The present scenario of soil health and its physicochemical properties could then be improved.

REFERENCES

1. Al-Rehiayani, S., Hafez, S.L., Thorton, M. and Sundararaj, P. (1999). Effects of *Pratylenchus neglectus, Bacillus megaterium* and oil radish or rapeseed green manure on reproductive potential of *Meloidogtne chitwodi* on potato. *Nematropica* 29, 37–49.
2. Anandraj, M. Venugopal, M.N., Veena, S.S., Kumar, A. and Sarma, Y.R. (2001): Ecofriendly management of disease of species. *Indian species.* 38 (3), 28–31.
3. Bagyaraj, D.J., Manjunath, A. and Redy, D.D.R. (1979). Interaction of vesicular arbuscular mycorrhizae with root knot nematode in tomato. *Plant and Soil* 51, 397–403.
4. Bevivino, A., Sarrocco, S., Dalmastri, C., Tabacchioni, S., Cantale, C. and Chiarini, L. (1998). Characterization of a free-living maize-rhizosphere population of *Burkholderia cepacia*: Effect of seed treatment on disease suppression and growth promotion of maize. *FEMS Microbiol. Ecol.* 27: 225–237.
5. Bhattacharyya, P. and Singh, T. (1992). Simplified methods of mass production and application of biofertilizer. *Proc. Natl. Sem. Organic Farming.* Edited by Rai MM, Verma LN, JNKVV and IGKVV, Indore, India, pp. 166–178.
6. Burton, J.C. (1981). Rhizobium inoculants for developing countries. *Trop. Agric.* 58: 291–295.
7. Bushby, H.V.A. and Marshall, K.C. (1977). Some factors affecting the survival of root- nodule bacteria on desiccation. *Soil Biol. Biochem.* 9, 143–147.
8. Cooper, R. (1959). Bacterial fertilisers in the Soviet Union. *Soils and Fert.* 22. 327–333.
9. Dubey, R.C. (2006). *A Textbook of Biotechnology.* 4th edn., S. Chand & Co. Ltd, New Delhi, ISBN 81-219-2608-4, p. 732.
10. Gulati, S.L. and Seth, V.K. (1973). An improvised portable fermentor for small-scale production of *Rhizobium* culture. *Curr. Sci.* 42, 394-395.
11. Hajra, N., Shahina, F. and Firoza, K. (2013). Biocontrol of root-knot nematode by arbuscular mycorrhizal fungi in *Lufa cylindrical. Pak. J. Nematol.* 31, 7–84.
12. I.S.I. (1986). Indian standard specification for *Rhizobium* inoculants. Standard: IS 8268–1986.
13. Jauhari, K.S. (1988). Dilution optima of broth cultures for maximizing production of carrier-based inoculants. *Curr. Sci.* 57. 507–509.
14. Jauhari, K.S. and Subba, Rao, N.S. (1984). Simple glass fermentor for *Rhizobium* culture production. *Curr. Sci.* 53, 984–985.
15. Khan, M.R., Khan, S.M., Mohidin, F.A. and Askary, T.H. (2007). Effect of certain Phosphate-solubilizing bacteria on root knot nematode disease of mung bean. Proc. First Intern. Meeting on Microbial Phosphate Solubilzation. *Develop. Plant and Soil Sci.* 102, 341–346.

16. Motsara, M.R. and Bhattacharyya, P. (1994). Status of biofertilizer and constraints in their large-scale commercialization. In: *Commercialization of Biotechnology for Agriculture and Aquaculture,* edited by. Srivasatava P., Chandrasekhar K.S., Oxford and IBH Publ. Co. pp. 285–298.
17. Mishra, B.K. and Dadhich, S.K. (2010). Methodology of nitrogen biofertilizer production. *J. Adv. Dev. Res.* 1, 3–6.
18. Padgham, J.L. and Sikora, R.L. (2007). Biological control potential and modes of action of *Bacillus megaterium* against *Meloidogyne graminicola* on rice. *Crop Prot.* 26, 971–97.
19. Sattar, M.A., Khanam, D., Ahmad, S., Haider, M.R., Podder, A.K. and Bhuiyan, M.A. (1997). In: *Extending Nitrogen Fixation Research to Farmers Fields,* edited by Rupela O.P., Johensen C.C., Herridge D.F., (ICRISAT, Pantencheru, India), pp. 227–234.
20. Sen, P. and Palit, P. (1988). *Biofertilizers – Potentialities* and *Problems.* Naya Prokash, Calcutta, pp. 51–56.
21. Sethi, S.K. and Adhikary, S.P. (2009a). Vegetative growth and yield of *Arachis hypogea* and *Vigna radiata* in response to region-specific *rhizobium* biofertilizer treatment. *J. Pure Appl. Microbiol.* 3, 295–300.
22. Somasegaran, P. and Hoben, H.J. (1985). *Methods in Legume Rhizobium Technology.* NifTAL, MIRCEN, Univ. Hawaii, Paia, USA, p. 367.
23. Srivastava, R., Khalid, A., Singh, U.S. and Sharma, A.K. (2010). Evaluation of arbuscular mycorrhizal fungus, fluorescent *Pseudomonas* and *Trichoderma harzianum* formulation against *Fusarium oxysporum* f. sp. *lycopersici* for the management of tomato wilt. *Biological Control,* 53, 24–31.
24. Subba Rao N.S. (1993). *Biofertilizers in Agriculture and Forestry,* edited by Subba Rao NS (Oxford and IBH Publishing Co, New Delhi), p. 242.
25. Wani, S.P. and Lee, K.K. (1992). Role of biofertilizers in upland crop production. In: *Fertilizers, Organic Manures, Recyclable wastes and Biofertilizers.* (ed.) H.L.S. Tandon, Fertilizers Development and Consultation Organisation, 204-204A Bhanot Corner, 1-2 Pamposh Enclave, New Delhi-110048, India. 91–112.

8

Application of Bioinformatics in Biofertilizer Technology

Asha Mishra, Sashwati Ghosh (Sachan), Vinod Kumar Nigam and Dev Mani Pandey*

Department of Bioengineering, Birla Institute of Technology Mesra, Ranchi – 835215 Jharkhand, India

***Corresponding Author E-mail:** dmpandey@bitmesra.ac.in*

ABSTRACT

Biofertilizers are microorganisms containing compound having potential to provide sustainable plant growth promotion and an increases in crop productivity. These are associated with the root of plant, which is an important soil ecological environment for plant-microbe interactions. Recent molecular biology and next generation sequencing approaches have produced enormous amount of data where use of various bioinformatics tools and software becomes very much important in their storage as well as providing information in public domain. Subsequent use of bioinformatics has also been extended in finding the information hidden in sequences as well as their biological significance, predicting the structures and functions, comparative analysis, phylogeny studies, interaction and networking studies and mechanism involved in symbiosis of plant and microbes resulting in enhanced crop productivity. Further, factors as well as mechanisms involved in solubilisation of minerals, phosphates and other nutrients, enhancing resistance to stresses, stabilizing soil aggregates, and improving soil structure and organic matter content could also be predicted by bioinformatics approaches. Therefore, recent development in bioinformatics tools and software will have tremendous applications in biofertilizers technology by enhancing agricultural productivity for human society welfare.

Keywords: Bioinformatics, Biofertilizers, Bacillus, Nitrogen Fixers, Phosphate Solubilizers, Rhizobacteria.

8.1 BIOFERTILIZERS

Agriculture contributes to a major share of national income and export earnings in many developing countries, while ensuring food security and employment. Sustainable agriculture is vitally important in today's world because it offers the potential to meet our future agricultural needs, something that conventional agriculture will not be able to do. Recently there has been a great interest in eco-friendly and sustainable agriculture. Plant growth promoting rhizobacteria (PGPR) are known to improve plant growth in many ways when compared to synthetic fertilizers, insecticides and pesticides. They enhance crop growth and can help in sustainability of safe environment and crop productivity. The rhizospheric soil contains diverse types of PGPR communities, which exhibit beneficial effects on crop productivity. PGPR as biofertilizers are well recognized as efficient soil microbes for sustainable agriculture and hold great promise in the improvement of agriculture yields.

PGPR has potential towards developing the sustainable systems. They have capacity to improve the fruitfulness of the soil for the growth of plants using biological wastes [1, 2]. Symbiotic (*Rhizobium* sp.) and non-symbiotic bacteria (*Azotobacter, Azospirillum, Bacillus,* and *Klebsiella* sp.) have also been used worldwide to achieve the enhanced plant productivity [3, 4]. Vacheron et al. [5] reported that PGPR colonize roots of monocots and dicots, enhancing plant growth by direct and indirect mechanisms. They do not carry any compound that causes harm to the soil; they rather enrich the soil with the existing micro-organism by producing organic nutrients. Therefore, enriched soils have capacity to tolerate biotic and abiotic stresses upto some extent.

Biofertilizers contain living microorganisms, which when applied to the seed, plant surface, or soil, colonize the rhizosphere and promote plant growth through increased supply of primary nutrients to the host plant [6]. Biofertilizers are important factors that are involved in reducing the cultivation cost and practising organic farming. The key function of biofertilizers is to provide nutrients to crop plants, thus enhancing plant growth with the production of phytohormones, regulating the activity of plant pathogens, maintaining soil structure and soil fertility as well as bioaccumulation of inorganic substances [7, 8]. Biofertilizers have capacity to enrich the nutritional content and quality of soil. The main living organisms used in biofertilizers are bacteria, fungi and blue-green algae (also known as cyanobacteria), etc. These living organisms have very much efficient mechanisms and symbiosis relationship with plants where both partners get benefits from each other. The nitrogen fixers, phosphate-solubilizing microbes, vesicular arbuscular mycorrhizas, plant growth promoting rhizobacteria and sulphur solubilising microbes are some of the examples of biofertilizers. In nitrogen fixers, bacteria (*Rhizobium, Azotobacter, Mycobacterium, Azospirillum, Bacillus,* etc.) and blue-green algae (*Anabaena, Nostoc, Tolypothrix* and *Anabaenopsis*) are very much important and used in biofertilizers. Similarly, *Azolla filiculoides* and *Azolla rubra* are also important and used in the preparation of biofertilizers.

In 1980s agriculture primarily depended upon the chemical fertilizers for enriching the nutritional status of soil and food production. Subsequently, after 1990s increased population and food demand, enhanced use of fertilizers became a major concern for environmental safety and human health. Therefore, the concept of microbial inoculants uses in agriculture originated. These microbial inoculants are environment-friendly and are ensuring sustainable agriculture since the past two decades. Singh et al. [9] proposed that inoculation of thermotolerant phosphate-solubilizing bacteria into agricultural and vegetable wastes will be an approachable and practical strategy for preparing multifunctional biofertilizers. Boraste et al. [10] has also studied the use of various microorganisms for the production of biofertilizers.

8.2 PGPR IN FIELD EXPERIMENT

Initiatives towards isolation of microbes from soil have already been taken by various research groups of different organizations in India and abroad. Subsequently, characterization and large-scale production of these microbes are also being performed and optimized protocol/method/technology has also been developed for some microorganisms like *Azotobacter, Azospirillum,* Blue-Green Algae (BGA), Mycorrhiza and *Rhizobium.* These technologies are being used in various industries for large-scale production and subsequent selling.

Pandey et al. [11] treated *Ocimum sanctum Linn.* with *Pseudomonas fluorescence* and *Glomus fasciculatum* and observed increased plant growth, yield and biomass by phosphate solubilization and phytohormone production. The effect of *Bacillus subtilis* Bs008 on cucumber plants was observed. *Bacillus subtilis* Bs008 increased growth of lateral roots, biomass, length of root and shoot of cucumber plants through stimulation of polysaccharide and production of IAA [12]. Sadiq and Ali [13] treated wheat plants with *Bacillus megaterium* ZmR-4, *Bacillus cereus* McR-3, *Bacillus megaterium* and *Acinetobacter calcoaceticus* McR-2 and observed increased yield and plant growth with the help of phosphate solubilization and ACC production. Kaur and Reddy [14] observed the effect of *Pantoea cypripedii* (PSB-3) and *Pseudomonas plecoglossicida* on maize crop. *Pantoea cypripedii* (PSB-3) and *Pseudomonas plecoglossicida* increased maize crop yield and soil fertility with the help of phosphate solubilization. Karnwal [15] used PGPR (*Bacillus* sp. AK21, *Bacillus subtilis* AK31, *Azotobacter diazotrophics* AK14, *Microbacterium* sp. AK19 and *Pseudomonas* AK32) as crop enhancer and biofertilizer for production of maize and wheat by phytohormone IAA production. Jahanian et al. [16] showed combined effect of *Pseudomonas putida, Azotobacter* and *Azospirillum* on artichoke plants. They noticed that this combination of phosphate solubilizer and nitrogen fixer increased radicle and shoots length, shoot weight, coefficient of velocity of germination, seedling vigourty index and decreased mean time of germination. It was suggested that PGPR isolates may be used as biofertilizer to improve the growth and production of medicinal and aromatic plants under local climatic conditions of Andhra Pradesh [17]. Naumavo et al. [18] suggested that specific combinations of PGPR can be considered most competent biofertilizers.

Biofertilizers are very much helpful in making the soil enriched with various nutrients that are ready to use by plants and increasing crop yield. In modern agriculture these chemical fertilizers (responsible for making environment and soil polluted by releasing some harmful compounds) are being replaced with biofertilizers. A study towards preparation of biofertilizers from agricultural waste materials using phosphate solubilizing bacteria *Bacillus megaterium* has been conducted [9]. Singh et al. [9] described that during the composting process inoculated biofertilizers having *Bacillus megatarium* resulted in higher soluble phosphorus content as well as germination rate as compared to noninoculated biofertilizers. Further, when these bacteria are added, they can shorten the maturity period, and improve the quality and increase the soluble phosphorus content.

It is thought that development of biofertilizers will be very much helpful in decreasing the production costs and environmental pollution. Studies are being conducted for the development of biofertilizers that will be alternate for nitrogenated chemical fertilizers for lupine. Recently a review describing the application of microbial fertilizer, laboratory settings and operations involved during process, current status of biofertilizers development with emphasis on the effect of inoculating associative and free living nitrogen-fixing bacteria on the growth of crops, farmer's acceptance and utilization of biofertilizers as well as future perspectives has been published [19]. Similarly, different types of biofertilizers and their application, potential role of biofertilizers in agriculture as well as constraints with biofertilizer in current scenario have also been discussed recently [20].

8.3 MECHANISM OF GROWTH PROMOTION BY PGPR

Generally, plant growth promoting rhizobacteria facilitate the plant growth directly by either assisting in resource acquisition (nitrogen, phosphorus and essential minerals) or modulating plant hormone levels, or indirectly by decreasing the inhibitory effects of various pathogens on plant growth and development in the form of biocontrol agents. Various studies have documented the increased health and productivity of different plant species by the application of plant growth promoting rhizobacteria under both normal and stressed conditions. The plant-beneficial rhizobacteria may decrease the global dependence on hazardous agricultural chemicals which destabilize the agro-ecosystems.

In addition to providing the mechanical support and facilitating water and nutrient uptake, plant roots also synthesize, accumulate, and secrete a diverse array of compounds (amino acids, sugars, organic acids, vitamins, enzymes and nucleosides, etc.). These compounds secreted by plant roots act as chemical attractants for a vast number of heterogeneous, diverse and actively metabolizing soil microbial communities. The chemicals which are secreted by roots into the soils are generally called root exudates. These chemical compounds modify the chemical and physical properties of the soil, and thus regulate the structure of soil microbial community in the immediate vicinity of root surface. In fact, some of the exudates act as repellants against microorganisms while others act as attractants to lodge the microbes. The composition of these exudates is dependent upon the physiological status and species of plants and microorganisms. Moreover, these exudates also promote the plant-beneficial symbiotic interactions and inhibit the growth of the competing plant species. Also, microbial activity in the rhizosphere affects rooting patterns and the supply of available nutrients to plants, thereby modifying the quality and quantity of root exudates. A fraction of these plant-derived small organic molecules is further metabolized by microorganisms in the vicinity as carbon and nitrogen sources, and some microbe-oriented molecules are subsequently re-taken up by plants for growth and development.

Various methods like seedling root dip (where a layer of water is spread on the soil area where the particular crop has to grow), seed treatment (where nitrogen and phosphorus fertilizers are mixed together in the water followed by dipping of seeds in the mixture) and soil treatment (where all the biofertilizers together with the compost fertilizers are mixed) are being followed for enhancing the crop productivity.

8.4 BIOINFORMATICS APPROACHES IN BIOFERTILIZERS AND ENHANCING CROP PRODUCTION

In current years bioinformatics is also playing important role in biofertilizers technology. Various PGPRs are being searched from different soil zones from different plants. These PGPRs might be characterized and their subsequent performance on plant growth and development may be studied. Plants showing good performance having individual PGPR or in combination may be used for further study. Some genes related to either enrichment of nutritional status by solubilizing mineral nutrients in soil or help in their uptake in plants have been reported. Sequence of these genes might be retrieved from various databases using online and offline computational approaches. These sequences may be used to design the primers and their subsequent amplification in different tissue of plant showing good performance. Amplified product may be submitted for sequencing followed by sequence analysis like gene structure, number of exons and introns, motifs, corresponding amino acid sequence, motif and domain, 3-D structure of protein, binding sites, as well as other analysis using bioinformatics approaches.

Now sequencing of genomes from plant and bacteria gave a platform where evolutionary studies can be studied with an aim to determine the phylogenetic tree and possibility of having common ancestor. Comparative analysis of plant genome with bacterial genome may give some clue about organization of their genes and whether they remained conserved during evolutionary time. The comparative analysis of these plant and bacterial genome will also reveal about symbiosis association between microorganism and plant. By comparative genomics, computational analysis, functional genomics and bioinformatics approaches information obtained from completely sequenced crop may be utilized to predict the structural and functional properties of partially sequenced crops as well as their improvement. Presently, complete genome sequences of some crops are available online. Similarly, some genes involved in symbiotic association have also been known. This new ability of the plants and microorganism to make symbiotic association resulted in increased nutritional quality of the crops. Significant progress has also been made in developing cereal varieties that have a greater potential in yield with the application of biofertilizers. These crops along with biofertilizers will be a boon for agriculture by improving poorer soil areas and, thus making more land to be available for agriculture globally. Studies are also in progress to produce technology having different crops and biofertilizers that are capable to resist various stresses, and enriching the nutritional status of soil. Now DNA-based profiling is studied for finding the novel microorganisms that might be used in biofertilizers. Recent recombinant DNA technology, functional genomics and bioinformatics approaches will further increase the directed production of biofertilizers. Indian government has also taken initiative in opening of bioinformatics centres in various parts of the country. These centres allow the study and exchange of biological information with efficient communication networking, bioinformatics software development, role of tolerant rhizobacteria in various crop cultivation and their *in-silico* studies involved in various processes including biofertilizer technology development.

Commonly used microbes that are being studied for biofertilizers include *Bacillus* spp., *Rhizobium* spp., *Pseudomonas* spp., *Bradyrhizobium* sp., and *Azospirillum* sp. It has been reported that *Bacillus vallismortis* Strain EXTN-1 is a commercial biofertilizer and/or PGPR that is very effective in increasing potato yield and chlorophyll content; besides it also provides protection to plant against potato virus Y and X [21]. It has also been reported that powdery and downy mildews are fungal diseases that infect vegetative portion of many organically-grown vegetables [22]. Some of the groups have also utilized the egg yolk and cooking oil to make the EYCO an effective biofertilizer for controlling pathogens related to powdery and downy mildew in organically-grown vegetables. These types of new discoveries will open avenues for the use of biofertilizers for more diversified, promising, eco-friendly and sustainable agriculture [23]. An effort towards development of biofertilizers (nitrogen as well as phosphorus biofertilizers) by Government of India has also been initiated. Detailed update about biofertilizers can be found at http://www. nationalfertilizers.com. Information about various companies that are involved in biofertilizers production and suppliers in India could be reached at http://dir.indiamart.com/impcat/biofertilizers.html. Similarly, information about potash mobilizing bacteria in India can be found at http://dir.indiamart.com/impcat/potash-mobilizing-bacteria.html.

8.5 RECENT BIOTECHNOLOGY AND BIOINFORMATICS APPROACHES IN BIOFERTILIZERS TECHNOLOGY

Bioinformatics has tremendous application in biofertilizer technology and has helped in the development as well as updating the databases, algorithm as well as computational tools and software used in the analysis of biological data. Recently development of advanced molecular biology, functional genomics and next generation sequencing technologies has generated huge amount of data. The new developments in bacterial study with involvement of recent next generation sequencing technologies, Real Time PCR and Digital PCR, highly efficient computational tools, and various omics technologies have scaled up their uses and application in agricultural plants. However, during this process huge data are generated that are being analysed by *in-silico* study using online and offline software. These data may be further used for comparative analysis with various data bases and predicting the changes that might occur during the developmental stage. Therefore, various computational approaches have been used to understand the information hidden in the biological data such as comparative analysis and alignment of nucleotide and protein sequences, 3-D structure prediction of protein and DNA, molecular modelling, protein-DNA interaction studies, etc. The involvement of various mechanisms that make the bacterial system and plant system mutually beneficial and provide resistance to survive under unfavourable condition are now been studied for their functional analyses. Figure 8.1 shows various steps involved in biofertilizers application in plants, use of molecular biology and bioinformatics approaches for various analysis and outcome.

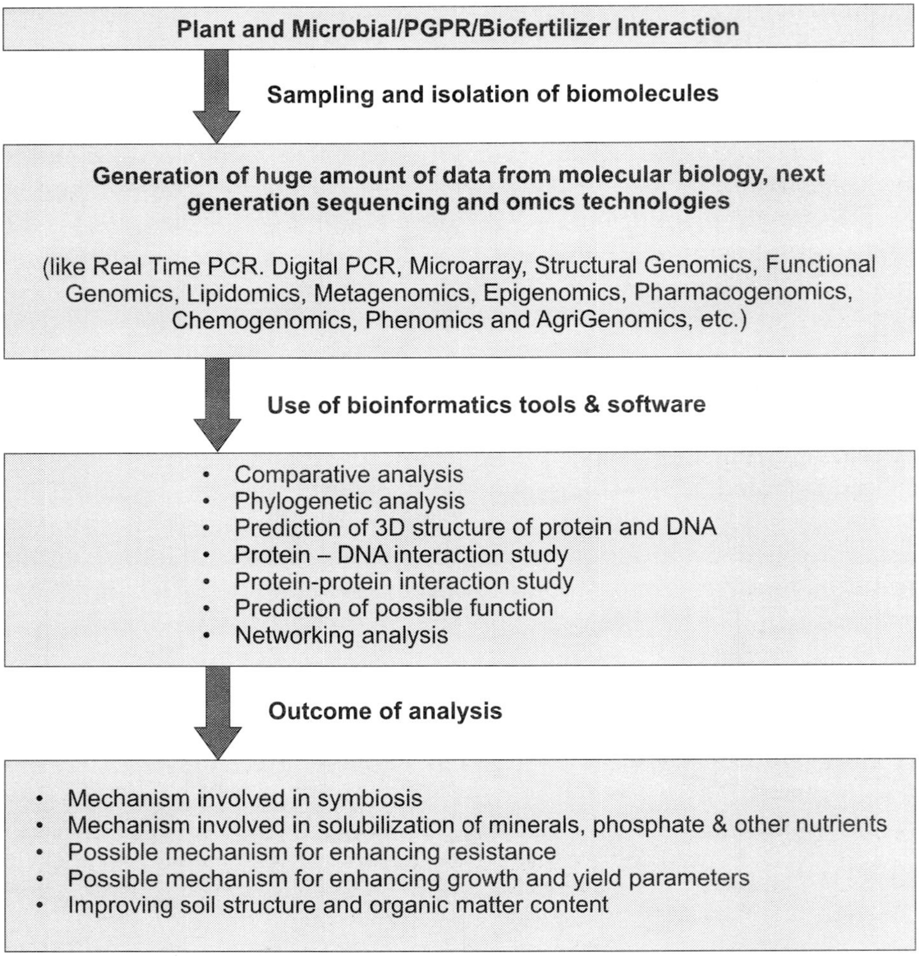

Fig. 8.1 Detailed diagram showing sequential steps involved in use of biofertilizers in plants, use of molecular biology and bioinformatics approaches for various analysis and outcome.

CONCLUSION

Recent innovations in bioinformatics tools and databases, biofertilizers, and agricultural biotechnology have tremendous implications in welfare of human society. Recent molecular biology techniques along with computational approaches will play a crucial role in analysing the competency of biofertilizers for increasing fertility of soils as well as increased crop productivity.

REFERENCES

1. Sturz, A.V. and Nowak, J. (2000). "Endophytic communities of rhizobacteria and the strategies required to create yield enhancing associations with crops". *Applied Soil Ecology* 15, 183–19.

2. Shoebitz, M., Ribaudo, C.M., Pardo, M.A., Cantore, M.L., Ciampi, L. and Cura, J.A. (2009). "Plant growth promoting properties of a strain of *Enterobacter ludwigii* isolated from Lolium perenne rhizosphere". *Soil Biology and Biochemistry* 41(9), 1768–1774.
3. Burd, G., Dixon, D.G. and Glick, B.R. (2000). "Plant growth-promoting bacteria that decrease heavy metal toxicity in plants". *Canadian Journal of Microbiology* l46, 237–245.
4. Cocking, E.C. (2003). "Endophytic colonization of plant roots by nitrogen-fixing bacteria". *Plant Soil* 252, 169–175.
5. Vacheron Jordan, Desbrosses Guilhem, Bouffaud Marie-Lara, Touraine Bruno, Moënne-Loccoz Yvan, Muller Daniel, Legendre Laurent, Wisniewski-Dyé Florence and Prigent-Combaret Claire (2013). "Plant growth-promoting rhizobacteria and root system functioning". *Frontiers in Plant Science* 4, 356.
6. Vessey, J.K. (2003). "Plant growth-promoting rhizobacteria as biofertilizers". *Plant Soil* 255, 571–586.
7. Brierley, J.A. (1985). "Use of microorganisms for mining metals" In: Halvorson H.O., Pramer D., Rogul M. (eds) *Engineered organisims in the Environment: Scientifc Issues*, ASM Press, Washington, pp. 141–146.
8. Ehrlich, H.L., (1990). *Geomicrobiology,* 2nd Edison. Dekker, New York, p. 646.
9. Singh, A.K., Masih, H., Nidhi, P., Kumar, Y., Peter, J.K. and Mishra, S.K. (2013). "Production of biofertilizer from agro-waste by using thermotolerant phosphate solubilising bacteria". *International Journal of Bioinformatics and Biological Science* 1(2), pp. 227–244.
10. Boraste, A., Vamsi, K.K., Jhadav, A., Khairnar, Y., Gupta, N., Trivedi, S., Patil, P., Gupta, G., Gupta, M., Mujapara, A.K. and Joshi, B. (2009). "Biofertilizers: *A Novel Tool for Agriculture*", *Interntional Journal of Microbiology Research.* 1(2), 23–31.
11. Pandey, P., Swami, A. and Bhatt, A.B. (2014). "Combined effect of *Pseudomonas fluorescence* and *Glomus fasciculatum* with zinc on growth of Indian Basil *(Ocimum sanctum Linn.)*". *Global Journal of Current Research.* 2(2), 30–35.
12. Buensanteai, N., Sompong, M., Saengchan, C. and Thumanu, K. (2014). "The cellular components of cucumber seedlings after primed with plant growth promoting Rhizobacteria, *Bacillus subtilis* Bs008. *African Journal of Microbiology Research.* 8(10), 1006–1011.
13. Sadiq, A. and Ali, B. (2013). "Growth and yield enhancement of *Triticum aestivum* L. by rhizobacteria isolated from agronomic plants". *Australian Journal of Crop Science* 7(10), 1544–1550.
14. Kaur, G. and Reddy, M.S. (2013). "Phosphate solubilizing rhizobacteria from an organic farm and their influence on the growth and yield of maize (*Zea mays* L.). *The Journal of General and Applied Microbiology* 59(4), 295–303.
15. *Karnwal A.* (2012). "Screening of plant growth-promoting rhizobacteria from maize (*Zea mays*) and wheat (*Triticum aestivum*)". *African Journal of Food, Agriculture, Nutrition and Development.* 12(3), 6170–6185.

16. Jahanian, A., Chaichi, M.R., Rezaei, K., Rezayazdi, K. and Khavazi, K. (2012). "The effect of plant growth promoting rhizobacteria (PGPR) on germination and primary growth of artichoke (*Cynara scolymus*)". *International Journal of Agriculture and Crop Sciences* 4(14), 923–929.
17. Malleswari, D. and Bagyanarayana, G. (2013). "In vitro screening of rhizobacteria isolated from the rhizosphere of medicinal and aromatic plants for multiple plant growth promoting activities". *Journal of Microbiology and Biotechnology Research,* **3**(1), 84–91.
18. Noumavo, P.A., Kochoni, E., Didagbé, Y.O., Adjanohoun, A., Sikirou, M.A.R., Gachomo, E.W., Kotchoni, S.O. and Baba-Moussa, L. (2013). "Effect of different plant growth promoting rhizobacteria on maize seed germination and seedling development". *American Journal of Plant Sciences 4*(5), 1013–1021.
19. Aggani, S.L. (2013). "Development of biofertilizers and their future perspective", *Scholar Academic Journal of Pharmacy,* 2(4), 327–332.
20. Rana R., Ramesh and Kapoor P. (2013). Biofertilizers and their role in agriculture. *Popular Kheti* 1(1), 56–61.
21. Park K., Paul D., Ryu K.R., Kim E.Y. and Kim Y.K. (2006). "*Bacillus vallismortis* strain EXTN-1 mediated systemic resistance against *Potato virus Y* and *X* in the field". *Plant Pathology Journal* 22(4), 360–363.
22. Callo, Jr., D.P., Medina J.R., Teofilo L.B., Tauli H.A. and Cuaterno W.R. (2002). "Manual of participatory technology development activities for IPM in vegetables. SEAMEO Regional Center for Graduate Study and Research in Agriculture (SEARCA), Los Baños, Laguna, Philippines pp. 380–386.
23. YAP Arthur C. (2009). "*Field Guide of Discovery-Based Exercises for Organic Vegetable Production*". Edited by Damaso P., Callo, Jr., Cesar A., Baniqued, Audy G., Maa Gad, Noelito C., Villa, Os Car T., Tobia *and* Kula Fu A., Seballosllos. Joinly published by Philippine Council for Agriculture, Forestry and Natural Recourses Research and Development, Philippines; The Philippine National IPM Programme, Phillippines; ASEAN IPM Knowledge Network, Philippines; and Food and Agriculture Organization Vegetable IPM Programme for Asia, FAO Regional Office for Asia-Pacific, Bangkok, Thaliand. pp. 307–310. ISBN: 978-971-94101-2-6.

9

nifH Database: Insights into Diversity and Evolution to Explore Next Generation Nitrogen Fixer

Partha Sarathi Das[1], Sunil Kumar[2], Keshab C. Mondal[1] and Bikas R. Pati[1]

[1]*Bioinformatics Infrastructure Facility, Department of Microbiology, Vidyasagar University, Midnapore – 721102, WB, India*

[2]*National Bureau of Agriculturally Important Microorganisms (NBAIM), Mau, UP, India*

E-mail: *parthadas77@gmail.com*

ABSTRACT

Microorganisms catalyse biological nitrogen fixation with the enzyme nitrogenase, which has remained highly conserved during evolution. Cloning, sequencing and metagenomics based studies of the nitrogenase expressing genes have provided a large, rapidly expanding database of sequences from diverse habitats. It has also been providing us important insights into possible ancient events of horizontal gene transfer. The distribution of nitrogen-fixing microorganisms, although not entirely dictated by the nitrogen availability in the environment, is non-random and can be predicted on the basis of habitat characteristics. The ability to assay for gene expression and investigate genome arrangements provides the promise of new tools for interrogating natural populations of diazotrophs. The broad analysis of nitrogenase genes provides a basis for developing molecular assays and bioinformatics approaches for the study of nitrogen fixation in the environment.

Keywords: Nitrogenase, *niH*, *nifD*, Phylogeny, Horizontal Gene Transfer, Evolutionary Biology.

9.1 INTRODUCTION

Microbial communities are a fundamental component of ecosystems playing critical roles in metabolism of organic matter and in biogeochemical transformations like N_2 fixation (Atlas and Bartha, 1998; 10). With recent advances in microbial techniques, the genetic diversity of the nitrogen fixers in natural communities are being thoroughly studied in recent days.The advanced techniques in molecular biology like analysis of ribosomal RNA (rRNA) phylotypes has even more enhanced this understanding. Now we have a considerable idea about the patterns of global distribution of major groups of certain phylotypes such as the beta proteobacteria in freshwater (Methe et al., 1998) and the distribution of individual genotypes [2].

These rRNA-based investigations made it clear that we do not have cultivated representatives of the numerically most abundant microorganisms in the environment [12]. So, new methods must be developed to study "the unculturables". Moreover, rRNA analysis alone is insufficient for a complete understanding of microbial ecology as it does not provide definitive information on biological, physiological or ecological. It is interesting to note that the nitrogenase gene is found among a variety of Archaea and Bacteria, but these are so sporadically distributed within clades in most cases [18] that phylogenetic closeness in the branch of evolutionary closeness cannot ensure an organism's capacity to have nitrogenase activity.

9.2 NITROGENASE

In the course of evolution, acquiring of certain proteins and enzymes to enable nitrogen fixation by certain organisms transformed them into beings of higher order. But it is surprising to note that most N_2-fixing microorganisms possess an evolutionarily conserved nitrogenase protein complex [7]. This high degree of similarity of protein sequences of nitrogenases among microorganisms suggests an early origin or lateral gene transfer among prokaryotes [13]. By studying the huge numbers of nitrogenase and 16S rRNA sequences uploaded to the biological databases [13] and also suggested by Young (1992) [19] that even among distantly diverge organisms the nitrogenase is highly conserved, it is believed that the nitrogenase evolution was a very ancient event and it must have happened in extinct ancestors. Recently, Doolittle [3], referring to the findings made through comparative genomics, however, again raised the speculation that nitrogenases could have been distributed in nature by lateral gene transfer mechanisms. Further detailed comparative phylogenetic analyses will ultimately help resolve the relative evolutionary histories of nitrogenases and rRNA genes. But this understanding and the fact that nitrogenese sequence has a high phyletic retention and conservation has raised hopes that through metagenomics approaches we will be able to classify and identify many uncultivated diazotrophic microorganisms.

The nitrogenase enzyme is a complex one. It is composed of two multisubunit metalloproteins. Component I contains the active site for N_2 reduction (molecular weight ~ 250 kDa) and is composed of two heterodimers, encoded by the *nifD* and *nifK* genes. Component II (~70 kDa) joins the events of ATP hydrolysis to interprotein electron transfer and is composed of two identical subunits encoded by the *nifH* gene. Thus, the interest to study *nifH* and *nifD* is ever growing.

The nitrogen fixation is a complex event as not only the structural genes for the conventional nitrogenase, but a number of other gene products are also required for regulation and assembly. The N_2 fixation event, apart from being metabolically expensive, also requires nearly 20 kb of DNA or more to encode all of the proteins involved in for assembly and function.Thus, it is also important to study the function and diversity of these proteins if further objective is to impart a nitrogen fixation capacity to transgenics.

In fact, as reported by Hoover, nitrogenase gene expression is highly regulated at levels ranging from transcription [1] to post-translational protein modification [9]. It has been observed that *nifHDK* operon is not constitutively expressed and is regulated in response to factors that control N_2-fixation. Hence, these days, *nifHDK* is being viewed as a good marker for nitrogenase activity. The high conservation of nitrogenase genes and wide availability of nitrogenase gene sequence data in databases, coupled with software tools customized for such searches (to be discussed later) has led us to various success stories of designing probes to ribosomal RNA phylotypes that have nitrogen-fixing capability [8, 11, 14, 15, 16, 18, 19]. These researchers investigated various diverse habitats ranging from soils, plants, bioreactors, invertebrate guts, lakes, and rivers to the open ocean, and have enriched the databases with a great diversity of nitrogenase sequences. It is interesting to note that the database for nitrogenase genes (specifically the *nifH* gene) has become one of the largest non-ribosomal gene datasets on uncultivated microorganisms developed so far. Apart from *nifH* and *nifD, nifK* also have been studied. But still now sparse amount of data on *nifD* and *nifK* are available in public databases. It is also not confirmed that these two genes are *phylogenetically* consistent. Thus still now, mainly *nifH* has been used to infer phylogenetically and development of markers. But in future if subsequently *nif K* and *D* are also used as markers, it is felt that they will also bring important insights to differentiate among closely related strains and better differentiate different *nif* gene family members such as the vanadium containing nitrogenases (which do not possess the traditional molybdenum in the protein molecule).

9.3 PHYLOGENETIC CLUSTERS ANALYSIS

The *nifH* sequence database is rapidly expanding. Gaby and Buckley [5] accomplished a colossal task of creating a nitrogenase gene sequence database which contained nearly 33000 aligned nitrogenase *nifH* sequences. The database includes 185 linked multigene entries including full-length *nifH, nifD, nifK* and 16S ribosomal RNA (rRNA) gene sequences. From the phylogenetic analysis, it is seen that homologs of the *nifH* gene can be divided into five main phylogenetic clusters.

Cluster I: It is composed of a diverse group of *nifH* genes from aerobic and facultatively anaerobic organisms that belong to the phyla including Proteobacteria, Cyanobacteria, Firmicutes and Actinobacteria.

Cluster II: Contains anfH, alternative nitrogenases that are paralogs of *nifH* and use an Fe–Fe co-factor in place of the Fe–Mo cofactor used by *nifH.*

Cluster III: Contains nifH genes that are almost exclusively found in obligate anaerobes including methanogenic archaea, *Treponema, Clostridium* and sulphate-reducing and sulphur-reducing species of *Deltaproteobacteria.*

Clusters IV and V: Contain paralogous genes that do not participate in nitrogen fixation.

For illustrative purpose, we had downloaded few sequence data of *nifH* genes and created a phylogenetic tree. In the gene tree, it is seen in some cases that are organisms are classified in some way as to defy the species tree, i.e., one organism is found to be closer to another distant genus in the clade. The same is confirmed by high bootstrap values seen in the cladogram. But these events are rare, compared to as commonly seen in cases of other proteins which are widely distributed among the taxa. This supports the view that *nifH* had been horizontally transferred in some cases, but that is a much ancient event.

Gaby et al. [5] has created a *nifH* database, also a multipurpose tool to facilitate the search through sequence-based techniques to study nitrogen-fixing bacteria. The database is implemented in the ARB software package. This package also enables the compilation and association of aligned sequence data, sequence metadata and phylogenetic trees.As discussed earlier, these new data layers facilitate improved evolutionary analysis and enable assessment of ancestral patterns of horizontal gene exchange.The dataset also displays a diversity of microbes, as there are 375 full-length *nifH* genes including 245 from sequenced genomes. The full-length *nifH* sequences are again from 294 different strains and 222 different species.

The speciality of this database can be highlighted based on another fact that the sequences were sourced from a wide array of environments and habitats, and not only the cultivated fields. Apart from the comprehensive database mentioned above, there are also other publicly available databases like the Functional Gene Pipeline (FunGene).

The FunGene Pipeline (FGP) [4] consists of a set of tools, along with reference files and parameters for each gene offered and Python scripts that coordinate running the individual tools that make up the pipeline. All of the tools which were developed and incorporated in the FGP are available under open source licences from RDPStaff GitHub page, http:// github.com/rdpstaff. The sequences in FunGene are pulled from public databases and aligned using the Hidden Markov Modelling algorithm HMMER3. The pipeline also uses the mcClust tool(http://github.com/rdpstaff/Clustering) for clustering similar genes. The FGP also offers diversity estimate tools for computing several alpha and beta diversity estimates and rarefaction curves from a cluster file. Thus, this database along with many useful tools is another comprehensive source for gathering information about the *nifH* and other *nif* genes.

Zehr et al. [19] has also built a database named ARBitrator, which can be accessed at http://pmc.ucsc.edu/~wwwzehr/research/database/ [6]. The inhererent problem of difficulty in finding all nif sequences by BLAST analysis has been addressed here. Zehr observes that the problem of misannotation, and homologous proteins in the database also makes a simple search for *nifH* sequence difficult and error prone in the standard public databases.

The procedure there uses representative *nifH* protein sequences to BLAST against GenBank to identify potential *nifH* and *nifH*-like genes. Then the results are screened for false positives, and those are eliminated from the database. As the identified *nifH* protein and the encoding DNA sequence are retrieved, the sequences are imported into ARB using the nucleotide GI (GenBank identifier) as the sequence name. Further, the DNA sequence is used to generate the amino acid sequence and it is checked if this sequence matches with reords in Genbank. Then the amino acid sequences are exported and aligned against a *nifH* PFAM using HMMR and the amino acid sequences re-imported into ARB. The aligned amino acid sequences are then used to align the DNA sequences using the "Backalign" feature of ARB. Thus, this databse provides a comprehensively checked and aligned sequences of *nifH* genes. The database contains all *nifH* amino acid and DNA sequences obtainable from BLAST analysis and it is regularly maintained and updated.

The clear idea about the phylogeny of diazotrophs and phyletic retention of nitrogenase has enabled us to create specific primers and paved the way for finding new nitrogenase producing genes by metagenomics approach. This is also facilitating the studies about regulations of *nif* gene clusters which may soon enable us to create transgenic plants with nitrogenase expression capacity. Recently, Qiong Wang et al. [17] had undertaken an extensive study of isolating *nifH* amplicon sequences from 222 soil samples collected from 4 National Ecological Observatory Network (NEON) sites in Alaska, Hawaii, Utah, and Florida. Total number of sequences were 1.1 million. For analyzing these ecofunctional genes they have also developed a software, Framebot, a tool for frameshift correction and nearest neighbour classification. The dataset of 1.1 million sequences was found to contain 278, 561 unique DNA reads. When they compared these sequences with the cluster groups I-V (as mentioned earlier), it was found that overall, 92.8%, 7.0%, 0.16%, and 0.004% of the reads belonged to *nifH* groups I, II, and III, and to groups IV and V, respectively, with some variation between sites. It is also interesting to note that the large number of sequences in the four sampling sites showed nearest matches to strains of *Rhizobium*, ranging from 2% to 10%, and to strains of *Bradyrhizobium* in the range of 6% and 13%. Thus, in near future, we may expect more information coming out from metagenomics-based laboratories.

CONCLUSION

There are distinct distribution patterns of diazotroph phylotypes across ecosystems. Different clusters of diazotrophs are found in different types of habitats. On the other hand, some habitats have unique and divergent populations. In some cases these diazotroph phylotypes are found even in the absence of strong N limitation. The expanding genomic database has provided information on genome context as well as sequence information that can be used in the next generation of environmental nitrogenase studies. The use of *nifDK* gene sequences can provide additional supporting or contrasting information for evaluating nifH phylotypes. The gene organization data provides context that can resolve phylogenies of some of the phylotypes that are difficult to identify conclusively. However, representation of N_2-fixing microorganisms in the database of sequenced genomes is still poor. Ultimately, expression analyses and possibly proteomics approaches will be critical for identifying the factors controlling N_2-fixation in the environment. Expression data is the key to identifying the microorganisms that are active in N_2 fixation, but more generally to understand the factors involved in controlling the distribution of nitrogenase gene-containing microorganisms in the environment. This information will help us to understand the links between diversity of functional genes and biogeochemical activity.

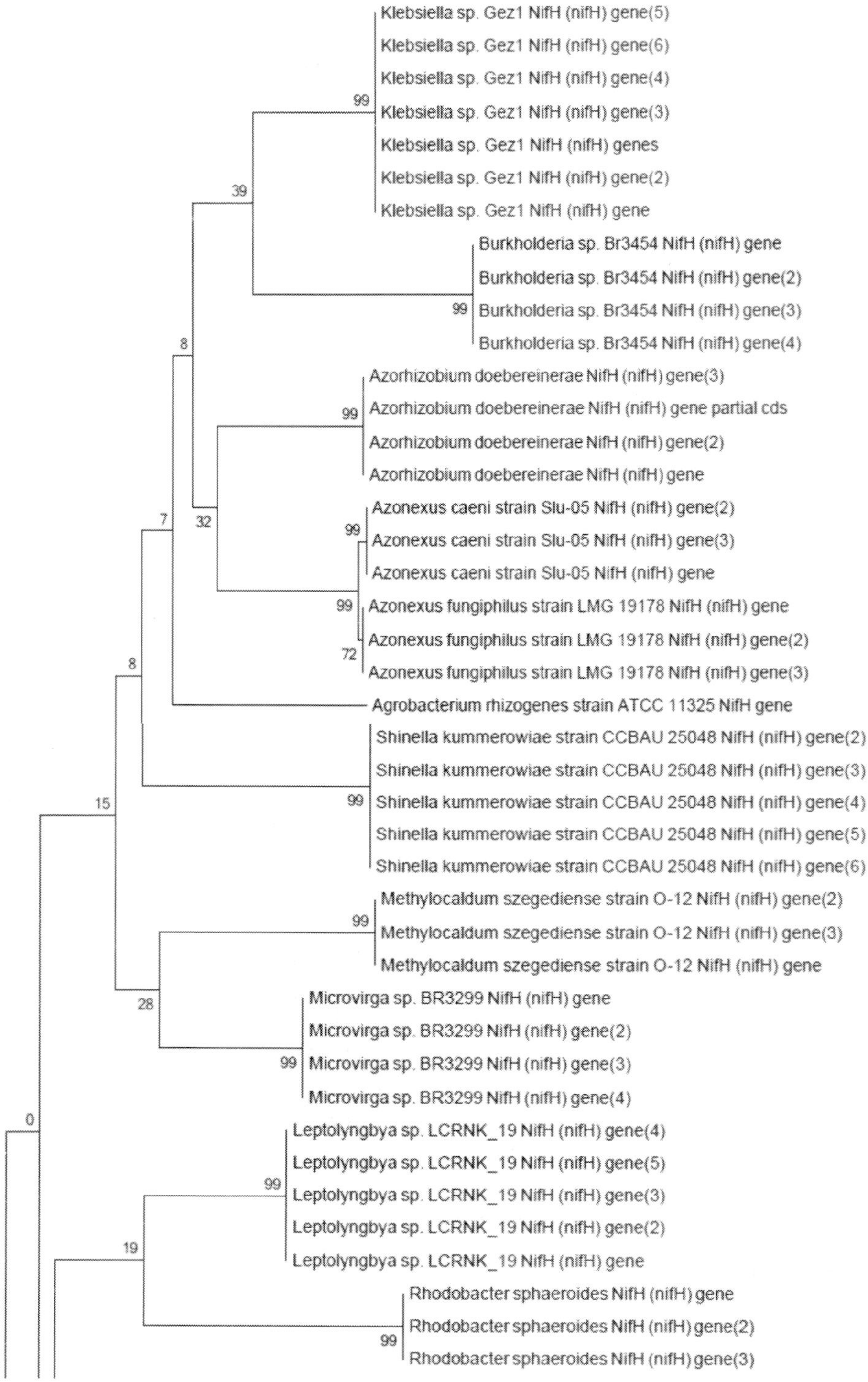
Klebsiella sp. Gez1 NifH (nifH) gene(5)
Klebsiella sp. Gez1 NifH (nifH) gene(6)
Klebsiella sp. Gez1 NifH (nifH) gene(4)
Klebsiella sp. Gez1 NifH (nifH) gene(3)
Klebsiella sp. Gez1 NifH (nifH) genes
Klebsiella sp. Gez1 NifH (nifH) gene(2)
Klebsiella sp. Gez1 NifH (nifH) gene
Burkholderia sp. Br3454 NifH (nifH) gene
Burkholderia sp. Br3454 NifH (nifH) gene(2)
Burkholderia sp. Br3454 NifH (nifH) gene(3)
Burkholderia sp. Br3454 NifH (nifH) gene(4)
Azorhizobium doebereinerae NifH (nifH) gene(3)
Azorhizobium doebereinerae NifH (nifH) gene partial cds
Azorhizobium doebereinerae NifH (nifH) gene(2)
Azorhizobium doebereinerae NifH (nifH) gene
Azonexus caeni strain Slu-05 NifH (nifH) gene(2)
Azonexus caeni strain Slu-05 NifH (nifH) gene(3)
Azonexus caeni strain Slu-05 NifH (nifH) gene
Azonexus fungiphilus strain LMG 19178 NifH (nifH) gene
Azonexus fungiphilus strain LMG 19178 NifH (nifH) gene(2)
Azonexus fungiphilus strain LMG 19178 NifH (nifH) gene(3)
Agrobacterium rhizogenes strain ATCC 11325 NifH gene
Shinella kummerowiae strain CCBAU 25048 NifH (nifH) gene(2)
Shinella kummerowiae strain CCBAU 25048 NifH (nifH) gene(3)
Shinella kummerowiae strain CCBAU 25048 NifH (nifH) gene(4)
Shinella kummerowiae strain CCBAU 25048 NifH (nifH) gene(5)
Shinella kummerowiae strain CCBAU 25048 NifH (nifH) gene(6)
Methylocaldum szegediense strain O-12 NifH (nifH) gene(2)
Methylocaldum szegediense strain O-12 NifH (nifH) gene(3)
Methylocaldum szegediense strain O-12 NifH (nifH) gene
Microvirga sp. BR3299 NifH (nifH) gene
Microvirga sp. BR3299 NifH (nifH) gene(2)
Microvirga sp. BR3299 NifH (nifH) gene(3)
Microvirga sp. BR3299 NifH (nifH) gene(4)
Leptolyngbya sp. LCRNK_19 NifH (nifH) gene(4)
Leptolyngbya sp. LCRNK_19 NifH (nifH) gene(5)
Leptolyngbya sp. LCRNK_19 NifH (nifH) gene(3)
Leptolyngbya sp. LCRNK_19 NifH (nifH) gene(2)
Leptolyngbya sp. LCRNK_19 NifH (nifH) gene
Rhodobacter sphaeroides NifH (nifH) gene
Rhodobacter sphaeroides NifH (nifH) gene(2)
Rhodobacter sphaeroides NifH (nifH) gene(3)

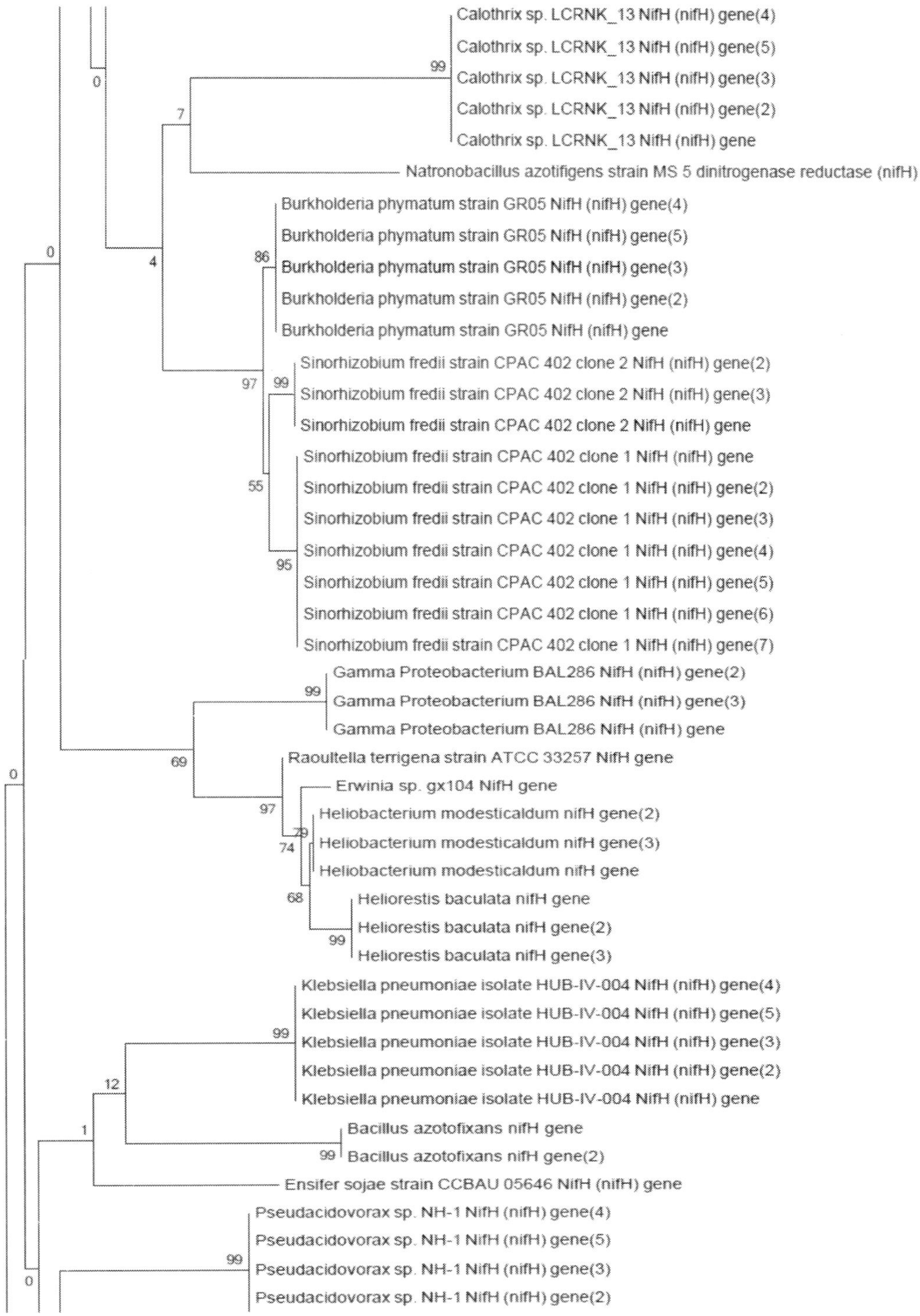
Calothrix sp. LCRNK_13 NifH (nifH) gene(4)
Calothrix sp. LCRNK_13 NifH (nifH) gene(5)
Calothrix sp. LCRNK_13 NifH (nifH) gene(3)
Calothrix sp. LCRNK_13 NifH (nifH) gene(2)
Calothrix sp. LCRNK_13 NifH (nifH) gene
Natronobacillus azotifigens strain MS 5 dinitrogenase reductase (nifH)
Burkholderia phymatum strain GR05 NifH (nifH) gene(4)
Burkholderia phymatum strain GR05 NifH (nifH) gene(5)
Burkholderia phymatum strain GR05 NifH (nifH) gene(3)
Burkholderia phymatum strain GR05 NifH (nifH) gene(2)
Burkholderia phymatum strain GR05 NifH (nifH) gene
Sinorhizobium fredii strain CPAC 402 clone 2 NifH (nifH) gene(2)
Sinorhizobium fredii strain CPAC 402 clone 2 NifH (nifH) gene(3)
Sinorhizobium fredii strain CPAC 402 clone 2 NifH (nifH) gene
Sinorhizobium fredii strain CPAC 402 clone 1 NifH (nifH) gene
Sinorhizobium fredii strain CPAC 402 clone 1 NifH (nifH) gene(2)
Sinorhizobium fredii strain CPAC 402 clone 1 NifH (nifH) gene(3)
Sinorhizobium fredii strain CPAC 402 clone 1 NifH (nifH) gene(4)
Sinorhizobium fredii strain CPAC 402 clone 1 NifH (nifH) gene(5)
Sinorhizobium fredii strain CPAC 402 clone 1 NifH (nifH) gene(6)
Sinorhizobium fredii strain CPAC 402 clone 1 NifH (nifH) gene(7)
Gamma Proteobacterium BAL286 NifH (nifH) gene(2)
Gamma Proteobacterium BAL286 NifH (nifH) gene(3)
Gamma Proteobacterium BAL286 NifH (nifH) gene
Raoultella terrigena strain ATCC 33257 NifH gene
Erwinia sp. gx104 NifH gene
Heliobacterium modesticaldum nifH gene(2)
Heliobacterium modesticaldum nifH gene(3)
Heliobacterium modesticaldum nifH gene
Heliorestis baculata nifH gene
Heliorestis baculata nifH gene(2)
Heliorestis baculata nifH gene(3)
Klebsiella pneumoniae isolate HUB-IV-004 NifH (nifH) gene(4)
Klebsiella pneumoniae isolate HUB-IV-004 NifH (nifH) gene(5)
Klebsiella pneumoniae isolate HUB-IV-004 NifH (nifH) gene(3)
Klebsiella pneumoniae isolate HUB-IV-004 NifH (nifH) gene(2)
Klebsiella pneumoniae isolate HUB-IV-004 NifH (nifH) gene
Bacillus azotofixans nifH gene
Bacillus azotofixans nifH gene(2)
Ensifer sojae strain CCBAU 05646 NifH (nifH) gene
Pseudacidovorax sp. NH-1 NifH (nifH) gene(4)
Pseudacidovorax sp. NH-1 NifH (nifH) gene(5)
Pseudacidovorax sp. NH-1 NifH (nifH) gene(3)
Pseudacidovorax sp. NH-1 NifH (nifH) gene(2)

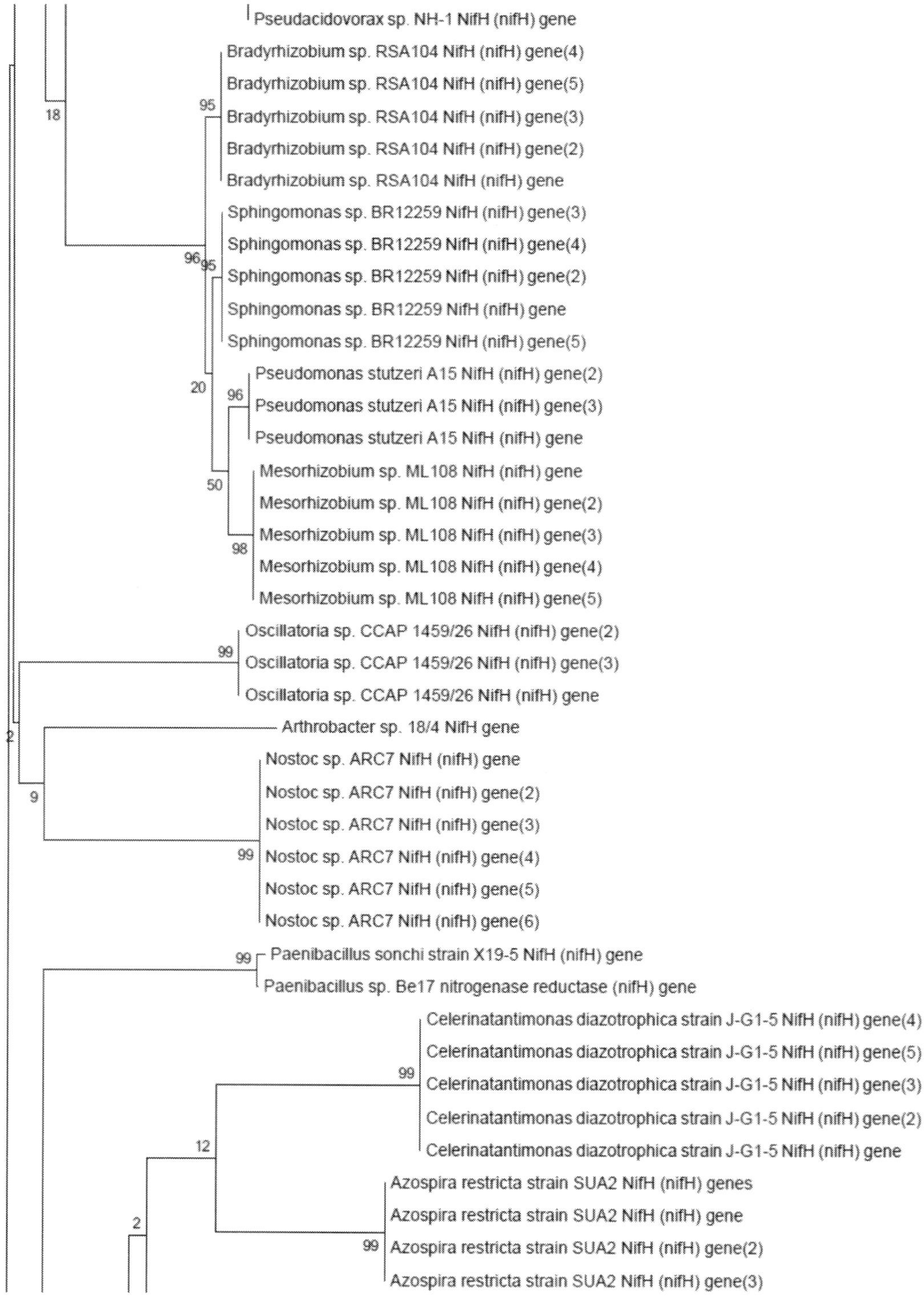
Pseudacidovorax sp. NH-1 NifH (nifH) gene
Bradyrhizobium sp. RSA104 NifH (nifH) gene(4)
Bradyrhizobium sp. RSA104 NifH (nifH) gene(5)
Bradyrhizobium sp. RSA104 NifH (nifH) gene(3)
Bradyrhizobium sp. RSA104 NifH (nifH) gene(2)
Bradyrhizobium sp. RSA104 NifH (nifH) gene
Sphingomonas sp. BR12259 NifH (nifH) gene(3)
Sphingomonas sp. BR12259 NifH (nifH) gene(4)
Sphingomonas sp. BR12259 NifH (nifH) gene(2)
Sphingomonas sp. BR12259 NifH (nifH) gene
Sphingomonas sp. BR12259 NifH (nifH) gene(5)
Pseudomonas stutzeri A15 NifH (nifH) gene(2)
Pseudomonas stutzeri A15 NifH (nifH) gene(3)
Pseudomonas stutzeri A15 NifH (nifH) gene
Mesorhizobium sp. ML108 NifH (nifH) gene
Mesorhizobium sp. ML108 NifH (nifH) gene(2)
Mesorhizobium sp. ML108 NifH (nifH) gene(3)
Mesorhizobium sp. ML108 NifH (nifH) gene(4)
Mesorhizobium sp. ML108 NifH (nifH) gene(5)
Oscillatoria sp. CCAP 1459/26 NifH (nifH) gene(2)
Oscillatoria sp. CCAP 1459/26 NifH (nifH) gene(3)
Oscillatoria sp. CCAP 1459/26 NifH (nifH) gene
Arthrobacter sp. 18/4 NifH gene
Nostoc sp. ARC7 NifH (nifH) gene
Nostoc sp. ARC7 NifH (nifH) gene(2)
Nostoc sp. ARC7 NifH (nifH) gene(3)
Nostoc sp. ARC7 NifH (nifH) gene(4)
Nostoc sp. ARC7 NifH (nifH) gene(5)
Nostoc sp. ARC7 NifH (nifH) gene(6)
Paenibacillus sonchi strain X19-5 NifH (nifH) gene
Paenibacillus sp. Be17 nitrogenase reductase (nifH) gene
Celerinatantimonas diazotrophica strain J-G1-5 NifH (nifH) gene(4)
Celerinatantimonas diazotrophica strain J-G1-5 NifH (nifH) gene(5)
Celerinatantimonas diazotrophica strain J-G1-5 NifH (nifH) gene(3)
Celerinatantimonas diazotrophica strain J-G1-5 NifH (nifH) gene(2)
Celerinatantimonas diazotrophica strain J-G1-5 NifH (nifH) gene
Azospira restricta strain SUA2 NifH (nifH) genes
Azospira restricta strain SUA2 NifH (nifH) gene
Azospira restricta strain SUA2 NifH (nifH) gene(2)
Azospira restricta strain SUA2 NifH (nifH) gene(3)

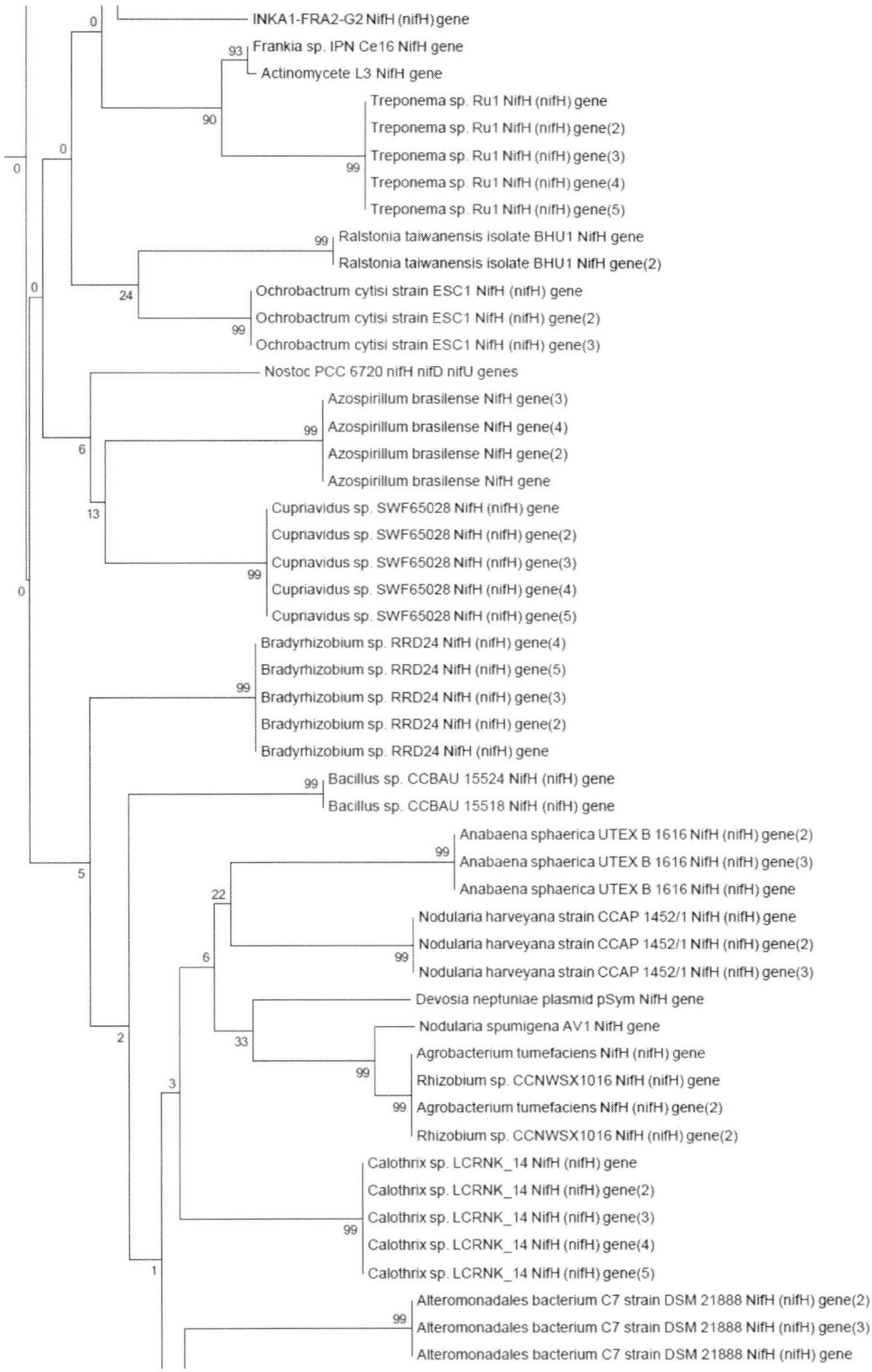
INKA1-FRA2-G2 NifH (nifH) gene
Frankia sp. IPN Ce16 NifH gene
Actinomycete L3 NifH gene
Treponema sp. Ru1 NifH (nifH) gene
Treponema sp. Ru1 NifH (nifH) gene(2)
Treponema sp. Ru1 NifH (nifH) gene(3)
Treponema sp. Ru1 NifH (nifH) gene(4)
Treponema sp. Ru1 NifH (nifH) gene(5)
Ralstonia taiwanensis isolate BHU1 NifH gene
Ralstonia taiwanensis isolate BHU1 NifH gene(2)
Ochrobactrum cytisi strain ESC1 NifH (nifH) gene
Ochrobactrum cytisi strain ESC1 NifH (nifH) gene(2)
Ochrobactrum cytisi strain ESC1 NifH (nifH) gene(3)
Nostoc PCC 6720 nifH nifD nifU genes
Azospirillum brasilense NifH gene(3)
Azospirillum brasilense NifH gene(4)
Azospirillum brasilense NifH gene(2)
Azospirillum brasilense NifH gene
Cupriavidus sp. SWF65028 NifH (nifH) gene
Cupriavidus sp. SWF65028 NifH (nifH) gene(2)
Cupriavidus sp. SWF65028 NifH (nifH) gene(3)
Cupriavidus sp. SWF65028 NifH (nifH) gene(4)
Cupriavidus sp. SWF65028 NifH (nifH) gene(5)
Bradyrhizobium sp. RRD24 NifH (nifH) gene(4)
Bradyrhizobium sp. RRD24 NifH (nifH) gene(5)
Bradyrhizobium sp. RRD24 NifH (nifH) gene(3)
Bradyrhizobium sp. RRD24 NifH (nifH) gene(2)
Bradyrhizobium sp. RRD24 NifH (nifH) gene
Bacillus sp. CCBAU 15524 NifH (nifH) gene
Bacillus sp. CCBAU 15518 NifH (nifH) gene
Anabaena sphaerica UTEX B 1616 NifH (nifH) gene(2)
Anabaena sphaerica UTEX B 1616 NifH (nifH) gene(3)
Anabaena sphaerica UTEX B 1616 NifH (nifH) gene
Nodularia harveyana strain CCAP 1452/1 NifH (nifH) gene
Nodularia harveyana strain CCAP 1452/1 NifH (nifH) gene(2)
Nodularia harveyana strain CCAP 1452/1 NifH (nifH) gene(3)
Devosia neptuniae plasmid pSym NifH gene
Nodularia spumigena AV1 NifH gene
Agrobacterium tumefaciens NifH (nifH) gene
Rhizobium sp. CCNWSX1016 NifH (nifH) gene
Agrobacterium tumefaciens NifH (nifH) gene(2)
Rhizobium sp. CCNWSX1016 NifH (nifH) gene(2)
Calothrix sp. LCRNK_14 NifH (nifH) gene
Calothrix sp. LCRNK_14 NifH (nifH) gene(2)
Calothrix sp. LCRNK_14 NifH (nifH) gene(3)
Calothrix sp. LCRNK_14 NifH (nifH) gene(4)
Calothrix sp. LCRNK_14 NifH (nifH) gene(5)
Alteromonadales bacterium C7 strain DSM 21888 NifH (nifH) gene(2)
Alteromonadales bacterium C7 strain DSM 21888 NifH (nifH) gene(3)
Alteromonadales bacterium C7 strain DSM 21888 NifH (nifH) gene

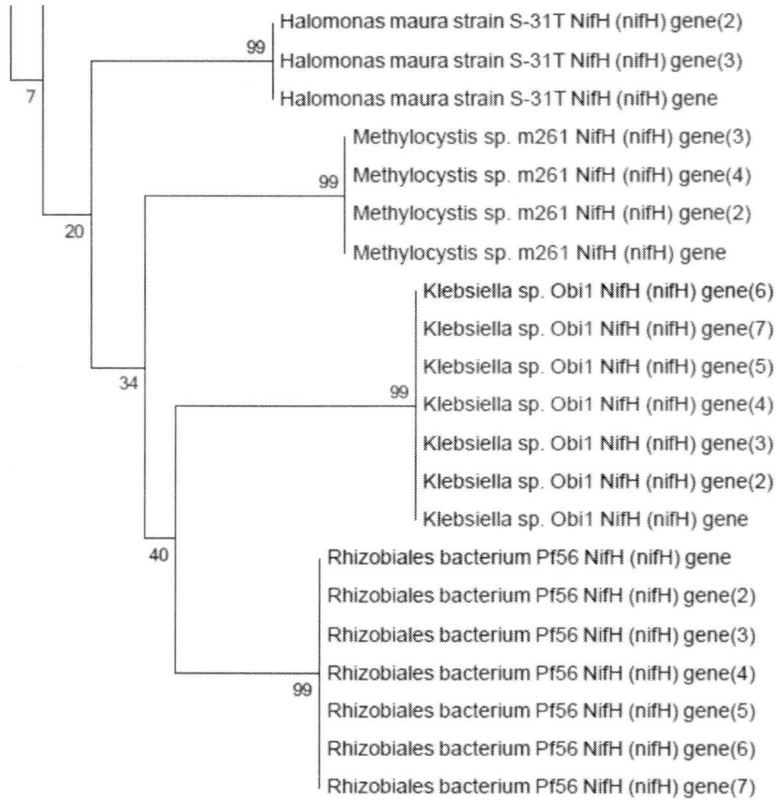

Fig. 9.1 Phylogenetic analysis of *infH* gene cluster.

REFERENCES

1. Chen, Y.B., Dominic, B., Mellon, M.T. and Zehr, J.P. (1998). Circadian rhythm of nitrogenase gene expression in the diazotrophic filamentous nonheterocystous cyanobacterium *Trichodesmium* sp. Strain IMS 101. *J Bacteriol* 180, 3598–3605.
2. Cho, J.C. and J.M., Tiedje (2000). Biogeography and degree of endemicity of fluorescent *Pseudomonas* strains in soil. *Appl. Environ. Microbiol.* 66, 5448–5456.
3. Doolittle, W.F., Phylogenetic classification and the universal tree, *Science.* 1999 Jun 25; 284(5423), 2124–9.
4. Fish, J.A., Chai, B., Wang, Q., Sun, Y., Brown, C.T., Tiedje, J.M. and Cole, J. (2013). FunGene: the Functional Gene Pipeline and Repository. *Front. Microbiol.* 4, 291.
5. Gaby J.C. and Buckley D.H. (2012). A comprehensive evaluation of PCR primers to amplify the *nifH* gene of nitrogenase. PLoS ONE 7(7): e42149. doi:10.1371/journal.pone.0042149.
6. Heller, P., Tripp, H.J., Turk-Kubo, K. and Zehr, J.P. (2014). ARBitrator: A software pipeline for on-demand retrieval of auto-curated *nifH* sequences from GenBank. *Bioinformatics,* btu417.

7. Howard, J.B. and Rees, D.C. (1996). Structural basis of biological nitrogen fixation. *Chem. Rev., 96*, 2965-2982.
8. Hurek, T., Handley, L.L., Reinhold-Hurek, B. and Piche, Y. (2002). Azoarcus grass endophytes contribute fixed nitrogen to the plant in an unculturable state. Mol Plant- Microbe Interacts 15, 233–242.
9. Kim, K., Zhang, Y.P. and Roberts, G.P. (1999). Correlation of activity regulation and substrate recognition of the ADP ribosyltransferase that regulates nitrogenase activity in *Rhodospirillum rubrum. J Bacteriol* 181, 1698–1702.
10. Madigan, M.T., J.M. Martinko and J. Parker (2000). *Brock Biology of Microorganism.* 8th edn., Prentice Hall, Upper Saddle River, pp: 891-921.
11. Ohkuma, M., Noda, S., Usami, R., Horikoshi, K. and Kudo, T. (1996). Diversity of nitrogen fixation genes in the symbiotic intestinal microflora of the termite *Reticulitermes speratus. Appl Environ Microbiol* 62, 2747–2752.
12. Pace, N.R. (1997). A molecular view of microbial diversity and the biosphere. *Science* 276, 734–740.
13. Postgate, J.R. and Eady, R.R. (1988). The evolution of biological nitrogen fixation. In: *Nitrogen Fixation: Hundred Years After*, pp. 31–40. Edited by H. Bothe., F.J. de Bruijn & W.E. Newton. Stuttgart: Gustav Fischer.
14. Simonet, P., Grosjean, M.C., Misra, A.K., Nazaret, S., Cournoyer, B. and Normand, P. (1991). *Frankia* genus-specific characterization by polymerase chain reaction. *Appl Environ Microbiol* 57,3278–3286.
15. Tan, Z.Y., Hurek, T., Vinuesa, P., Muller, P., Ladha, J.K. and Reinhold-Hurek, B. (2001). Specific detection of *Bradyrhizobium* and *Rhizobium* strains colonizing rice (Oryza sativa) roots by 16S-23S ribosomal DNA intergenic spacer-targeted PCR. *Appl Environ Microbiol* 67, 3655–3664.
16. Ueda, T., Suga, Y., Yahiro, N. and Matsuguchi, T. (1995a). Remarkable N_2-fixing bacterial diversity detected in rice roots by molecular evolutionary analysis of *nifH* gene sequences. J. Bacteriol 177, 1414–1417.
17. Wang Q., Quensen J.F., III, Fish J.A., Lee T.K., Sun Y., Tiedje J.M. and Cole J.R. (2013). Ecological patterns of *nifH* genes in four terrestrial climatic zones explored with targeted metagenomics using FrameBot, a new informatics tool. mBio 4(5), e00592-13. doi:10.1128/mBio.00592-13.
18. Young, J.P.W. (1992). Phylogenetic classification of nitrogen-fixing organisms. In: *Biological Nitrogen Fixation.* Stacey, G., Evans, H.J., and Burris, R.H., (eds). New York: Chapman & Hall, pp. 43–86.
19. Zehr, J.P., Mellon, M.T. and Zani, S. (1998). New nitrogen-fixing microorganisms detected in oligotrophic oceans by the amplification of nitrogenase (*nifH*) genes. *Appl Environ Microbiol* 64, 3444–3450.

10

Arbuscular Mycorrhizal Dependency of Plants

Somdatta Ghosh[1], Debashis Kuila[1] and Nagendra Kumar Verma[2]

[1]*Department of Botany, Midnapore College (Autonomous), Midnapore – 721101*

[2]*Department of Botany & Forestry, Vidyasagar University, Midnapore – 721102, WB*

E-mail: *somdattaghosh@yahoo.co.in*

ABSTRACT

Arbuscular mycorrhizal (AM) symbiosis occur with more than 80% of land plants and are widely distributed in almost all the ecological range where plants grow. Mycorrhizal fungi as such are non-host specific but as reported, the preference and mycorrhizal dependency of plants may vary with the soil. It is known that phosphorus is a relatively immobile element in soil and its demand is much higher than its mobility. Other than phosphorus, AM fungi also help plants to capture micronutrients such as copper, zinc, boron, etc. AM is said to be a key factor of plant establishment in different plant communities; plant growth and soil sustainability in nutrient poor soils. Several workers have worked on the mycorrhizal dependency of different plants which include weeds, agricultural, horticultural and forest crops. Parameters mainly used to determine MD include biomass, leaf phosphorus, phosphorus utilization of plants, root morphology, etc. Mycorrhizal dependency is the degree to which a plant is dependent on the mycorrhizal condition to produce its maximum growth or yield, at a given level of soil fertility. Even same plant with same VA fungi may vary with MD, according to some reports. Majority of agricultural species are either obligate or facultative mycorrhizal. Most of the mycorrhiza dependent plant species show better result at lower P content. Increased soil P concentration depresses the mycorrhizal infection and also the intensity of infection which is reflected in the degree of MD of such plant species. However, totally dependent species may continue to derive benefit from mycorrhizal symbiosis irrespective of concentration of soil P.

Keywords: Vesicular Arbuscular Mycorrhiza, Arbuscular Mycorrhiza, Soil P, Mycorrhizal Dependency.

10.1 INTRODUCTION

The mycorrhizal symbiosis is a key stone to the productivity and diversity of natural plant ecosystems [72]. It is believed that the development of VAM symbiosis played a crucial role in the initial colonization of the land by plants and in evolution of the vascular plants [27]. It has been said that it is quicker to list the plants that do not form mycorrhizae than those that do [59]. The symbiosis is a highly evolved mutualistic relationship found between fungi and plants, the most prevalent plant symbiosis known [122], and as a result VAM symbiosis is found in more than 80% of vascular plant families of today [119].

AM are well-known to enhance the P-uptake of seedlings growing under P-poor conditions [26]. The altered P nutrition alone has a substantial impact on the host carbon balance. It is likely that the formation of AM symbiosis is an important factor in the successful establishment of different mycorrhiza dependent plants [69] [70] in the nutrient deficient soil of the degraded lands. Some leguminous and non-leguminous plant species depend on AM for uptake of phosphorus when grown in a nutrient-deficient soil.

It is known that phosphorus is a relatively immobile element in soil and its demand is much higher than its mobility. Mycorrhizal plants can explore great volume of soil resulting in the increased flow of phosphorus from soil resulting in the increased flow of phosphorus from soil to plants [136]. Non-mycorrhizal plants growing in low to moderately fertile soils frequently require a supply of soluble phosphate in order to improve their P nutrition and growth. Such improvement in phosphorus uptake by roots occurs through improving physical exploration of the soil. VAM fungi also help plants to capture micronutrients such as copper and zinc, compared to non-mycorrhizal plants [126].

Many researchers have shown how AM fungi can benefit plants growth and develop essential conditions to sustain it. This growth stimulating effect, referred to as mycorrhizal dependency [46], is highly variable [115] and influenced by several factors such as plant species [45] and soil fertilization [61]. Gerdemann [46] defined the mycorrhizal dependency (MD) as the degree to which a plant is dependent on the mycorrhizal condition to produce its maximum growth or yield, at a given level of soil fertility.

Mycorrhizal inoculation results in an appreciable increase in the growth of seedlings. An increase in growth response due to AM inoculation has been earlier reported by several workers using different tests [94] [21] [73] [118] [120] [119] [48] [114].

10.2 OCCURENCE OF MYCORRHIZAL INFECTION

Mycorrhizal fungus penetrates the cortical cells of the roots of a plants. Round-shaped, thick cell walled mycorrhizal spores contain very large number of nuclei, up to 2000 per spore [20]. Nuclei from the spore move into the extending mycelium and some of them undergo mitosis [110]. AM spores germinate under appropriate water and temperature conditions. After germination, hyphae are always coenocytic [42]. Hyphae grow for about 2–3 weeks. During this time the fungal colonies extend a few centimetres, and show the characteristic growth pattern of hyphal branching. After attachment of a hypha to the root surface by means of an appressorium, the fungus penetrates into the cortex and forms distinct morphologically specialized structures: inter- and intra-cellular hyphae, coils and arbuscules. After host colonization, the fungal mycelium grows out of the root exploring the soil in search of mineral nutrients, and it can colonize other susceptible roots.

10.3 AVAILABLE P UPTAKE

Many researchers have shown that VAM infection can improve the phosphorus nutrition of plants grown in soils of low fertility [97]. In phosphate-deficient soils plants inoculated with vesicular-arbuscular (AM) mycorrhizal fungi could mobilize P from rock phosphate [13] [14] and utilize or take up more phosphorus and grow better than non-mycorrhizal plants [90] [7] [19] [45] [61] [99]. Mycorrhizas develop extramatrical mycelium which increases the root P absorbing sites [24] and these VA mycorrhizal roots take up more phosphorus than non-mycorrhizal roots [25] [51]. How the mycorrhizal plants obtain the extra phosphorus is not clear. But it was shown that mycorrhizae can hydrolyze insoluble P and insoluble inorganic phosphates (phytates)[53]. Hyphae of mycorrhizal roots extend into the soil and absorb soluble phosphate beyond the phosphate depletion zone and can also absorb phosphate from insoluble forms not available to non-mycorrhizal roots [62]. The intra- and extra-radical hyphae act as a link between soil nutrient pool and roots [37]. The mechanism of nutrient uptake by AM is mainly by increasing the absorptive surface of roots [91] [57], procuring nutrients from the zone beyond the reach of roots and exploiting larger area [18] [85] [112] and increasing affectivity of absorption by finer hyphae narrower in diameter than roots from a low diffusion gradient [9]. AM hyphae also increase availability of nutrients from locked source by decomposing large organic molecules [44], increasing mobility of nutrients by aggregation of soil particles [44] [67]. Physiochemical release of inorganic and organic phosphate by organic acids also occurs by the action of organic anions such as oxalate [41], replacing P absorbed at metal hydroxide surface through legand exchange reactions. The AM fungi may enhance plant P nutrition and increase the plant growth by diluting metal effect in host plant or by binding of the metal to the fungal mycelium and immobilizing them in rhizosphere or roots [31]. Thus, mycorrhizal plants take up more phosphate from the soil than non-mycorrhizal plants.

10.4 ABSORPTION OF OTHER NUTRIENTS

In addition to phosphorus, AM fungi are also able to uptake and transport organic and inorganic nitrogen [16] [63] [87], potassium [10] and micronutrients like S, Mg, Zn, Cu, Ca and Na [29] [126] [36]. Manganese and iron uptake [86] have also been reported. Marschner and Dell [92] reported that AMF could deliver up to 80% of plant P, 25% plant N, 10% of plant K, 25% of plant Zn and 60% of plant Cu; P uptake is strain dependent [1].

Mycorrhizal fungi can solubilize surrounding weatherable minerals through excretion of organic acids such as α-ketoglutaric acid. This organic compound could exert a selective influence on soil microbial communities [39]. Arbuscular mycorrhizal fungi (AMF) stimulate revegetation by supplementing the nutrient absorption capacity of the plant root systems, resulting in increased seedling survival and growth of the host plant [104].

AM fungal hyphae networks step up the polymerization of plant sugars and also help to form humus, a high molecular weight gel-like substance that holds four to twenty times its own weight in water. Humic substances significantly improve soil structure, porosity, cation exchange capacity and plant growth [76].

AM acts as biofertilizers for agricultural, horticultural and silvicultural plant species in polluted area [83]. AMF can also benefit plants by stimulating the production of growth regulating substances, increasing photosynthesis, improving osmotic adjustment under drought and salinity stresses and increasing resistance to pests and soil-borne diseases [6]. They have been known to enhance crop growth and yield [38] through increased water and nutrient uptake, as well as alteration of some physiological processes in the plants that result in increased yield [103]. They act as antagonist against some plant pathogens [34]. Mycorrhizal fungi, upon root colonization, develop an external mycelium which is a bridge connecting the root with the surrounding soil microhabitats. Therefore, the mycorrhizal symbiosis, by linking the biotic and geochemical portions of the ecosystem, can contribute to nutrient capture and supply [71]. Arbuscular mycorrhizal symbiosis play direct role in nutrient cycling rates and patterns in both agro-ecosystem and natural environments [23] [75].

10.5 MYCORRHIZA CAN IMPROVE WATER HOLDING CAPACITY OF THE PLANT

Mycorrhizal association can enhance water absorption [43] and help in water retention as well. Water uptake may be carried out by the external hyphae of AM fungi [39]. Mycorrhizal fungi play an important role in plant-water dynamics. The hyphal tips of mycorrhizal fungi are hydrophilic at both the ends, i.e., internally in root and externally in the soil, enabling both water and nutrients to diffuse from one end to the other along a moisture gradient [76].

10.6 IMPROVEMENT OF SOIL PROPERTIES

AM fungi accumulate heavy metals from soil and also from fly ash. AM fungi help in binding the fine particles and reducing heavy metal phytotoxicity [30] [8] in which they arrest the movement of heavy metals and also help in uptake of micronutrients [2]. AM fungi have been used as bioremediation agents [84]. They may suppress the uptake of Al, Fe, and Mn that may be present at toxic levels in some soils [100]. They can improve soil texture by producing humus. AM fungi improved soil properties in stressed environments [116] [102]. They supply phosphorus, enabling plants to grow under salt stress conditions [105] [17]. Many researchers have shown that these mycorrhizal fungi play a significant role in re-vegetation of disturbed and toxic environments by contributing nutrient availability, immobilizing heavy metals in the soil, and binding soil particles into stable aggregates that improve soil structure.

10.7 MYCORRHIZAL DEPENDENCY

Mycorrhizal workers defined mycorrhizal dependency (MD) of plants variously. Generally when a particular plant species or variety or genotype produces greater biomass in mycorrhizal inoculation than the non-mycorrhizal at any soil phosphorus level, the species or variety or genotype is considered mycorrhiza dependent [123]. For quantifying MD, selection of most efficient mycorrhiza for the plant is necessary.

Mycorrhizal dependency is mostly related to soil fertility. According to Gerdemann [46], mycorrhizal dependency is the degree to which a plant relies on mycorrhizal condition to produce maximum yield at a given level of fertility. MD has been quantified by various formula based on dry mass of mycorrhizal and non-mycorrhizal plants [93] [105] [56] [102].

Parameters mainly used to determine MD are leaf phosphorus [3], phosphorus utilization of plants [88] [64], root morphology [88] and mineral contents [80], etc.

The term mycorrhizal dependency (MD) proposed by Plenchette et. al., [105] was used to express the degree of growth change associated with arbuscular mycorrhizal colonization and is estimated by the following formula:

Mycorrhizal dependency (%) = [(dry weight of mycorrhizal plant dry weight of non-mycorrhizal plant)/dry weight of mycorrhizal plant] × 100.

Values calculated from this equation were sometimes referred to as mycorrhizal responsiveness [138]. In some another reports, dry weight of non-mycorrhizal plant instead of that of mycorrhizal plant was used in a denominator, referred to as growth response [65].

Growth response (%) = [(dry weight of mycorrhizal plant - dry weight of non-mycorrhizal plant)/dry weight of non-mycorrhizal plant] × 100.

However, the physiological and genetical basis of MD has not yet been elucidated [74]. A selection of optimal relationships between mycorrhizal fungi and host plants could result in more effective symbiosis and better growth of the host plants [108]. The degree of plant nutrient acquisition depending on fungi is defined as mycotrophy.

Cruz et al. [32] suggested three MD classes:

(i) 41% or above as highly dependent;

(ii) 10%-40% as intermediate; and

(iii) less than 10% as non-dependent.

In spite of apparent lack of host specificity [97] [46], some degree of preference has been reported [111] [132]. Some evidences have also indicated the presence of genetic factors that limit or promote colonization [82]. Better colonization has been reported in several families like Leguminosae, Asteraceae and Graminae [107] [68] [47]. Members of same family may also show variation to AM response [47]. None or rarely mycotrophic fruit trees are dominant in *Rosaceae*, and several AM species have shown specificity in association with them [135] [133]. Mycorrhizal dependencey of plant species may vary [113] [135] [137]. Majority of agricultural species are either obligate or facultative mycorrhizal [126].

Most of the mycorrhiza dependent plant species showed better result at lower P content [55] [125] [121]. But in some highly mycorrhiza dependent species, like *Leucaena leucocephala*, biomass increase correlated with soil P concentration [54]. They have opined that species that differ in mycorrhizal dependency differ in the degree to which increased soil P concentration depressed the infection. Totally dependent species continue to derive benefit from mycorrhiza irrespective of concentration of soil P [88]. Information related to MD of plant species is presented in Table 10.1.

Table 10.1 Mycorrhizal dependency (MD) of different plant species

Plant	*AM species*	*Dependency*	*Worker(s)/Year*
Wheat cultivars-Tormes, Anza, Negrillo, Cerros, Bastion, Pane 247, Lozano, Cocorit, Champlein, Castan, Tajo, Boulmiche and Jupateco	*Glomus mossea*	100 to 217	Azcon and Ocampo, 1981. [11]
Avena sativa	*Indigenous*	- 22	Plenchette et a1., 1983. [105]
Lycopersicon esculentum	*Indigenous*	59	
Solanum tuberosum	*Indigenous*	42	
Liquidambar styraciflua	*Glomus macrocarpum*	88	Pope et al., 1983. [106]
Platanus occidentalis	*Glomus macrocarpum*	38	
Citrus reshni	*Glomus intraradices*	91	Graham and Syvertsen, 1985. [50]
P. trifoliata X C. sinensis	*Glomus intraradices*	80	
Centrosema pascuorum	*Glomus manzhotis*	69	Saif, 1987. [112]
Desmodium canum	*Glomus manzhotis*	50	
Pueraria maximum	*Glomus manzhotis*	27	
Stylosanthes capitata	*Glomus manzhotis*	67	
Arabis hirsuta	*Glomus constrictum*	7	Grime et al., 1987. [52]
Briza media	*Glomus constrictum*	18	
Hieracium pilosella	*Glomus constrictum*	62	
Cotoneaster dammeri	*Glomus intraradices*	48	Nelson, 1987. [98]
Euonymus fortunei	*Glomus intraradices*	37	
Allium cepa	*Indigenous*	46	Krikun et al., 1990. [81]
Capsicum annuum	*Indigenous*	90	
Cucumis melo	*Indigenous*	- 23	
Lycopersicon esculentum *No. of cultivars = 10*	*Glomus etunicatum*	9 to 61	Bryla and Koide, 1990. [28]
Leucaena diversifolia	*Glomus aggregatum*	73	Manjunath and Habte, 1991. [89]
Leucaena retusa	*Glomus aggregatum*	35	
Sesbania formosa	*Glomus aggregatum*	24	
Sesbania grandiflora	*Glomus aggregatum*	35	
Sesbania seban	*Glomus aggregatum*	14	
Oryza sativa *No. of cultivars = 3*	*Glomus etunicatum*	8-33	Dhillion, 1992. [35]
Triticum aestivum *No. of cultivars = 23*	*Glomus etunicatum*	- 15 to 50	Hetrick et al., 1992. [68]

Contd...

Contd...

Agropyron smithii	*Glomus etunicatum*	27	Bentivenga and Hetrick, 1992. [22]
Bromus inermis	*Glomus etunicatum*	32	
Poa pratensis	*Glomus etunicatum*	- 20	
Sorghastrum nutans	*Glomus etunicatum*	99	
Acacia koa	*Not stated*	37	Miyasaka, Habte and Matsuyama, 1993. [95]
Hordeum vulgare *No. of cultivars = 8*	*Glomus etunicatum*	- 8 to 36	Baon et al., 1993. [15]
Triticum paleocolichium	*Glomus etunicatum*	- 5	Hetrick et al., 1993. [66]
Triticum polonicum	*Glomus etunicatum*	0	
Triticum speltoides	*Glomus etunicatum*	6	
Triticum tauschii *var. typical*	*Glomus etunicatum*	63	
Triticum carthlicum	*Glomus etunicatum*	- 13	
Triticum dicoccum	*Glomus etunicatum*	- 34	
Glycine max *No. of cultivars = 5*	*Glomus intraradices*	53 to 94	Khalil et al., 1994. [79]
	Gigaspora margarita	40 to 94	
Arachis hypogaea	*Glomus etunicatum*	93	Ahiabor and Hirata, 1995. [5]
Cajanus cajan	*Glomus etunicatum*	100	
Vigna unguiculata	*Glomus etunicatum*	99	
Musa *No. of cultivars = 7*	*Glomus macrocarpum*	13-54	Declerck et al., 1995. [33]
Eucalyptus bosistoana	*Glomus intraradices*	27	Adjoud et al., 1996. [3]
Eucalyptus delegatensis	*Glomus intraradices*	20	
Eucalyptus dumosa	*Glomus intraradices*	14	
Eucalyptus macanhu	*Glomus intraradices*	14	
Achillea millefolium	*Glomus etunicatum*	23	Wilson and Hartnett, 1998. [138]
Allium canadense	*Glomus etunicatum*	86	
Amaranthus spinosus	*Glomus etunicatum*	2	
Amorpha canescens	*Glomus etunicatum*	98	
Andropogon scoparius	*Glomus etunicatum*	99	
Andropogon virginicus	*Glomus etunicatum*	96	
Angropogon bladhii	*Glomus etunicatum*	99	
Artemisia ludoviciana	*Glomus etunicatum*	44	

Contd...

Contd...

Asclepias tuberosa	*Glomus etunicatum*	91	
Aster laevis	*Glomus etunicatum*	97	
Aster sericeus	*Glomus etunicatum*	77	
Astragalus crassicarpus	*Glomus etunicatum*	84	
Baptisia bracteata	*Glomus etunicatum*	83	
Baptista lactea	*Glomus etunicatum*	94	
Bouteloua curtipendula	*Glomus etunicatum*	87	
Bouteloua gracilis	*Glomus etunicatum*	33	Wilson and Hartnett, 1998. [138]
Bromus inermis	*Glomus etunicatum*	- 33	
Bromus japonicus	*Glomus etunicatum*	- 5	
Bromus tectorum	*Glomus etunicatum*	4	
Buchloe dactyloides	*Glomus etunicatum*	85	
Capselia bursa-pastoris	*Glomus etunicatum*	10	
Cassia Chamaecrista	*Glomus etunicatum*	- 12	
Cirsium vulgare	*Glomus etunicatum*	94	
Cynodon dactylon	*Glomus etunicatum*	25	
Dalea candida	*Glomus etunicatum*	96	
Dalea purpurea	*Glomus etunicatum*	94	
Daucus carota	*Glomus etunicatum*	30	
Desmanthus illinois	*Glomus etunicatum*	76	
Desmodium illinoense	*Glomus etunicatum*	78	
Dichanthelium clandestinum	*Glomus etunicatum*	51	
Digitaria sangumalis	*Glomus etunicatum*	26	
Echeinacea angustifolia	*Glomus etunicatum*	89	
Eragrostis spectabilis	*Glomus etunicatum*	65	
Eragrostis curvula	*Glomus etunicatum*	82	
Erigeron annuus	*Glomus etunicatum*	- 15	
Euphorbia corollata	*Glomus etunicatum*	96	
Festuca arundinacea	*Glomus etunicatum*	56	
Helianthus annuus	*Glomus etunicatum*	2	
Helianthus maximiliani	*Glomus etunicatum*	93	
Hieracium longipilum	*Glomus etunicatum*	98	

Contd...

Contd...

Hordeum jubatum	*Glomus etunicatum*	- 9	Wilson and Hartnett, 1998. [138]
Hordeum pusillum	*Glomus etunicatum*	- 17	
Koeleria pyramidata	*Glomus etunicatum*	- 17	
Linum sulcatum	*Glomus etunicatum*	26	
Lolzum perenne	*Glomus etunicatum*	16	
Mimosa biuncifera	*Glomus etunicatum*	44	
Oenothera biennis	*Glomus etunicatum*	30	
Oenothera speciosa	*Glomus etunicatum*	88	
Oxalis stricta	*Glomus etunicatum*	75	
Panicum virgatum	*Glomus etunicatum*	98	
Plantago lanceolata	*Glomus etunicatum*	- 6	
Poa pratensis	*Glomus etunicatum*	- 19	
Polygonum lapathifolium	*Glomus etunicatum*	- 19	
Psoralea esculenta	*Glomus etunicatum*	4	
Rudbeckia hirta	*Glomus etunicatum*	98	
Rumex crispus	*Glomus etunicatum*	98	
Salvia azurea	*Glomus etunicatum*	88	
Senecio plattensis	*Glomus etunicatum*	92	
Setaria glauca	*Glomus etunicatum*	13	
Setaria viridis	*Glomus etunicatum*	18	
Solidago canadensis	*Glomus etunicatum*	94	
Solidago rigida	*Glomus etunicatum*	36	
Sorghastrum nutans	*Glomus etunicatum*	100	
Sporobolus airoides	*Glomus etunicatum*	91	
Sporobolus heterolepis	*Glomus etunicatum*	98	
Tnpsacum dactyloides	*Glomus etunicatum*	70	
Tridens flavus	*Glomus etunicatum*	98	
Vernonia fasciculata	*Glomus etunicatum*	4	
Yucca glauca	*Glomus etunicatum*	87	Wilson and Hartnett, 1998. [138]
Zizyphus mauritiana	*Acaulospora spinosa*	91	Guissou et al., 1998. [53]
Allium fistulosum No. of cultivars = 16	*Glomus fasciculatu*	- 30 to 78	Tawaraya et al., 1999. [129]

Contd...

Contd...

Zea mays No. of cultivars = 6	*Glomus etunieatum*	42 to 83	Kaeppler et al., 2000. [77]
Adansonia digitata	*Glomus aggregatum*	7	Ba et al., 2000. [12]
Anacardium occidentale	*Glomus aggregatum*	23	
Balanites aegyptiaca	*Glomus aggregatum*	0	
Cordyla pinnata	*Glomus aggregatum*	28	
Dialium guineense	*Glomus aggregatum*	45	
Parkia biglobosa	*Glomus aggregatum*	33	
Sclerocarya birrea	*Glomus aggregatum*	12	
Tamarindus indica	*Glomus aggregatum*	53	
Zizyphus mauritiana	*Glomus aggregatum*	98	
Acacia holosericea	*Glomus aggregatum*	42	Duponnois et al., 2001. [40]
Acacia raddiana	*Glomus aggregatum*	63	
Crotalaria goreensis	*Glomus aggregatum*	91	
Indigofera senegalensis	*Glomus aggregatum*	83	
Indigofera stenophylla	*Glomus aggregatum*	93	
Indigofera tinctoria	*Glomus aggregatum*	49	
Prosopis juliflora	*Glomus aggregatum*	22	
Rhynchosia minima	*Glomus aggregatum*	63	
Sesbania grandiflora	*Glomus aggregatum*	- 4	
Sesbania nudica	*Glomus aggregatum*	- 32	
Sesbania pubescens	*Glomus aggregatum*	- 69	
Stylosanthes sp.	*Glomus aggregatum*	75	
Aspidosperma parvifolium	*Glomus etunicatum*	68	Siqueira and Saggin-Junior, 2001. [124]
Cedrela fissilis	*Glomus etunicatum*	95	
Copaifera langsdorffii	*Glomus etunicatum*	50	
Cordia trichotoma	*Glomus etunicatum*	98	
Leucaena leucocephala	*Glomus etunicatum*	92	
Luehea grandiflora	*Glomus etunicatum*	98	
Myrsine umbellata	*Glomus etunicatum*	98	
Sapindus saponaria	*Glomus etunicatum*	36	
Senna macranthera	*Glomus etunicatum*	87	
Solanum granuloso leprosum	*Glomus etunicatum*	99	
Tabebuia impetiginosa	*Glomus etunicatum*	58	
Tabebuia serratifolia	*Glomus etunicatum*	89	

Contd...

Contd...

Tibouchina granulosa	*Glomus etunicatum*	100	
Trema micrantha	*Glomus etunicatum*	98	
Glycine max	*Glomus clarum*	36	Kelly et al., 2001. [78]
Saccharum officinarum	*Glomus clarum*	9	
Zea mays	*Glomus clarum*	54	
Alnus cordata	*Glomus mosseae*	93	Monzon and Azcon, 2001. [96]
Alnus glutinosa	*Glomus mosseae*	99	
Alnus incana	*Glomus mosseae*	96	
Allium fistulosum No. of cultivars = 27	*Glomus fasciculatum*	73 to 95	Tawaraya et al., 2001. [130]
Triticum aestivum No. of cultivars = 3	*Glomus versiforme*	5 to 12	Yao et al., 2001. [139]
Triticum aestivum No. of cultivars = 6	*Glomus intraradices*	– 83 to – 43	Zhu et al., 2001. [140]
Acacia mangium	*Acaulospora delicate*	51	Ghosh and Verma, 2006. [48]
	Glomus aggregatum	46	
	Glomus mosseae	47	
	Paraglomus occultum	57	
Acacia auriculiformis	*Acaulospora delicata*	69	Ghosh, 2007. [49]
	Glomus aggregatum	65	
	G. mosseae	64	
	Paraglomus occultum	60	
Eucalyptus auriculiformis	*Acaulospora delicata* *Glomus aggregatum* *G. mosseae* *Paraglomu socccultum*	64 62 61 59	Ghosh, 2007. [49]
Albizia lebbeck	*Acaulospora delicate* *Glomus aggregatum* *G. mosseae* *Para glomus occultum*	58 51 57 54	Ghosh, 2007. [49]
Liriodendron tulipifera (Magnoliaceae)	*G. fasciculatum* *G. macrocarpum* *G. etunicatum*	65 35 33	Aggangan et al., 2011. [4]
Acacia melanoxylon	Not stated	30 to 62	Renuka et al., 2012. [109]
Pepper cultivars	*Glomus etunicatum*	20 to 39	Hassan and Ibrhahim, 2012. [60]

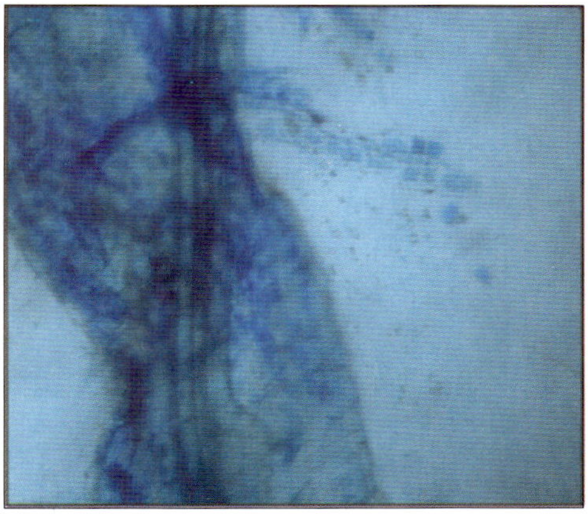

Arbuscles transformed to sporangioles

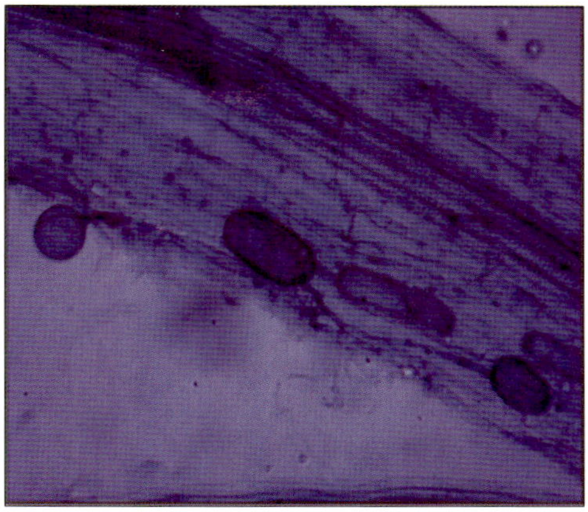

Spore and vesicle

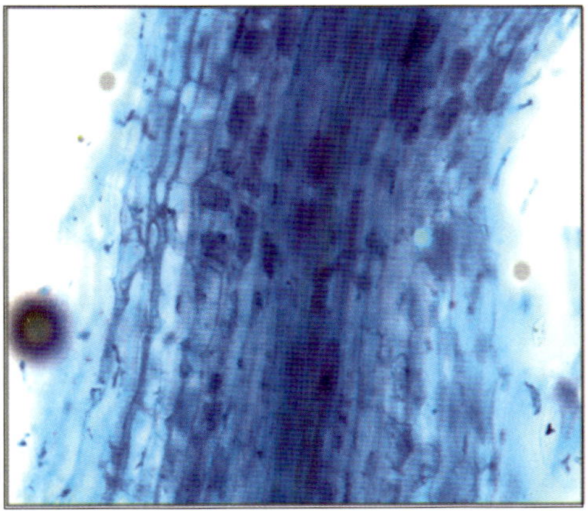

Arbuscle

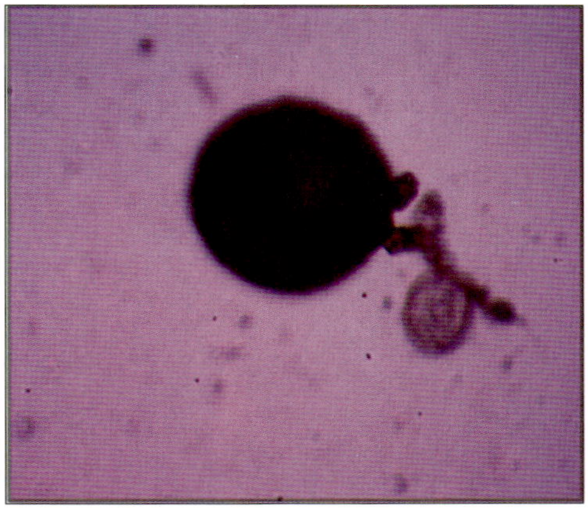

Isolated and pure cultured spores

Hyphae and arbuscle

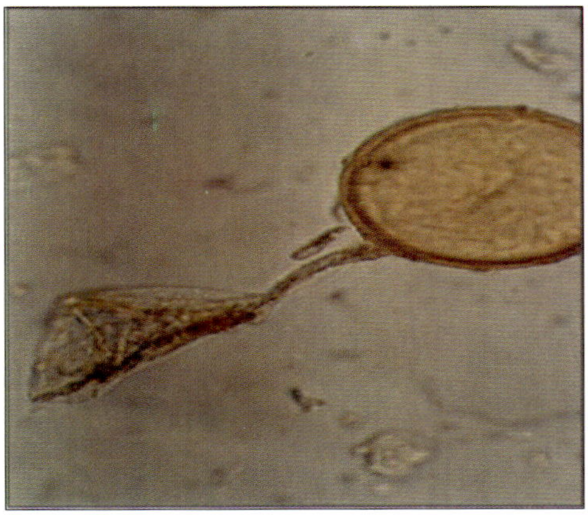

Acaulospora

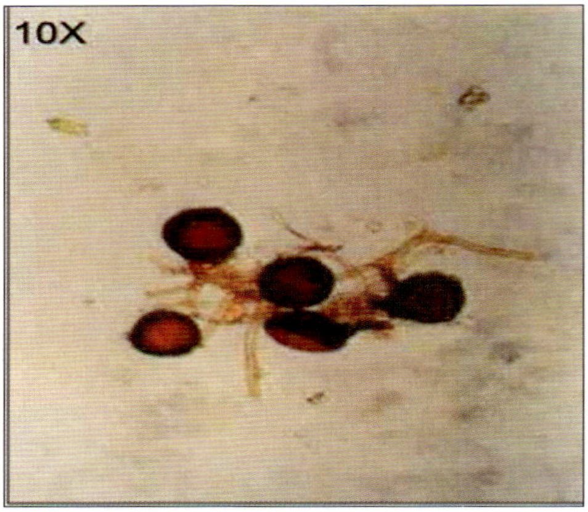

Glomus aggregatum

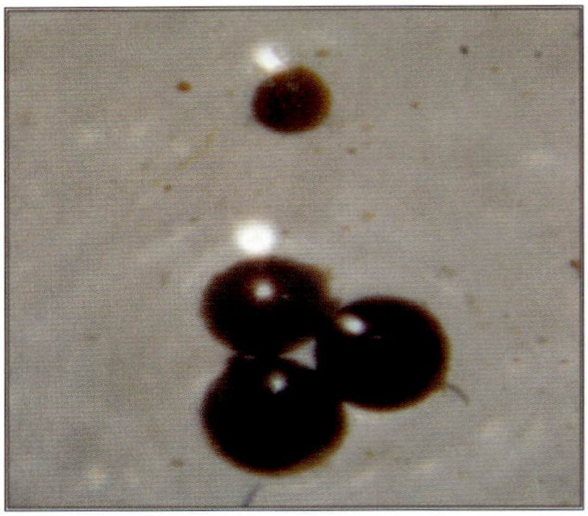

Scutellospora nigra

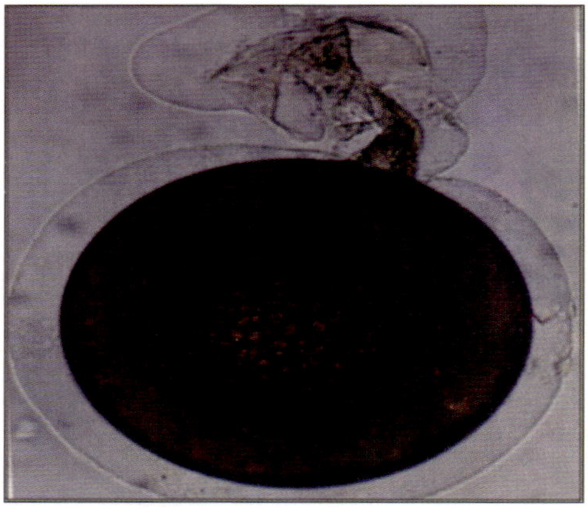

Glomus mosseae

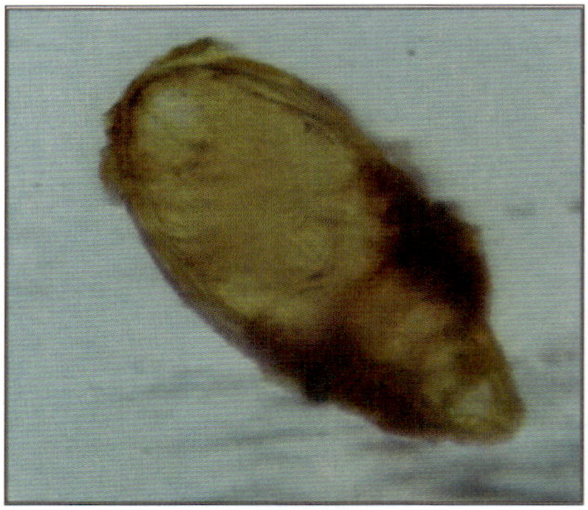

Sclerocystis

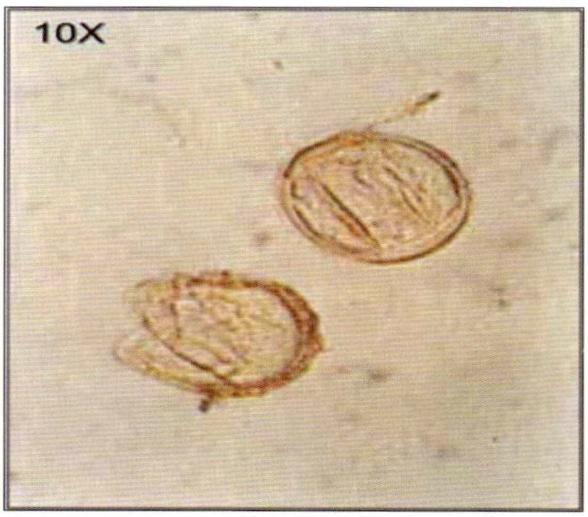

Paraglomus occultum

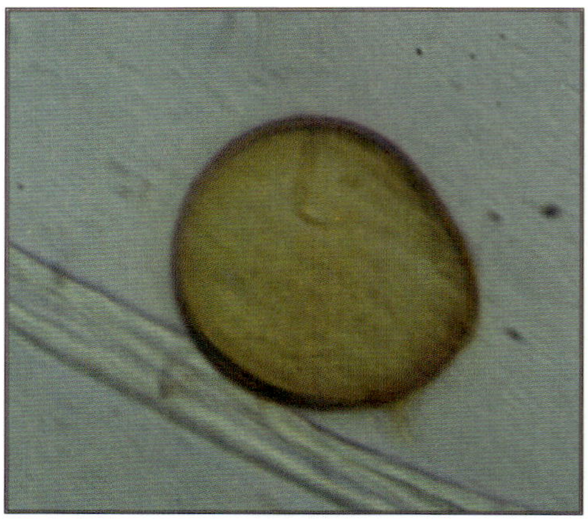

Acaulospora

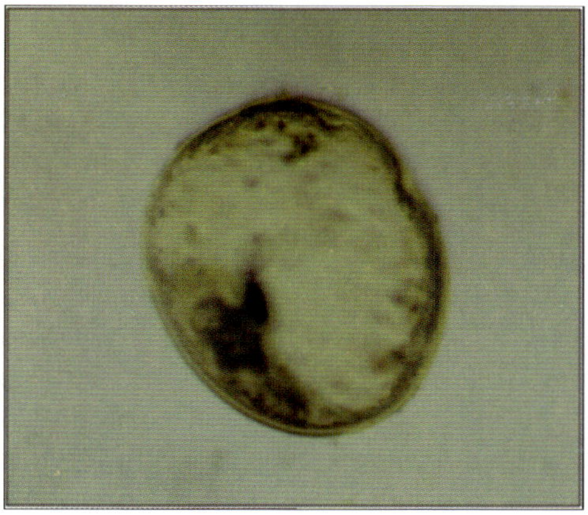

Acaulosora

Mass culture

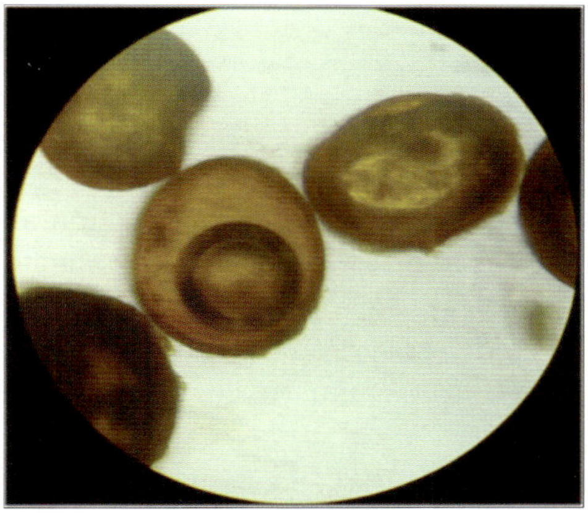

Acaulospora

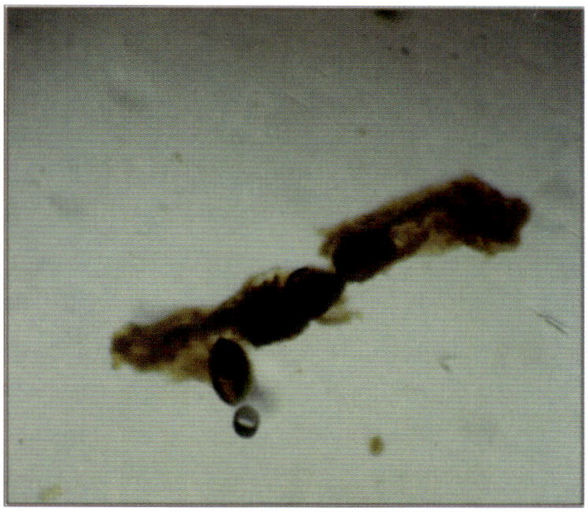

Spores in root

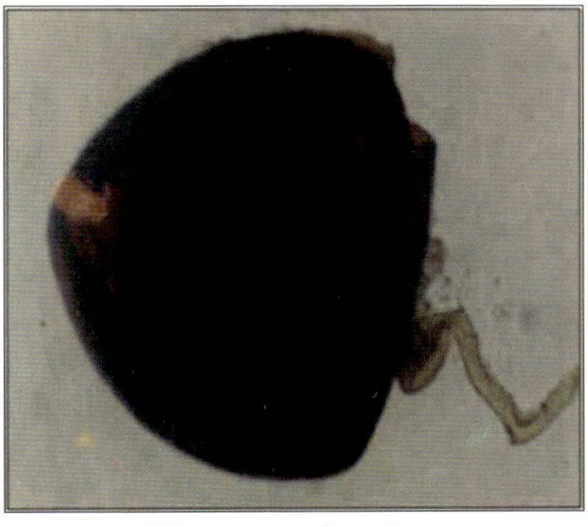

Gigaspora gigantean

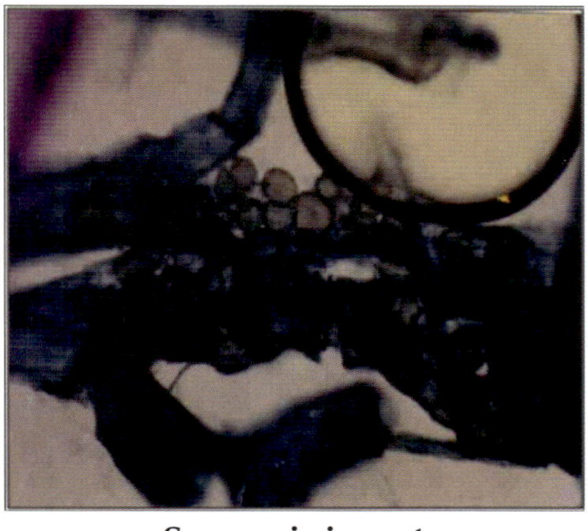

Sporangia in root

CONCLUSION

Various levels of mycorrhizal dependency of plants, no doubt, give some indication to design the artificial inoculation regime in the selected crop for better growth, quality and quantity of yield. However, the desired result may or may not replicate in field, because in field, there is succession of VA-mycorrhizal species in seasonal and annual cycle [134], and as a rule there is no host specificity in mycorrhizal association. Also, the synergistic effect of beneficial microbial population influenced by the dominant mycorrhizal fungi active in particular season can never be managed artificially. Therefore, if the mycorrhizal dependency of the host plant is more than 40% at field level trial, then only the artificial inoculation of field level would be economically viable. However, if there is integration of dominant VA-mycorrhizal fungi in a particular season, when the less costly inoculum and easy and standardized methods of inoculation is available, the dependency based inoculation will always prove cheaper and eco-friendly because the build-up inoculum of mycorrhizal fungal species in soil will maintain the diversity of associated microflora on sustainable basis.

REFERENCES

1. Abbot, L.K. and Robson, A.D. (1991). "Factors influencing the occurrence of vesicular arbuscular mycorrhizae". *Agr. Ecosys and Env.*, 35, 121–128.
2. Adholeya, A. (2000). "Utilization of fly ash for commercial plant production and environmental protection using microbes". *In Proceedings of Second International Conference on Fly Ash Disposal and Utilization*, vol. II, 32–35.
3. Adjoud, D., Plenchette, C., Halli Hargas, R. and Lapeyrie, F. (1996). "Response of *Eucalyptus* species to inoculation with three arbuscular mycorrhizal fungi". *Mycorrhiza*, 6(2), 129–135.
4. Aggangan, Nelly S., Kim, Yong-Wook., Han, Sim-Hee., Park, So-Young. and Moon, Heung-Kyu. (2011). "Selection of arbuscular mycorrhizal inoculants for improved growth of somatic embryo-derived Liriodendron tulipifera (Magnoliaceae) Plantlets". *Journal of Environmental Science and Management*, 15(1), 59–71.
5. Ahiabor, B.D. and Hirata, H. (1995). "Influence of growth stage on the association between some tropical legumes and two variant species of Glomus in an Andosol". *Soil Sci. Plant Nutr.*, 41, 481–496.
6. Al-karaki, G.N. (2006). "Nursery inoculation of tomato with arbuscular mycorrhizal fungi and subsequent performance under irrigation with saline water". *Scientia Horticulture*, 109, 1–7.
7. Antunes, V. and Cardoso, E.J.B.E., (1991). "Growth and nutrient status of citrus plants as influenced by mycorrhiza and phosphorus application". *Plant and Soil*, 131, 11–19.
8. Arriagada, C.A., Herrera, M.A., Garcia-romera, I. and Ocampo, J.A. (2004). "Tolerance to Cd of soybean (Glycine max) and eucalyptus (Eucalyptus globulus) inoculated with water air soil pollutant arbuscular mycorrhizal and saprobe fungi". *Symbiosis.*, 36, 285–299.

9. Aúge, R.M., Duan, X., Etel, R.C. and Stodola, A.J.W. (1994). "Non-hydraulic signalling of soil drying in mycorrhizal maize". *Planta,* 193, 74–82.
10. Azcón, R. and Barea, J.M. (1997). "Nodulation, nitrogen fixation (Nitrogen[15]) and nitrogen nutrition relationships in mycorrhizal and phosphate amended alfalfa plants". *Symbiosis,* 12(1), 33–42.
11. Azcón, R. and Ocampo, J.A. (1981). "Factors affecting the vesicular-arbuscular infection and mycorrhizal dependency of thirteen wheat cultivars". *New Phytol.,* 87, 677–685.
12. Ba, A.M., Plenchelle, C., Danthu, P., Duponnois, R. and Guissou, T. (2000). "Functional compatibility of two arbuscular mycorrhizae with thirteen fruit trees in Senegal". *Agroforest. Syst.,* 50, 95–105.
13. Bâ, A.M. and Guissou. T. (1996a). "Rock phosphate and vesicular-arbuscular mycorrhiza effects on growth and nutrient uptake of Faidherbia albida (Del.) seedlings in an alkaline sandy soil". *Agroforestry System,* 34, 129–137.
14. Bâ, A.M., Bazie´, M. and Guissou, T. (1996b). "Effet du phosphate naturel sur de jeunes Acacia albida Del. Enpre´senceou non de mycorhizes."*Les parcsa` Faidherbia,* 12, 237–244.
15. Baon, J.B., Smith, S.E. and Alston, A.M. (1993). "Mycorrhizal responses of barley cultivars differing in P efficiency". *Plant Soil,* 157, 97–105.
16. Barea, J.M. (1991). "Vesicular arbuscular mycorrhizae as modifiers of soil fertility". In: B.A. Stewart (ed.), *Advances in Soil Science.* Springer-Verlag, New York, 1–40.
17. Barea, J.M., Azcon, R. and Azcon-Aguliar, C. (1992). "Vesicular arbuscular mycorrhizal fungi in nitrogen-fixing systems". In: J.R. Norris, D.J. Read and A.K. Varma (eds.), *Methods in Microbiology.* Academic Press, London, 167–189.
18. Barrow, N.J., Malajczuck, N. and Shaw, T.C. (1977). "A direct test of the ability of vesicular arbuscular mycorrhizal to help plants take up fixed soil phosphate". *New Phytol.,* 78, 269–276.
19. Baylis, G.T.S. (1967). "Experiments on the ecological significance of phycomycetous mycorrhizas". *New Phytol.,* 66, 231.
20. Bécard, G. and Pfeffer, P.E. (1993). "Status of nuclear division in arbuscular mycorrhizal fungi during in vitro development". *Protoplasma.,* 174, 62–68.
21. Becard, G. and Piche, Y. (1989). "Fungal growth stimulation by CO_2 and root exudates in vesicular arbuscular mycorrhizal symbiosis". *Appl. Environ. Microbial.,* 55, 2320–2325.
22. Bentivenga, S.P. and Hetrick, B.A.D. (1992). "Seasonal and temperature effects on mycorrhizal activity and dependence of cool and warm-season tall grass prairie grasses". *Can. 1. Bot.,* 70, 1596–1602.
23. Bethlenfalvay, G.J. and Linderman, R.G. (1992). "Mycorrhizae in Sustainable agriculture". *ASA Spec Publ Madison,* WI.
24. Bolan, N.S. (1991). "A critical review on the role of mycorrhizal fungi in the uptake of phosphorus by plant". *Plant and Soil,* 134, 189–207.

25. Bowen, G.D. and Theodorou, C. (1967). "Studies on phosphate uptake by mycorrhizas". *Proc. int. Un. for. Res. Orgn., 14th,* Munich, 5, 116.
26. Bowen, G.D. (1973). "Mineral nutrition in mycorrhizas". In: *Ectomycorrhizae* (ed. G.C. Marks and T.T. Kozlowski), Academic Press, New York, 151–201.
27. Brundrett, M.C. (2004). "Diversity and classification of mycorrhizal association". *Biol. Rev.,* 79, 473–495.
28. Bryla, D.R. and Koide, R.T. (1990). "Role of mycorrhizal infection in the growth and reproduction of wild vs. cultivated plants in eight wild accessions and two cultivars of *Lycopersicon esculentum* Mill". *Oecologia,* 84, 82–92.
29. Burkert, B. and Robson, A.D. (1994). "^{65}Zn uptake in subterranean clover (*Trifolium subterraneum*) by three vesicular arbuscular mycorrhizal fungi in a root-free sandy soil". *Soil biology and Biochemistry.,* 26, 1117–1124.
30. Chen, B.D., Zhu, Y.G., Duan J., Xiao, X.Y. and Smith, S.E. (2007). "Effects of the arbuscular mycorrhizal fungus *Glomus mosseae* on growth and metal uptake by four plant species in copper mine tailings". *Environmental Pollution.,* 147, 374–380.
31. Chen, B.D., Christie, P. and Li X.L. (2001). "A modified glass bead compartment cultivation system for studies on nutrient uptake by arbuscular mycorrhiza". *Chemosphere.,* 42, 185–192.
32. Cruz, R.E. de la, Zarate, J.T., Aggangan, N.S. and Lorilla, E.B. (1992). "Differential mycorrhizal dependency of some agricultural, horticulture and forestry crops on inoculation with arbuscular mycorrhizal fungi". In: *Proceedings of the International Symposium on Management of Mycorrhizas in Agriculture, Horticulture and Forestry* (ed. Jasper), Australia, 54.
33. Declerck, S., Plenchette, C. and Strullu, D.G. (1995). "Mycorrhizal dependency of banana (Musa acuminata, AAA group) cultivar". *Plant Soil,* 176, 183-187.
34. Dehne, H.W. (1982). "Interaction between vesicular arbuscular mycorrhizal fungi and plant pathogens". *Phytopathology,* 72, 1115–1119.
35. Dhillion, S.S. (1992). "Host-endophyte specificity of vesicular arbuscular mycorrhizal colonization of *Oryza sativa* at the pretransplant stage in low or high phosphorus soil". *Soil Bioi. Biochem.,* 24, 405–411.
36. Diaz, G., Azcon-aguliar, C. and Honnubia, L. (1996). "Influence of vesicular arbuscular mycorrhizae on heavy metal (Zn and Pb) uptake on growth of *Lygeum spartum* and *Anthylis cystoids*". *Plant Soil.,* 180, 241–249.
37. Dodd, J.C. (1994). "Approaches to the study of the extra-radical mycelium of arbuscular mycorrhizal fungi". In: *Impact of Arbuscular Mycorrhizal on Sustainable agriculture and natural ecosystems* (Ed. - Gianinazzi, S and Scheiepp, H.), Birkhauser Verlag., 147–166.
38. Douds, D.D., Nagahashi, G., Pfeffer, P.E., Kayser, W.M. and Reider, C. (2005). "On-farm production and utilization of arbuscular mycorrhizal fungus inoculum". *Can. J. Plant Sci.* 85, 15–21.
39. Duponnois, R., Colombet A., Hien V. and Thioulouse J. (2005). The mycorrhizal fungus *Glomus intraradices* and rock phosphate amendment influence plant growth

and microbial activity in the rhizosphere of *Acacia holosericea*. *Soil Biology & Biochemistry*, 37, 1460–1468.

40. Duponnois, R., Plenchette C. and Ba, A.M. (2001). "Growth stimulation of seventeen fallow leguminous plants inoculated with *Glomus aggregatum* in Senegal". *Euro. Soil Bioi.*, 37, 181–186.
41. Fox, T.R., Comeford, N.B. and McFee, W.W. (1990). "Kinetics of phosphorus release from spodosols: Effects of oxalate and formate". *Soil Sci. Soc. Am. J.*, 54, 1441–1447
42. Garg, N. and Chandel, S. (2010). "Arbuscular mycorrhizal networks: process and functions - A review". *Agron. Sustain. Dev.* 30, 581–599.
43. George, E., Haussler, K., Vetterlein, G., Gorgus, E. and Marschner, H. (1992). "Water and nutrient translocation by hyphae of *Glomus mosseae*". *Canadian Journal of Botany*, 70, 2130–2137.
44. George, E., Marschner, H. and Jackobson, I. (1995). "Role of arbuscular mycorrhizal fungi in uptake of phosphorus and nitrogen from soil". *Critical Rioiass in Biotechnology*, 15, 257–270.
45. Gerdemann, J.W. (1968). "Vesicular arbuscular mycorrhiza and plant growth: A review". *Phytopath.*, 6, 3–7.
46. Gerdemann, J.W. (1975). "Vesicular arbuscular mycorrhiza. In: *The Development and Function of Roots*", ed. JG Torrey and DT Clarkson. Academic Press, London, 575–592.
47. Ghosh, S. and Verma, N.K. (2002). "A preliminary report on VA mycorrhizal status of some common weeds, trees and aquatic plants of south West Bengal". *Sci Cul.*, 68(5-6), 155–157.
48. Ghosh, S. and Verma, N.K. (2006). "Growth and mycorrhizal dependency of *Acacia mangium*. Willd, inoculated with three vesicular arbuscular mycorrhizal fungi in lateritic soil". *New Forests*, 31, 75–81.
49. Ghosh, S. (2007). "Studies related to the effect of arbuscular mycorrhizae and plant growth promoting rhizobacteria (PGPR) on the growth and nutrition of forest crops". *Ph.D dissertation*. Vidyasagar University.
50. Graham, J.H. and Syvertsen, J.P. (1985). "Host determinants of mycorrhizal dependency of *Citrus* rootstock seedlings". *New Phytol.*, 101, 667–676.
51. Gray, L.E. and Gerdemann, J.W. (1969). "Uptake of phosphorus - 32 by vesicular-arbuscular mycorrhiza". *Pl. Soil*, 30, 415.
52. Grime, J.P., Mackey, J.M.L., Hillier, S.H. and Read, D.J. (1987). "Floristic diversity in a model system using experimental microcosms". *Nature*, 328, 420–422.
53. Guissou, T., Ba A.M., Ouadba, J.M., Guinko, S. and Duponnois, R. (1998). "Responses of *Parkia biglobosa* (Jacq.) Benth, *Tamarindus indica* L. and *Ziziphus mauritiana* Lam. to arbuscular mycorrhizal fungi in a phosphorus-deficient sandy soil". *Biol. Fertil. Salls*, 26, 194–198.
54. Habte, M. and Manjunath, A. (1987). "Soil solution phosphorus status and Mycorrhizal dependency in *Leucena lucocephala*". *Appl. and Envirn. Microbiol. (Apl).*, 797–801.

55. Habte, M. and Musoko, M. (1994). "Changes in the vesicular arbuscular mycorrhizal dependency of *Albizzia ferruginea* and *Enterolobium cyclocarpum*in response to soil phosphorus concentration". *Jr. Plant Nutrition.*, 17(10), 1769–1780.
56. Habte, M. and Menge, A. (1991). "Categories of vesicular arbuscular mycorrhizal dependency". *Mycorrhiza.* , 1, 3–12.
57. Hampp, R., Nehls, U. and Wallenda, T. (2000). "Physiology of mycorrhiza". In: K. Esser, J.W. Kadereit, U. huttge & M. Runge (eds.), Genetics Physiology Systemetes, *Ecology Springer Verlag,* Berlin. *Progress in Botany,* 61, 223–254.
58. Harley, J.L. (1969). "*The Biology of Mycorrhiza*". Leonard Hill, London, 334.
59. Harley, J.L. and Smith, S.E. (1983). "*Mycorrhizal Symbiosis.*" Academic Press, New York, 306.
60. Hasan, P. and Ibrahim, O. (2012). "Screening of mycorrhizal dependency of different pepper genotypes". *Ç.Ü Fen ve Mühendislik Bilimleri Dergisi Yıl,* 27(3), 35–44.
61. Hayman, D.S. and Mosse, B. (1971). "Plant growth responses to vesicular arbuscular mycorrhiza". I. Growth of Endogone inoculated plants in phosphate-deficient soils. *New Phytol.*, 70, 19.
62. Hayman, D.S. and Mosse, B. (1972). "Plant growth responses to vesicular arbuscular mycorrhiza- increased uptake of labile P from soil". *New Phytol.*, 71, 41–47.
63. Heidi, J.H., Johansen, A. and George, E. (2000). "Uptake and transport of organic and inorganic nitrogen by arbuscular mycorrhizal fungi". *Plant and Soil,* 226, 275–285.
64. Hetrick, B.A.D, Wilson, G.W.T. and Todd, T.C. (1996). "Mycorrhizal response in wheat cultivers, relationship to phosphorus". *Jr of Bot.*, 1, 19–25.
65. Hetrick, B.A.D., Wilson, G.W.T. and Cox, T.S. (1992). "Mycorrhizal dependence of modern wheat varieties Landraces and Ancestors". *Can. I Bot.*, 70, 2032–2040.
66. Hetrick, B.A.D., Wilson, G.W.T. and Cox, T.S. (1993). "Mycorrhizal dependence of modern wheat cultivars and ancestors: A synthesis". *Can. Bot.*, 71, 512–518.
67. Hooker, J.E. and Black, J.E. (1995). "Arbuscular mycorrhizal fungi as a component of sustainable soil plant systems". *Critical Reviews in Biotechnology,* 15, 201–212.
68. Isobe, K. and Tsuboki, Y. (1998). "The relationship between growth promotion by arbuscular mycorrhizal fungi and root morphology and phosphorus absorption in graminaceous and leguminous crops". *Japan. J. Crop. Sc.*, 67(3), 347–352.
69. Jasper, D.A., Abbott, L.K. and Robson, A.D. (1989a). "Hyphae of a VA mycorrhizal fungus maintains infectivity in dry soil except when the soil is disturbed". *New Phytol.*, 112, 101–107.
70. Jasper, D.A., Abbott, L.K. and Robson, A.D. (1989b). "Soil disturbance reduces the infectivity of external hyphae of VA mycorrhizal fungi". *New Phytol.*, 112, 93–99.
71. Jeffries, P., and Barea, J.M. (1994). "Biogeochemical cycling and arbuscular mycorrhizas in the sustainability of plant-soil systems". In: *Impact of Arbuscular Mycorrhizas on Sustainable Agriculture and Natural Ecosystem,* Gianinazzi, S. and Schüepp, H. (eds.), Birkhäuser Verlag Basel, Switzerland, 101–115.

72. Jeffries, P., Gianinazzi, S., Perotto, S., Turnau, K. and Barea, J.M. (2003). "The contribution of arbuscular mycorrhizal fungi in sustainable maintenance of plant health and soil fertility". *Biology and Fertility of Soils*, 37, 1-16.
73. Johanssen, A., Jakobsen, I. and Jensen, E.S. (1994). "Hyphal N transport by a vesicular-arbuscular mycorrhizal fungus associated with cucumber grown at three nitrogen levels". *Plant and Soil*, 160, 1-9.
74. Johnson, N.C., Graham, J.H. and Smith, F.A. (1997). "Functioning of mycorrhizal associations along the mutualism-parasitism continuum". *New Phytol.*, 135, 575-585.
75. Joner, E.J. and Jakobsen, I. (1994). "Contribution by two arbuscular mycorrhizal fungi to P uptake by cucumber (*Cucumissativus L*) from 32P-labelled organic matter during mineralization in soil". *Plant Soil*, 163, 203-209.
76. Jones, C. (2014). "The Natural Farmer", B-14, www.amazingcarbon.com
77. Kaeppler, S.M., Parke, J.L., Mueller, S.M., Senior, L., Stuber, C. and Tracy, W.F. (2000). "Variation among maize inbred lines and detection of quantitative trait loci for growth at low phosphorus and responsiveness to arbuscular mycorrhizal fungi". *Crop Sci.*, 40, 358-364.
78. Kelly, R.M., Edwards, D.G., Thompson, J.P. and Magarey, R.C. (2001). "Responses of sugarcane, maize and soybean to phosphorus and vesicular arbuscular mycorrhizal fungi". *Aust. J. Agric.* Res., 52, 731-743.
79. Khalil, S., Loynachan, T.E. and Tabatabai, M.A. (1994). "Mycorrhizal dependency and nutrient uptake by improved and unimproved corn and soybean cultivars". *Agro. J.*, 86, 949-958.
80. Kim, K.Y., Jordon, D. and McDonald, G.A. (1998). "Effect of phosphate solubilizing bacteria and vesicular arbuscular mycorrhizal fungus on tomato growth and soil microbial activity". *Biol. Fert Soils*. 26, 79-87.
81. Krikun, J., Haas, J.H., Dodd, J.C. and Kmsbursky, R. (1990). "Mycorrhizal dependence of four crops in a P-sorbing soil". *Plant Soil*, 122, 213-217.
82. Krishna, K.R., Shetty, K.G., Dart, P.J. and Andrews, D.J. (1985). "Genotype dependent variation in mycorrhizal colonization and response to inoculation as pearl millet". *Plant Soil.*, 86, 113-125.
83. Lakshman, H.C. (2009). "Importance of AM fungal technology for sustainable agriculture". In the proceedings of *NAS - Bangalore ICAR - National Conference* April 14-16., 19-23.
84. Leyval, C., Turnau, K. and Haselwandler, K. (1997). "Effect of heavy metal pollution on mycorrhizal colonization and function: physiological, ecological and applied aspect". *Mycorrhiza*, 7, 139-153.
85. Li, Xia - Lin, George, E., and Marschiner H. (1991). "Extension of the phosphorus depletion zone in VA mycorrhizal white clover in calcareous soil". *Plant Soil*, 136 (1), 41-48.
86. Liu, A., Hamel, C., Hamilton, R.I., Ma, B.L. and Smith, D.L. (2000). "Acquisition of Cu, Zn, Mn, and Fe by mycorrhizal maize (*Zea mays* L.) grown in soil at different P and micronutrients levels". *Mycorrhiza*, 9(6), 331-336.

87. Madder, P., Vierheileg, H., Boller, T., Streitwalf, B., Freg, E., Chritie, P. and Wietnken, A. (2000). "Transport of ^{15}N from a soil compartment separated by a polytetra flouroethylene membrane to plant roots via the hyphae of arbuscular mycorrhizal fungi". *New Phytologist.*, 146, 155–161.
88. Manjunath, A. and Habte, M. (1991a). "Relationship between mycorrhizal dependency and rate variable associated with P uptake, utilization and growth". *Communications in Soil Science and Plant Analysis*, 22 (13/14), 1423–1437.
89. Manjunath, A. and Habte, M. (1991b). "Root morphological characteristics of host species having distinct mycorrhizal dependency". *Can. Bot.*, 69, 671–676.
90. Manjunath, A., Hue, N.V. and Habte, M. (1989). "Response of *Leuceana leucocephala* to vesicular arbuscular mycorrhizal colonization and rock phosphate fertilization in an Oxisol". *Plant and Soil*, 114, 127–133.
91. Marschner, H. (1998). "Role of root growth, arbuscular mycorrhiza and root exudates for the efficiency in nutrient acquisition". *Field Crop. Res.*, 56(1-2), 203–207.
92. Marschner, H. and Dell, B. (1994). "Nutrient uptake in mycorrhizal symbiosis". *Plant and Soil*, 159, 89–102.
93. Menge, J.A., Stierle, D., Bagyaraj, D.J., Johnson, E.L.V. and Leonard, R.T. (1978). "Phosphorus concentrations in plants responsible for inhibition of mycorrhizal infection". *New Phytol.*, 80, 575–578.
94. Mishra, R.R. and Verma, N.K. (1982). "Effect of different mycorrhizal treatments on the growth of onion". *Acta Botanica India.* 11, 49–52.
95. Miyasaka, S.C., Habte, M. and Matsuyama, D.T. (1993). "Mycorrhizal dependency of two Hawaiian endemic tree species: koa and mamane". *Journal of Plant Nutrition*, 16(7), 1339–1356.
96. Monzon, A. and Azcon, R. (2001). "Growth responses and N and P use efficiency of three Alnus species as affected by arbuscular- mycorrhizal colonisation". *Plant Growth Regul.*, 35, 97–104.
97. Mosse B. (1973). "Advances in the study of vesicular arbuscular mycorrhiza". *Ann. Rev. Phytopath.* 11, 171–196.
98. Nelson, S.D. (1987). "Rooting and subsequent growth of woody ornamental softwood cuttings treated with endomycorrhizal inoculum". *Am. Soc. Hortie. Sci.*, 112, 263–266.
99. Nicolson, T.H. (1967). "Vesicular arbuscular mycorrhiza—a universal plant symbiosis". *Sci. Prog. Ox*, 55, 561.
100. Ning J. (2000). "Mycorrhizal roles in broomsedge plants under phosphorus limitation and aluminum toxicity". *PhD Dissertation*, West Virginia University, Morgantown WV, USA.
101. Nouiam, R. and Choussad, R. (1994). "Mycorrhizal dependency of micropropagated argan tiree (*Argamoa spinosa*): 1. Growth and biomass production". *Agroforestry System*, 27(1), 53–65.
102. Ortega-Larrocea, M.P., Xoconostle-Cázares, B., Maldonado-Mendoza, I., Carrillo-González, R., Hernández-Hernández, J. and DíazGardunño, M., et al., (2010). "Plant

and fungal biodiversity from metal mine wastes under remediation at Zimapan, Hidalgo, Mexico". *Environmental Pollution*. 158, 1922–1931.

103. Oyetunji, O.J., Ekanayeke, I.J. and Osonubi, O. (2003). "The influence of arbuscular mycorrhizae fungus, mulch and fertilizer application on the yield of yams in an agroforestry system in south western Nigeria". *Maurik Bull.* 6, 75–82.

104. Perry, D.A. and Amaranthus, M.P. (1990). "The plant-soil bootstrap: Microorganisms and reclamation of degraded ecosystems". Berger J.J., (Ed.), *Environmental Restoration: Science and Strategies for Restoring the Earth*. Island Press, Washington, DC, 94–102.

105. Plenchette, C., Fortm, J.A. and Furlan, V. (1983). "Growth response of several plant species to mycorrhizae in a soil of moderate P fertility. I. Mycorrhizal dependency under field conditions". *Plant Soil*, 70, 199–209.

106. Pope, P.E., Chaney, W.R., Rhodes, J.D. and Woodhead, S.H. (1983). "The mycorrhizal dependency of four hardwood tree species. *Can. J. Bot.*, 61, 412-417.

107. Ragupati, S. and Mahadevan, A. (1993). "Distribution of vesicular arbuscular mycorrhiza in the plants and rhizosphere soils of the tropical plains of Tamilnadu, India". *Mycorrhiza*, 3, 123–136.

108. Rengel, Z. (2002). "Breeding for better symbiosis". *Plant Soil*, 245, 147–162.

109. Renuka, G., Rao, M.S., Kumar, Praveen V., Ramesh, M. and Reddy, M.S. (2012). "Arbuscular mycorrhizal dependency of *Acacia melanoxylon* R." Br. Proc. Natl. Acad. Sci., India, Sect. *B Biol. Sci.*, 82(3), 441–446.

110. Requena, N., Mann, P. and Franken, C. (2000). "A homologue of the cell-cycle check-point TOR2 from yeast exists in the arbuscular mycorrhizal fungus *Glomus mosseae*".*Protoplasma*, 211, 89–98.

111. Rosendahl, S., Rosendahl, C. and Sochting, U. (1990). "Distribution of VA mycorrhizal endophytes amongst plants from a Danish grassland community". *Agri. Gco. Environ.*, 29, 329–336.

112. Saif, S.R. (1987). "Growth responses of tropical forage plant species to vesicular arbuscular mycorrhizae: I. Growth, mineral uptake and mycorrhizal dependency". *Plant Soli*, 97, 25–35.

113. Salamanca, C.P., Herrera, M.A., Barea and J.M. (1992). "Mycorrhizal inoculation in micropropagated woody legumes used in revegetation programmes for desertified Mediterranean ecosystems". *Agronomic*, 12, 869–972.

114. Samanta, S. and Verma, N.K. (2006). "Effect of VA mycorrhiza on the growth and protein content in fruits of *Capsicum annuum* grown in acid lateritic soil". *J. Mycopathol. Res.*, 44 (2), 197–200.

115. Sanders, F.E., Tinker, P.B., Black, R.L. and Palmerley, S.M. (1977). "The development of endomycorrhizal root systems: I. Spread of infection and growth promoting effects in four species of vesicular arbuscular mycorrhizae". *New Phytologist*, 78, 257–268.

116. Sarangi, P.K. and Mishra, P.C. (1998). "Soil metabolic activities and yield in ground nut, ladies finger and radish in fly ash amended soil". *Journal of Chemistry and Environmental Research*, 2, 7–13.
117. Schüßler, A., Schwarziff, D. and Walker, C. (2001). "A new fungal phylum, the Glomeromycota: Phylogeny and evolution". *Mycological Research*, 105, 1413–1421.
118. Sengupta, D. (2000). "VAM – *Rhizobium* dual culture as biofertilizer for increasing the productivity of soybean and groundnut in acid lateritic soil of South West Bengal". *PhD Dissertations*. Vidyasagar University. W.B., India.
119. Sengupta, D., Verma, N.K. and Ghosh, B.C. (2006). "Effect of vesicular arbuscular mycorrhiza and *Rhizobium* on field grown groundnut in acid lateritic soil". In: *Mycorrhiza* (eds. Anil Prakash and V.S. Mehrotra) Scientific Publishers (India). Jodhpur, 215–218.
120. Sengupta, D., Verma, N.K. and Ghosh, B.C. (2001). "Effect of lime on VAM-*Rhizobium-Arachis hypogaea* tripartite association". *Natl. Bot. Soc.* 55 (5-6), 155–57.
121. Sharma, A.K., Srivastava, P.C. and Johri, B.N., (1999). "Effect of vesicular arbuscular mycorrhization on uptake and translocation of phosphorus in *Sesbania aculeate*". *Curr. Sci.* 77(10), 1351–1355.
122. Simon, L., Bousquet, J., Levesque, R.C. and Lalonde, M. (1993). "Origin and diversification of endomycorrhizal fungi and coincidence with vascular plants". *Nature*, 363, 67–69.
123. Singh, S. (2001). "Mycorrhizal dependency, Part. 1: Selection of efficient mycorrhizal fungi". *Mycorrhiza News*, 11(1), 2–9.
124. Siqueira, J.O. and Saggin-Junior, O.J. (2001). "Dependency on arbuscular mycorrhizal fungi and responsiveness of some Brazilian native woody species". *Mycorrhiza*, 11, 245–255.
125. Siqueira, J.O., Saggin-Junior, Flores, Aylas W.W. and Guimares, P.T.G. (1998). "Arbuscular mycorrhizal inoculants and superphosphate application influenced plant development and yield of coffee in Brazil". *Mycorrhiza*, 7, 293–300.
126. Smith, S.E. and Read, DJ (1997). "*Mycorrhizal Symbiosis*", Academic Press, London.
127. Smith, S.E., Dickson, S., Morris, C. and Smith, F.A. (1994). "Transfer of phosphate from fungus to plant in VA mycorrhizas: calculation of the area of symbiotic interface and fluxes of P from two different fungi to *Alhium porum* L". *New Phytol.* 127, 93–99.
128. Tawaraya, K. (2003). "Arbuscular mycorrhizal dependency of different plant species and cultivars". *Soil Sci. Plant Nutl.*, 49 (5), 655–668.
129. Tawaraya, K., Imai, T. and Wagatsuma, T. (1999). "Importance of root length in mycorrhizal colonisation of Welsh onion". *J. Plant Nutr.*, 22, 589–596.
130. Tawaraya, K., Tokairin, K. and Wagatsuma, T. (2001). "Dependence of *Allium fistulosum* cultivars on the arbuscular mycorrhizal fungus, *Glomus fasciculatum*". *Appl. Soil Ecol.*, 17: 119–124.

131. Torres - Barragan, A., Zavalta-Meja, E., Chavez, G. and Ferrea - Cerrato, R. (1996). "The use of arbuscular mycorrhizae to control onion white rot (*Sclerotium cepivolium)* under field conditions". *Mycorrhiza*. 6, 253–257.

132. Van der, Heijden V., Boller, G.A.T., Wiemken, A. and Sanders, I.R. (1998). "Different arbuscular mycorrhizal fungal species are potential determinants of plant community structure". *Ecology*. , 79, 2082–2091.

133. Varma, A. and Schüepp, H. (1994). "Infectivity and effectiveness of *Glomus intraradices* on micropropagated plants". *Mycorrhiza*, 5, 29–38.

134. Verma, N.K. (2000). "Seasonal influence on type and succession of VA-mycorrhizal association in two species of *Eupatorium* (*Asteraceae*) of Meghalaya". *Sci. & Cult.*, 66 (1-2), 97-98.

135. Vestburg, M. (1992). "Arbuscular mycorrhizal inoculation of micropropagated strawberry and field observation in Finland". *Agronomie.*, 12, 865–867.

136. Weber, E., George, E., Beck, D.P., Saxena, M.C. and Marschner, H. (1992). "Vesicular arbuscular mycorrhizal and phosphorus uptake in chickpea grown in Northern Syria". *Expl. Agric*, 28, 433–442.

137. Williums, S.C.K., Vestburg, M., Uosukainen, M., Dodd, J.C. and Jeffris, P. (1992). "Effects of fertilizers and arbuscular mycorrhizal fungi on the *post - vitro* growth of micropropagated strawberry". *Agronomie.*, 12, 851–857.

138. Wilson, G.W.T. and Hartnett, D.C. (1998). "Interspecific variation in plant responses to mycorrhizal colonization in tallgrass prairie". *Am. J. Bot.*, 85, 1732–1738.

139. Yao, Q., Li, X.L., Feng, G. and Christie, P. (2001). "Influence of extramatrical hyphae on mycorrhizal dependency of wheat genotypes *Commun*". *Soil Sci. Plant Anal.*, 32, 3307–3317.

140. Zhu, Y.G., Smith, S.E., Barritt, A.R. and Smith, F.A. (2001). "Phosphorus (P) efficiencies and mycorrhizal responsiveness of old and modern wheat cultivars". *Plant Soil*, 237, 249–255.

11

Vermicomposting to Modified Vermicomposting: Technological Upgradation for Improving Quality and Nutrient Status

Debabrata Das[1], Anupam Roy[1], Bijoy Chandra Ghosh[1], Shanker Lal Shrivastava[1] and Sreyajit Saha[2]

[1]*Agricultural & Food Engineering Department. Indian Institute of Technology Kharagpur, Kharagpur – 721302, India*

[2]*Amity Institute of Food Technology, Amity University, Noida, Uttar Pradesh, India*

E-mail: *bcg@agfe .iitkgp.ernet.in*

ABSTRACT

Vermicomposting systems are a consortium of complex microbial and invertebrate food web that results in the recycling of organic matter and release of nutrients. Biotic interactions between decomposers (i.e., bacteria and fungi) and microfauna result in competition, mutualism, predation, followed by rapid changes in the substrate quality. For sustainable farming, nutrient quantity of vermicompost is not sufficient for plant growth and yields. So, vermicompost is required to be enriched or modified with technological upgradation for value addition and higher nutrient quality. The "enriched vermicompost" or "modified vermicompost" is a mixture of vermicompost, nutrient-rich material (i.e. minerals) and microorganisms. Enriched vermicompost not only contains additional nutrients, but it also minimizes conventional vermicompost preparation time. Depending upon the specific requirements of different plants and soil, the ingredient and the concentration of nutrients in modified vermicompost can be formulated. This requirement can be easily achieved from agro wastes i.e. farm manure and legume residues. The nutrient status is further modified using natural minerals i.e. rock phosphate, dolomite and mica powder. Besides, microorganisms are also included in the vermicompost preparation, which are effective at providing and protecting fertilizers and also help to mineralize elements that are used by the crops. This chapter deals with the process of vermicomposting and its technological upgradation as modified vermicomposting.

Keywords: Modified Vermicomposting, Microbial Enrichment, Rock Mineral.

11.1 INTRODUCTION

The beneficial effects of earthworms in improving soil nutrition have long been known [1] but the better efficiency of some earthworm species over others for their ability to degrade organic matter was known much later [2]. Agriculture is facing a challenge to develop strategies for sustainability. The prime focus is on conserving non-renewable natural resources such as soil and enhancing the use of renewable resources such as organic wastes. More than 3000 million tonnes of organic wastes are generated per year in India [3]. The disposal of ever increasing amounts of organic wastes is becoming a serious problem in India. The hygienic disposal of organic wastes by composting is an environmentally sound and economically viable technology. This can effectively increase the production of organic fertilizer, which is a basic and valuable input in organic farming. In India, about 350 million tonnes of agricultural wastes are generated annually, of which vegetable waste alone is a major portion [4]. The biological treatment of these wastes appears to be most cost-effective and carry a less negative environmental impact. Among the different options for recycling wastes, application of wastes to agricultural land is probably the most reliable and cost-effective technique to supply organic matter to field crops [5]. Wastes of animal and plant origin are one of the major underutilized resources in many countries. These wastes could not be fully exploited due to the non-availability of a viable technology for their economic recycling. Composting, the controlled biological oxidative decomposition of organic matter [6], is an important process for recycling waste materials. However, composting has some disadvantages including nutrient loss during composting, the cost of land, equipment and labour requirement for composting and odor associated with composting [7]. The success of sustainable agriculture is very much dependent upon the availability of cheap and good quality organic manures. Under such perspective, vermicomposting has been recognized as an eco-friendly technology for converting organic wastes into high-value organic manure offering bio-chemical restoration [8], enhancing soil fertility and soil health.

11.2 SCIENCE BEHIND VERMICOMPOSTING

Vermicomposting is a biotechnological process which transforms organic wastes into stabilized humus-like product. Products have relatively high contents of readily available plant nutrients, growth enhancing substances, enzymes and a number of beneficial and active microorganisms like N_2-fixing, P-solubilizing and cellulose decomposing organisms. A variety of biochemical processes are involved in the biological decomposition of organic matter. In vermicomposting process, during the passage through the worm's gut, organic matter undergoes biochemical changes by the combined effect of earthworms and microbial activities. The major constituents like cellulose, hemicellulose, lignin, starch and different protein compounds present in waste are degraded by specific enzymes. Earthworms are divided into three categories based on their ecological existence with respect to soil levels, viz. endogeics, anecics and epigeics [9]. Endogeics reside deep inside the soil, while anecics are found just below the surface, whereas epigeics are the ones on the surface. Among these three classes, epigeics are the most voracious eaters and they consume soil at a very high rate although they utilize the least organic part of the soil they consume and excrete about 90-95% of the soil as vermicast. Vermicasts undergo rapid aerobic decomposition promoted by a number of growth promoting substances. The intestines of the worms have a large number of symbionts which are excreted along with the vermicast [10]. Quantification of enzyme activity during composting offers composting dynamics i.e., the decomposition pattern of organic matter and nitrogen transformation [11].

Vermicomposting relates to the bio-oxidation and stabilization of organic matter by the joint action of earthworms and microorganisms. In the process of vermicomposting, microorganisms, biochemically degrade the organic matter, whereas earthworms play a crucial role by aerating, conditioning, and fragmenting the substrate that facilitates proper environment for microbial growth and activity. Besides, it also moves fragments and bacteria-rich excrement through the waste profile, effectively homogenizing the organic material. The main role of earthworms is to act as mechanical blenders, and by comminuting the organic matter, they modify its physical and chemical status by gradually reducing the C:N ratio, thus increasing the surface area of the substrate exposed to microorganisms, making it much more favorable for microbial activity and further decomposition [12]. In vermicomposting, initially earthworms process the waste, modify its physical state and it eventually goes through microbial degradation [13]. After maturation of waste to vermicompost, the earthworms move toward fresher layers of undigested waste. Vermicomposting is a mesophilic, completely aerobic decomposition process and is the result of microbial activities. Earthworms and other soil animals do not have cellulolytic systems to digest plant material. Therefore, much of the animal nutrition depends on the action of microbes, either free-living or associated with their guts. But as mentioned earlier, the role of earthworms is also mostly important in conditioning the substrate and promoting microbial activity. In general, it can be concluded that the vermicomposting process is microbial mediated, with the earthworms also having a strong influence. But the whole soil fauna community also plays an important role in these processes through its interactions with soil microbes, and hence it must be considered in a holistic perspective.

11.3 VERMICOMPOSTING—ECO-FRIENDLY ADVANTAGEOUS APPROACH IMPROVING SOIL STATUS

Vermicompost, a potent biofertilizer, brings out several benefits to soil. It solubilizes essential minerals, gets hold of nutrients, offers micronutrients in a more utilizable form for plants, and takes part in biological nitrogen fixation. As mentioned earlier, earthworms can break down organic matter rapidly, resulting in stable, nontoxic vermicomposts with better structure, microbial content, and available nutrient content than general composts. The final compost is a finely divided, peat-like material with a low C:N ratio, excellent structure, aeration, porosity drainage and moisture-holding capacity. Besides, it also provides a suitable mineral balance, improves plant nutrient availability, and could act as complex-nutrient source granules.

Besides, vermicomposting reduces waste bulk density. Quite interestingly, it also greatly reduces the population of pathogenic microorganisms. It is generally accepted that the thermophilic stage of composting eliminates human pathogens, but it has been shown that human pathogens are also eliminated during vermicomposting, probably by means of an antagonism mechanism.

Vermicomposting leads to N mineralization, but the presence of earthworms in vermicomposting increases and accelerates the N mineralization rate. Moreover, the humification rates that take place during the maturation stage are higher and faster during vermicomposting, resulting in a greater decrease in bioavailable heavy metals. There is circumstantial evidence that the final product may contain hormone-like compounds or plant growth regulators that accelerate plant growth and crop yield.

Thus, these have a potentially high economic value as soil conditioners or media for plant growth. Quite interesting is that the best final products and the shortest residence times can be obtained by high-technology systems, whereas low technology ones can be easily adapted and managed on small farms or livestock operations.

Vermicomposting is not yet adopted fully in the larger industrial scale [12]. The temperature in the vermicomposting always stays in the mesophilic range. Thus, pathogen removal is not completely ensured, although, some study has provided good evidences of suppression of pathogens [14] during vermicomposting. In some cases, before vermicomposting, pre-treatment of organic residues is necessary as they may contain substances that are toxic to earthworms, i.e. acidic compounds, NH_3 and salts [15].

11.4 MODIFIED VERMICOMPOSTING—A NEW DIRECTION OF TECHNOLOGICAL UPGRADATION

For sustainable farming, vermicompost is not sufficient to fulfill the deficiency of soil fertility. So enrichment of waste recycling product is of major interest. Research fortifying natural waste with different nutrient rich waste substances is opening new direction of technological upgradation improving the quality and nutrient status of vermicompost. In general, modification of vermicompost is done either by microbial enrichment or fortifying with nutrient rich base material like mineral rock and agricultural waste etc.

11.5 MODIFICATION OF VERMICOMPOST BY MICROBIAL ENRICHMENT

Microbial enrichment of vermicompost serves multidimensional benefits. Treating the waste with certain efficient microflora at the pre-decomposition state helps in reducing the time period of vermicomposting processes. The addition of N_2-fixing bacteria, phosphorus solubilizing bacteria and potassium releasing bacteria in the vermicomposting process might be increasing the quality and also produces enriched vermicompost by increasing nitrogen, phosphorus, and potassium content [16] (Fig. 11.1). Microorganisms of this group are generally known as plant growth promoting microorganisms (PGPMs) which include *Azospirillum, Azotobacter, Phosphobacteria, Rhizobia,* and cyanobacteria.

The PGPMs are capable of putting forth advantageous properties on growth and yield characteristics of several cultivable crops in different parts of the world. Rhizosphere bacteria promote plant growth by improving the availability of nutrients by suppressing the growth of plant pathogens or by production of hormones such as auxin, gibberellins and cytokinin, and also fulvic acid and humic acids functioning as plant growth regulators. Plant growth promoting organisms like *Azospirillum*, VAM, *Phosphobacteria, Rhizobium, Azotobacter,* etc. which live in close association with the roots

Fig. 11.1 Experimentally modified vermicomposting process showing 1. Natural Mineral, 1.(a) Rock Mineral: 1.(a) Mineral Powder 2. Mixing: 2.(a) Mixing of rock mineral and microbial culture, 2.(b) Mixing of straw and 2.(c) Mixing of earth worm 3. Maturation of vermicompost 4. Final vermicompost.

of many cultivated plants, are the main types of organisms widely recommended for many field crops. Significant improvement in growth yield and quality of vegetables with different biofertilizer applications has been reported in various crops. Phosphorus is the second major nutrient required by plants. Phosphorus exists in nature in a variety of organic and inorganic forms which vary from insoluble form to very poorly soluble form with varying solubility [17]. Phosphate solubilising bacteria have the ability to solubilise insoluble mineral phosphate by producing various organic acids, mineral acids, proteins, humic substances etc. Phosphorus solubilising microorganisms solubilise different forms of insoluble phosphate by producing either chelating organic acids such as citric acid, succinic acid, fumaric acid and malic acid or even mineral acids such as sulphuric acid and nitric acid. *Bacillus megatherium, Pseudomonas liquifacians, Bacillus megaterium var. phosphaticum, Bacillus polymyxa, Pseudomonas striata, Agrobacterium, Aspergillus fumigates, Aspergillus niger, Penicillium digitatum,* etc. [18] are efficient Phosphorus solubilising micro organisms having the capacity to solubilise phosphorus content supplied through fertilizer application in soil. These organisms, in the form of culture broth or water suspension of biofertilizer product were added in the vermicomposting process which improves the quality of vermicompost and reduces the composting time also.

11.6 MODIFICATION OF VERMICOMPOST BY FORTIFYING WITH NUTRIENT RICH BASE MATERIAL

For the development of sustainable farming and livelihood of farmers, waste recycling is of main interest from the view point of ecological, environmental or economic aspects. To improve the nutrient status of vermicompost, several nutrient rich materials were tried to incorporate into the time course of vermicomposting [19-20]. Oil cakes (e.g. mustard cake, groundnut cake, karanj cake etc), poultry manure, green manure crops or waste from leguminous crops can be named as popular N-rich natural products. Rock phosphate, bone meal, basic slag, horn and hoof meal, wastes from fish and animal processing are used as P-rich natural products. K-rich natural products are wood ash (casuarina wood, babul wood etc.). Limestone, gypsum and dolomites are natural sources of Mg and Ca respectively. To improve the productivity, balanced plant nutrition has an eminent role, for which use of organic resources can be a good option. The addition of rock materials and microbes into the composting processes improves the quality of vermicompost in terms of macro and micronutrient contents, as well as helps to enhance the mobilization of unavailable form to available form of nutrients. Thus organic wastes may be recycled back to field by converting them into value-added product through composting technology.

A Special Focus on the Enrichment using Rock Mineral

The search for low-cost and easy-to manage techniques that increase mineral solubilisation and improve the fertilization of agro-ecosystems, such as vermicomposting, could transform the use of inorganic residues such as rock powder into environmentally and economically sound alternatives for farmers. Therefore, fertilization with rock powder could become an environmentally and economically viable option to fertilize tropical soils that are generally highly leached and poor in plant available nutrients [21-22]. Different vermicompost enrichment strategies from natural sources are summarized in Table 11.1.

Table 11.1 Different vermicompost enrichment strategies from natural sources

Enrichment	*Natural product used*
Nitrogen enrichment	Common oilcake like Mustard cake, Groundnut cake, Neem cake, Karanj cake, Sesame cake, Castor cake etc, OR Poultry manure or mushroom compost, Aquatic fern like *Azolla*, Green manure crops or waste from leguminous crops.
Phosphorus enrichment	Rock phosphate, Bone meal, Basic slag, Horn and hoof meal, Wastes from fish and animal processing.
Potassium enrichment	Sylvite, Wood ash [Eucalyptus wood, Casuarina wood , Tobacco stem, Babool wood , etc].
Calcium enrichment	Limestone, gypsum
Magnesium enrichment	Dolomite

Rock minerals and waste mixture can be combined at the time of vermicomposting process to give sound enrichment. Vermicomposting may accelerate the solubilization of the inorganic residues because the minerals undergo chemical and physical weathering due to enzymatic action and grinding of the materials as they pass through the earthworm's intestine.

Involvement of earthworms (*Eisenia fetida*) for the degradation of organic wastes and production of vermicompost is near commercialization because loss of nitrogen from agricultural wastes and dung is minimized when the vermicompost is the source of organic matter. There is the possibility of increasing the nitrogen content of compost by inoculation with nitrogen-fixing organisms and the phosphorus content by the addition of rock phosphate and then inoculation with phosphate solubilizing bacteria, since direct application of rock phosphate is not useful, particularly in neutral and alkaline soils [16]. Vermicomposting organic refuse with rock phosphate and microbial activities may help to solubilize phosphorus and increase phosphorus availability to plants. Nitrogen-fixing bacteria, besides fixing N, solubilize P due to production of organic acids and enzymes [23]. In case of potassium enrichment, direct application of clay mineral like illite, mica (rock K) may be agronomically more useful and environmentally more feasible [24]. Rock potassium is a cheaper source of potassium that is not readily available to a plant because the minerals are released slowly and their use as fertilizer often causes insignificant yield increase in current crop [25]. Potassium solubilizing bacteria are able to solubilize rock K mineral powder, such as micas, illite and orthoclases, through production and excretion of organic acids [26, 27].

CONCLUSION

An increase in the use of chemical fertilizers for agricultural processes has created both ecological and economical hazards. Vermicomposts thus can provide a sustainable alternative for their chemical counterparts. Earthworms can act as viable sources of fertilizers by the use of their ability to degrade various organic materials. All earthworms do not have the ability to degrade any organic material, thus purpose-specific selection of earthworms is utmost necessary. To improve the productivity, balanced plant nutrition has an eminent role, for which the use of organic resources are thus a good option. The addition of rock materials and microbes into the composting processes improves the quality of vermicompost in terms of macro- and micronutrient contents, as well as helps to enhance the mobilization of unavailable form to available form of nutrients. Microflora of the gut of earthworms also have beneficial effects, thus they can also provide microbial enrichment when they excrete the ingested soil. Thus organic wastes may be recycled back to field by getting converted into value-added product through vermicomposting technology. Besides humus-like substances are known to act as plant growth regulators with the ability to control root initiation, seed germination, development and plant height, nutrient uptake, photosynthesis and energy metabolism [28]. Thus modified vermicompost can serve as a better future alternative for improving soil health.

REFERENCES

1. Darwin C. (1837). On the Formation of Vegetable Mould, Vol. 2, series 5, Trans Geol Soc, London, pp. 505–509.
2. Senapati B.A. (1993). "Earthworm gut contents and its significance", In: Ghosh A.K., (ed.) *Earthworm Resources and Vermiculture*. Zoological Society of India, Kolkata, pp. 97–100.

3. Alok B. (2010). "Management of kitchen waste material through vermicomposting", *Asian Journal of Experimental Biol. Sci.*, 1 (1): 175–177.
4. Jadia C.D. and Fulekar M.H. (2008). "Vermicomposting of vegetable wastes: A bio physicochemical process based on hydro-operating bioreactor", Afr. J. Biotechnol. 7 (20), 3723–3730.
5. Coker, E.G., Hall, J.E., Carlton-Smith, C.H. and Davis, R.D. (1987). "Field investigations into the manorial value of lagoon-matured digested sewage sludge", The Journal of Agricultural Science 109, 467–478.
6. Hoitinik, H.A. and Kuter, G.A. (1986). "Effects of compost in growth media on soil-borne pathogens", In. Chen, Y., Avnimelech, Y., (eds.), *The Role of Organic Matter in Modern Agriculture.* Martinus Nijhoff, Dordrecht, 289-306.
7. Eghbal, B. and J.F., Power (1999). "Phosphorus and nitrogen-based manure and compost applications: corn production and soil phosphorus", Soil Sci. Soc. Am. J., 63: 895–901.
8. Garcia, C., Herná ndez, T., Costa, F. and Ceccanti, B. (1995). "Phosphatase and β-glucosidase activities in humic substances from animal wastes", *Bioresource Technology* 53, 79–87.
9. Bouche M.B. (1977). "Strategies lombriciennes. In: Lohm U., Persson T., (eds.) *Soil Organisms as Components of Ecosystems*", Vol. 25. Biol Bull, Stockholm, pp. 122–132.
10. Altavinyte O. and Vanagas J. (1982). "Effect of earthworms on the quality of barley and rye grain", *Pedobiologia*, 23, 255–362.
11. Dash M.C. and Senapati B.K. (1986). "Vermitechnology: an option for organic waste management in India", In: Dash M.C., Senapati B.K., Mishra P.C., (eds.) *Verms and Vermicomposting.* Sambalpur University, Orissa, School of Life Sciences, pp. 157–172.
12. Dominguez J. and Edwards C.A. (1997). " Effects of stocking rate and moisture content on the growth and maturation of *Eisenia Andrei* (Oligochaeta, Lumbricidae)", *Pedobiologia*, 42, 566–576.
13. Lores, M., Gómez, M., Pérez, D. and Domínguez, J. (2006). "Using FAME profiles for the characterization of animal wastes and vermicomposts", *Soil Biology and Biochemistry*, 38, 2993–2996.
14. Monroy, F., Aira M. and Domínguez J. (2009). "Reduction of total coliform numbers during vermicomposting is caused by short-term direct effects of earthworms on microorganisms and depends on the dose of application of pig slurry", *Sci. Tot Environ*, 407, 5411–5416.
15. Nair, J., Sekiozoic V. and Anda M. (2006). "Effect of pre-composting on vermicomposting of kitchen waste", *Bioresource Technology*, 97, 2091–2095.
16. Vivek, K. and Singh, K.P. (2001). "Enriching vermicompost by nitrogen fixing and phosphate solubilizing bacteria", *Bioresource Technology*, 76, 173–175.
17. Premono, Edi. M., Moawad, M.A. and Vlek, P.L.G. (1996). "Effect of phosphate-solubilizing *Pseudomonas putida* on the growth of maize and its survival in the rhizosphere", *Indonesian Journal of Crop Science*, 11, 13–23.

18. Rajasekar, K., Daniel T. and Karmegam, N. (2012). International Scholarly Research Network (ISRN) *Soil Science* Volume 2012, Article ID 946079.
19. Singh., A. and S., Sharma, (2002). "Composting of a crop residue through treatment with microorganisms and subsequent vermicomposting", *Bioresource Technology,* 85(2), 107–111.
20. Kumar, R., Verma, D., Singh, B.L., Kumar, U. and Shweta (2010). "Composting of sugar cane waste by-products through treatment with microorganisms and subsequent vermicomposting", *Bioresource Technology,* 101(7), 6707–6711.
21. Kumar R. and Shweta (2011). "Enhancement of wood waste decomposition by microbial inoculation prior to vermicomposting", Bioresource Technology, vol. 102(2), pp. 1475–1480.
22. Leonardos, O.H., Theodoro, S.C.H. and Assad, M.L. (2000). Remineralization for sustainable agriculture: A tropical perspective from a Brasiliam viewpoint *Nutr. Cycl. Agroecosyst.,* 56, 3–9.
23. Kumar, V. and Narula, N. (1999). "Solubilization of inorganic phosphates and growth emergence of wheat as affected by *Azotobacter chroococcum", Biology and Fertility of Soils* 28, 301–305.
24. Rajan S.S.S., Watkinson J.H. and Sinclair A.G. (1996). "Phosphate rock for direct application to soils", *Adv. Agron.,* 57, 77–159.
25. Zapata F. and Roy, R.N. (2004). Use of Phosphate rock for sustainable agriculture. FAO and IAEA, Rome, Italy.
26. Friedrich S., Platonova N.P., Karavaiko G.I., Stichel E. and Glombitza F. (1991). "Chemical and microbiological solubilization of silicates", *Acta Biotechnol.,* 11, 187–196.
27. Ullman W.J., Kirchman D.L. and Welch S.A. (1996). "Laboratory evidence for microbially mediated silicate mineral dissolution in nature", *Chem. Geol.,* 132, 11–17.
28. Zandonadi D.B and Busato J.G. (2012). "Vermicompost humic substances: technology for converting pollution into plant growth regulators", *International Journal of Environmental Science and Engineering Research (IJESER),* 3(2), 73–84.

12

Biofertilizer for Bioremediation

Subash Chandra Santra, Anusaya Mallick and Alok Chandra Samal
Department of Environmental Science, University of Kalyani, West Bengal, India
E-mail: *scsantra@yahoo.com*

ABSTRACT

Industrial revolution is the mother of environmental pollution. This affects the quality of life and ecosystem. To reduce the environmental pollution, various attempts are being made. Among these attempts, bioremediation is one of the good techniques. In this technique microorganisms are used to reduce the toxicity of harmful substances from contaminated environments. Bioremediation solves the problem in the field of solid and liquid wastes by the process of detoxification. It is an attractive clean-up technology due to its cost effectiveness and environmental-friendly nature. Biofertilizer is a large population of a specific or a group of beneficial microorganisms incorporated aseptically into sterile carrier materials. Microorganisms perform several important processes such as cycling of nutrients and degradation of various compounds. In sustainable agriculture, algae, bacteria and fungi are being utilized as biofertilizers. Plant growth promoting bacteria (PGPB) play important role in soil fertility and can be used as biofertilizers that stop deterioration of soil caused by excessive application of chemical fertilizers. Different microbes are utilized in degradation of different environmental contaminants like hydrocarbons, heavy metals, and toxic substances. Recently, research is being carried out to study the effect of bioremediation by using biofertilizer in reducing the environmental hazards.

Keywords: Microorganisms, Biofertilizer, Bioremediation, Biodegradation.

12.1 INTRODUCTION

Biofertilizers are the best modern tools for sustainable agriculture. Biofertilizer is an important component of integrated nutrient management in agricultural system. It is a cost effective and renewable source of plant nutrients to supplement the chemical fertilizers. The commercial history of biofertilizers began with the launch of 'Nitragin' by Nobbe and Hiltner, a laboratory culture of *Rhizobia* in 1895, followed by the discovery of *Azotobacter* and then the blue green algae and a host of other microorganisms. In India the first study on legume *Rhizobium* symbiosis was conducted by N.V. Joshi, and the first commercial production started in middle of 1950s.

Biofertilizers are the most advanced biotechnology tools that support developing organic agriculture, sustainable agriculture, green agriculture, and non-polluting agriculture. This bioorganic fertilizer can increase the output, and improve the quality of agriculture. In recent years the biofertilizer has been widely used with excellent results in agriculture all over the world. It is well known that continuous and excessive use of chemical fertilizers and pesticides has a detrimental effect on our environment and health. The alternate sources of environment-friendly plant nutrients are the immediate need for the recent times. Biofertilizers have become a hope for the future world as far as economical and environmental viewpoints are concerned. Especially in developing countries like India it can solve the problem of high cost of chemical fertilizers and help in saving the economy of the country [1, 2]. The disposal of biodigested slurry after biogas production is a major concern for the environment [3-8]. It contains considerable amount of plant nutrients and helps to improve crop production, apart from preventing adverse environmental impacts of waste disposal [9, 10]. The application of biodigested slurry in agricultural field increases the soil fertility by fixing atmospheric nitrogen, solubilizing soil phosphates and producing plant growth substances in the soil. The role and importance of biofertilizers in sustainable crop production has been reviewed by several authors [11-13]. The degradation of organophosphorus insecticides, chlorpyrifos – methyl, cyanophos and malathion in mineral salts media were studied by several researchers [14-18]. The effect of additional biofertilizers, individually or combined with organic amendments, on chlorpyrifos and cyanophos degrading activity in soil was investigated. *Paenibacillus polymyxa* (*Bacillus polymyxa*) and *Azospirillum lipoferum* (Beijerinck) were found to degrade the organophosphorus insecticides as a carbon and phosphorus source. *Bacillus polymyxa* appeared to be more effective than *Azospirillum lipoferum* in degrading all the tested organophosphate pesticides in mineral salts media.

There is a continuous interest in algal-based waste stabilization pond systems that are inexpensive and are known for their ability to achieve good removal of pathogens and organic pollutants all over the world [19]. Cyanobacteria and microalgae play an important role by supplying molecular oxygen to heterotrophic partners and thus support the initial steps of degradation [20]. Nutrient removal with the aid of algae compares favourably with other conventional technologies [21-23]. It was also found that some cyanobacteria and algae might remove xenobiotics from the environment by sorption, transformation and degradation [24]. Several attempts have been made to explore the efficiency of microalgae for metal removal [25-27]. Increasing the light intensity can lead to a higher microalgal activity and an increased removal of nutrients from wastewater [28].

Phycoremediation is the process of employing algae, for improving water quality which can fix carbon dioxide by photosynthesis and remove excess nutrients effectively at minimal cost [29-31]. The use of algae in purification of wastewater and to eliminate the nutrients was studied widely [32, 33]. Additionally, these photosynthetic microorganisms are also useful in bioremediation applications and as nitrogen-fixing biofertilizers [34-36]. *Anabaena* is capable of evolving hydrogen through the process of indirect biophotolysis of water, using nitrogenase as a catalyst. Hydrogen is a promising alternative to fossil fuels, being renewable, eco-friendly and efficient [37]. Photobiological hydrogen production by cyanobacteria is an ideal process, owing to their simple nutritional requirements from air, water, mineral salts, and light as the main energy source [38, 39]. There are several reports concluding that the metal-binding capacities of several biomasses including marine algae, fungi, and yeast are very high [40-44].

12.2 MICROBES IN BIOFERTILIZER PRODUCTION

Biofertilizers are products containing living cells of different types of microorganisms which, when applied to seed, plant surface or soil, colonize the rhizosphere or the interior of the plant and promote growth by converting nutritionally important elements (nitrogen, phosphorus) from unavailable to available form through biological process such as nitrogen fixation and solubilization of rock minerals [45]. Beneficial microorganisms in biofertilizers accelerate and improve plant growth and protect plants from pests and diseases [46]. Soil microorganisms have an important role in sustainable development of agriculture [47-48]. Organisms that are commonly used as biofertilizers component are nitrogen (N) fixers, potassium (K) solubilizers and phosphorus (P) solubilizers, or with the combination of molds or fungi. Most of the bacteria included in biofertilizer have close relationship with plant roots. *Rhizobium* has symbiotic interaction with legume roots, and Rhizobacteria inhabit root surface or rhizosphere soil.

12.2.1 Nitrogen-Fixing Microorganisms

Nitrogen is one of the major important nutrients essential for plant growth. Nitrogen fixation in soil by different microorganisms is a complex biochemical process by which atmospheric elemental nitrogen (unavailable to plants) is transformed into organic nitrogenous compound, which is available to plants. Different species of bacteria, some actinomycetes and blue-green algae are agriculturally important nitrogen fixers. The value of nitrogen-fixing legumes is improving and higher yield of legumes and other crops can be achieved by the application of biofertilizers [49]. Nitrogen-fixing bacteria (NFB) fix atmospheric N_2 usable for plants [50-51]. Nitrogen fixers are used in biofertilizer as a living fertilizer composed of microbial inoculants or groups of microorganisms which are able to fix atmospheric nitrogen. They are grouped into free-living bacteria (*Azotobacter* and *Azospirillum*) and the blue-green algae and symbionts such as *Rhizobium*, *Frankia* and *Azolla* [52]. The N_2-fixing bacteria associated with non-legumes include species of *Achromobacter, Alcaligenes, Arthrobacter, Acetobacter, Azomonas, Beijerinckia, Bacillus, Clostridium, Enterobacter, Erwinia, Derxia, Desulfovibrio, Corynebacterium, Campylobacter, Herbaspirillum, Klebsiella, Lignobacter, Mycobacterium, Rhodospirillum, Rhodopseudomonas, Xanthobacter,* and *Methylosinus* [53].

12.2.1.1 Symbiotic nitrogen-fixing microorganisms for leguminous plants

Symbiotic bacteria infecting plant roots, and numerous taxa of less intimately associated N_2-fixing bacteria can be considered for improvement of crop yield. Examples of such bacteria include *Acetobacter diazotrophicus* and *Herbaspirillum* spp. associated with sugar cane, sorghum, and maize [54-56]; *Azoarcus* spp. associated with Kallar grass *(Leptochloa fusca)* [57]; and *Alcaligenes, Azospirillum, Bacillus, Enterobacter, Herbaspirillum, Klebsiella, Pseudomonas,* and *Rhizobium* associated with rice and maize [58]. The genus *Azospirillum* colonizes a great variety of annual and perennial plants, many of which have never been reported to be colonized by N_2-fixing bacteria. Accordingly, *Azospirillum* possesses a great potential as a general root colonizer, whose use is not limited by host specificity [59]. Several studies indicate that *Azospirillum* can increase the growth of various crops, including sunflower, carrot, oak, sugar beet, tomato, eggplant, pepper, and cotton in addition to wheat and rice [59, 60].

Some species of *Rhizobium* bacteria form nodules by colonizing the roots of different leguminous plants. They derive carbohydrates (energy) and water from the root tissues of the host plants, which enables them to directly fix atmospheric nitrogen for the production of amino acids and protein to make cell materials. In turn, the host plants receives the mentioned nitrogenous compounds synthesized by the bacteria for their own growth and development.

12.2.1.2 Free-living or non-symbiotic nitrogen-fixing microorganisms

Many free-living bacteria also fix atmospheric N_2 like *Azotobacter, Beijerinckia,* and *Clostridium.* During favourable conditions nodulating bacterial symbionts (e.g., *Frankia*) of plant roots can also fix N_2 without symbiotic association (free-living) with their plant host [61]. More interestingly, it has been found that *Frankia* can occur and possibly fixes atmospheric N_2 in the rhizosphere of non host plants. *Frankia* has been recorded in the rhizosphere of *Betula pendula* and in soil where actinorhizal plants were not present [61, 62]. Certain free-living microorganisms exist in soil and water and are able to fix atmospheric nitrogen without symbiosis; these organisms are mainly bacteria (i.e., *Azotobacter, Clostridium, Verxia, Azospirillum,* etc.). Among the free-living nitrogen-fixing bacteria, *Azotobacter* is most important. They colonize rhizosphere and derived their energy by oxidation of soil organic matter and fix atmospheric nitrogen into the soil and subsequently increase the soil fertility.

12.2.1.3 Blue-green algae and azolla

N_2-fixing blue-green algae (BGA) or cyanobacteria and *Azolla* (free floating water-borne fern) have been shown to be the most important in maintaining and improving the productivity of rice fields [63]. It has been demonstrated that the N fertility of soil is sustained better under flooded conditions than under dryland conditions [64]. Favourable conditions for biological N_2 fixation by such BGA is considered to be one of the reasons for the relatively stable yield of rice under flooded condition. Unlike chemical N fertilizers, BGA and *Azolla* neither contaminate the environment nor consume the photosynthates of rice plants [65]. The importance of N_2- fixing BGA was first recognized by De [66, 67], who attributed the self-maintenance of the N status of tropical rice-field soils to the growth of N_2-fixing BGA. Similarly, the fertilizing value of *Azolla* in rice fields is well-known and has been utilized over year old in China and Vietnam [68, 69]. The plant-available N of rice soils is increased considerably by the growth of N_2-fixing BGA and *Azolla* [70-80]. BGA liberate extracellular organic compounds and photosynthetic O_2 during their growth, while *Azolla* prevent a rise in the pH, reduce water temperature, curb NH_3 volatilization and suppress weeds; and both of them contribute towards biomass.

The symbiotic association of *Anabaena* and *Azolla* produces 40-60 tonnes organic matter and at the same time they are capable of fixing 100 to 150 kg of atmospheric nitrogen per hectare per year. In India, their use as biofertilizer is yet to be popularized and their location-specific and thermo-sensitive cultures are not easily available in the market. Another difficulty in mass production of these biofertilizer is their high cost.

12.2.2 Phosphate Solubilizing Microorganisms

Phosphorus (P) is one of the least mobile elements in soil. In most soils (especially in acid soil), soluble phosphatic fertilizers locked in the soil react with calcium, ferrous and aluminium and are converted into insoluble phosphate. Phosphorus is the second most limiting plant nutrient after nitrogen [81]. Total P content in soil is usually high, but most of this soil P pool is not in forms available for plant uptake. Bacteria that can mobilize P from unavailable soil pools and increase P availability to plants are of great importance. Most predominant phosphorus-solubilizing bacteria (PSB) belong to the genera *Bacillus* and *Pseudomonas* [82]. The fixed phosphorus in the soil can be solubilized by phosphate solubilizing bacteria, which have the capacity to convert inorganic unavailable phosphorus to available for the plants. Bacteria are more effective in phosphorus solubilization than fungi [83]. Among the whole microbial population in soil, phosphate solubilizing bacteria (PSB) constitute 1 to 50%, while phosphorus solubilizing fungi (PSF) are only 0.1 to 0.5% in P solubilization potential [84].

Microorganisms involved in phosphorus acquisition include mycorrhizal fungi and phosphate solubilizing microorganisms (PSMs) [85]. Among the soil bacterial communities, ectorhizospheric strains from *Pseudomonas* and *Bacilli*, and endosymbiotic rhizobia have been described as effective phosphate solubilizers [86]. Strains from bacterial genera *Pseudomonas, Bacillus, Rhizobium* and *Enterobacter* along with *Penicillium* and *Aspergillus* fungi are the most powerful P solubilizers [87]. *Bacillus megaterium, B. circulans, B. subtilis, B. polymyxa, B. sircalmous, Pseudomonas striata,* and *Enterobacter* could be referred as the most important strains for phosphate solubilizer [88, 89]. A nematofungus *Arthrobotrys oligospora* also has the ability to solubilize the phosphate rocks [90].

12.2.3 Potassium Accumulating Organisms

Potassium (K) is one of the important nutrients for plants. There are a number of microorganisms which can also accumulate potassium and thus help in enhancing soil fertility. Some microorganisms in the soil are able to solubilize unavailable forms of K-bearing minerals, such as micas, iolite and orthoclase, by excreting organic acids which either directly dissolve rock K or chelate silicon ions to bring the K into solution [91,92]. These microorganisms are commonly known as potassium solubilizing bacteria (KSB) or potassium dissolving bacteria or silicate dissolving bacteria. Some research has been done regarding the use of potassium dissolving bacteria, known as "biological potassium biofertilizer (BPF)", particularly in China and South Korea to investigate the bio-activation of soil K-reserves so as to alleviate the shortage of K-fertilizer. It was shown that KSB increased K availability in soils and increased mineral uptake by plant [93, 94]. Therefore, application of KSB holds a promising approach for increasing K availability in soils.

12.2.4 Sulphur Solubilizing Organisms

Sulphur is generally regarded as trace element in majority of crop plants. But this is one of the major elements in oilseed crops, some important vegetables (onion, oat, cauliflower, etc.), and in some spices (ginger, garlic, etc.). Sulphur is essential for biochemical synthesis of some important glycosides, pungent compounds and disease resistant properties. Deficiency of sulphur in agricultural soil could be corrected using *Azotobacter pasturianam* as boifertilizer.

Sulphur is essential for all plants and indispensable for the growth and metabolism. Khandkar et al., observed that the nodule in blackgram increased due to sulphur application [95]. Sulphur is involved in the formation of nitrogenase enzyme and is known to promote nitrogen fixation in legumes [96]. Sulphur is also involved in the formation of chlorophyll [97]. Hanesklaus and Schnug reported that sulphur is associated with the production of crops of superior nutritional and market quality [98]. The sulphur application increased the total chlorophyll content of green gram [99]. Sulphur plays a dominant role in improving the quality of pulses [100].

12.2.5 Mycorrhiza

Mycorrhizal fungi form mutualistic symbiosis with a vast majority of land plants [101]. Possibly more than 80% of all land plants form mycorrhizal symbiosis. The extent of mycorrhizal symbiosis emphasizes the ancient evolutionary history and potential importance of fungal symbiosis for plant production and physiology. The association between plants and their root-colonizing mycorrhizal fungi is a functional symbiosis in which the mycorrhizal fungus is obligately or facultatively dependent on host photosynthates and energy. Mycorrhizal fungi may also improve soil quality by having a direct influence on soil aggregation and therefore aeration and water dynamics [102]. An interesting potential application of mycorrhizal fungi is their ability to allow access of plants to nutrient sources generally unavailable to the host plants. For example, crop plants may be able to use insoluble sources of phosphorus when inoculated with mycorrhizal fungi, but not in the absence of inoculation [101].

12.2.6 Cellulose Decomposing Organisms

Cellulases are especially common in soil and plant inhabiting fungi. Many fungi in the Ascomycotina and Basidiomycotina are able to digest cellulose. The necessary enzymes are less common in the members of Zygomycotina. Presence of cellulase enzymes in moist soil may indicate release of glucose dimers available for absorption by all microbes in the near vicinity, indicating a continuing and community-based process of cellulose degradation by microbes. Cellulose is the most abundant constituent of plant residues. It has been reported that cellulose decay would be enhanced by nitrogen addition due to the increased cellobiases [103,104]. *Serpula lacrimans*, a cellulose degrading fungus, produced more mycelium and resulted in faster decay of cellulose under nitrogen fertilization [105].

Excess doses of manure or sludge fertilizers can harm mesofauna because of toxicity (e.g., anhydrous ammonia) and high osmotic pressure created due to salt [106] or heavy metal accumulation [107]. The repellent nature of ammonium can affect soil invertebrates adversely [108]. The potential of toxic effects can be decreased by applying composted manure or sludge [109]. The time period for material composted prior to incorporating it into soil must be considered. For example, the plant-pathogenic fungus *Rhizoctonia solani* may cause damping-off disease in soil when fresh or immature compost material of high content of cellulose is added. However, in aged compost, cellulose is degraded and the biocontrol fungus *Trichoderma* spp. can grow and parasitise the pathogen effectively. Many soil-borne fungal species like *Aspergillus, Penicillium, Trichoderma, Chaetomium,* etc., act as activators in the decomposition process of plant bodies containing cellulose or lignin.

12.3 CONSTRAINTS AND SUCCESS OF BIOFERTILIZER TECHNOLOGY

Though the biofertilizer technology is a low cost, eco-friendly technology, several constraints limit the application or implementation of the technology. The constraints may be environmental, technological, infrastructural, financial, human resources, unawareness, quality, marketing, etc.

12.3.1 Technological Constraints

- Use of improper, less efficient strains for production.
- Lack of qualified technical personnel in production units.
- Unavailability of good quality carrier material or use of different carrier materials by different producers without knowing the quality of the materials.
- Production of poor quality inoculants without understanding the basic microbiological techniques.
- Short shelf life of inoculants.

12.3.2 Infrastructural Constraints

- Non-availability of suitable facilities for production.
- Lack of essential equipment power supply etc.
- Space availability for laboratory, production, storage etc.
- Lack of facility for cold storage of inoculant packets.
- Financial constraints.
- Non-availability of sufficient funds and problems in getting bank loans.
- Less return by sale of products in smaller production units.

12.3.3 Environmental Constraints

- Seasonal demand for biofertilizers.
- Simultaneous cropping operations and short span of sowing/planting in a particular locality.

- Soil characteristics like salinity, acidity, drought, waterlogging, etc.

12.3.4 Human Resources and Quality Constraints

- Lack of technically qualified staff in the production units.
- Lack of suitable training in the production techniques.
- Ignorance on the quality of the product by the manufacturer.
- Non-availability of quality specifications and quick quality control methods.
- No regulation or act on the quality of products.
- Awareness about the technology.
- Unawareness regarding the benefits of the technology.
- Problem in the adoption of the technology by the farmers due to different methods of inoculation.
- No visual difference in the crop growth immediately as that of inorganic fertilizers.

12.3.5 Awareness about the Technology

- Unawareness about the benefits of the technology.
- Problem in the adoption of the technology by the farmers due to different methods of inoculation.
- No visual difference in the crop growth immediately as that of inorganic fertilizers.
- Unawareness about the damages caused to the ecosystem by continuous application of inorganic fertilizer.

12.3.6 Marketing Constraints

- Non-availability of right inoculant at the right place in right time.
- Lack of retain outlets or the market network for the producers.

12.4 BIOREMEDIATION PROCESS: AN OVERVIEW

Bioremediation is defined as a process whereby hazardous wastes are biologically degraded under controlled conditions to an innocuous state, or to levels below concentration limits established by regulatory authorities [110]. This techniques uses living organisms, especially plants and microorganisms, to reduce, eliminate, transform, and detoxify the unwanted products present in soils, sediments, water, and air. Phytoremediation technology, one of its many approaches, uses plants as filters for accumulating, immobilizing, and transforming the contaminants to less harmful forms [111]. More specifically, it is the utilization of vascular plants, algae, and fungi to control, breakdown, remove wastes, or to encourage degradation of contaminants in the rhizosphere [112]. Phytoremediation has recently become a tangible alternative to the traditional methodology in restoring the polluted sites [113].

Bioremediation technologies currently used in many countries are mostly directed to the remediation of oil spillages on land and include *in-situ* biotreatment of contamination, for example, the addition of bacterial fertilizers, mineral, and organic nutrients to the oil-contaminated soil [114-116]. However, these technologies are not acceptable for the treatment of oil wastes, as the high concentration of toxic contaminants and anaerobic conditions in the pit content prevent the development of an active oil-oxidizing microbial consortium.

Rhodococcus biosurfactants have been used for the bioremediation of oil-contaminated agricultural soils after an accidental oil spill [117, 118]. The application of composting systems enhanced by nutrient addition, bulking with straw and inoculation of *Rhodococcus*-biosurfactant complexes provided a 57% decrease in oil contamination during a 3-month treatment. An *ex-situ* biotechnology has been developed, which employs a *Rhodococcus* biosurfactant-based biofertilizer for the decontamination of heavily oil-polluted soil [119].

12.5 BIOFERTILIZER TECHNOLOGY IN BIOREMEDIATION

Success or failure of bioremediation depends on several factors such as the competitive ability of the bioremedial agents and biotic factors such as soil moisture, pH, and temperature [120, 121]. Inoculated biofertilizers (Phosphoren, Microbien, Cerealin and *Azospirillum*) may act as potential agents for soil inoculation to bioremediate pesticides contaminated soil [122]. Successful removal of pesticides by the addition of bacteria (bioaugmentation) has been reported for many compounds including coumaphos (ethoprophos, dicofol and malathion) [123-126].

12.5.1 Use of Microorganisms in Bioremediation

Plant growth-promoting bacteria (PGPB) are mainly used to improve agricultural yields along with solving environmental problems [127]. A number of microorganisms (bacteria and fungi) have been used in soil inoculations intended to improve the supply of nutrients to crop plants, to stimulate plant growth, to control or inhibit the activity of plant pathogens and to improve soil structure (Table 12.1). Other, more recent, objectives for the introduction of microorganisms into soil are the mineralization of organic pollutants [121]. Phosphoren, micrebien, cerealin and azospirillum are the biofertilizers (derived from the living biomass of different microbial species) produced by general organization belonging to the Egyptian Ministry of Agriculture. These biofertilizers, as mixed inoculums are beneficially applied to many fields and horticultural crops [128].

Table 12.1 Plant growth promoting bacteria (PGPB) applied in bioremediation strategies

Characteristics of PGPB	*Microorganism*	*Plants associated*	*References*
Nitrogen fixation			
Freely associated bacteria	*Azotobacter chroococcum*	*Brassica juncea*	Wu et al., (2006) [129]
Symbiotic bacteria	*Sinorhizobium meliloti*	*Medicago truncatula*	Bianco and Defez (2009) [130]

Contd...

Contd...

Enhanced the nodule numbers, seed yield, grain protein, root and shoot	*Bradyrhizobium* sp. (vigna)	Greengram (*Vigna radiate*)	Wain et al., (2007) [131]
Phosphate mobilization			
Inorganic P source	*Pseudomonas aeruginosa*	*Vigna mungo*	Ganesan (2008) [132]
Organic P source	*Bacillus amyloliquefaciens*	*Zea mays*	Idriss et al., (2002) [133]
Phosphate solubilizing bacteria	*Pseudomonas putida, Azospirillum, Azotobacter*	Artichoke (*Cynara scolymus)*	Jahanian et al., (2012) [134]
Siderophore release			
Hydroxamates	*Streptomyces acidiscabies*	*Cicer arietinum*	Dimkpa et al., (2008) [135]
Phenol catecholates	*Rhizobium* sp.	*Sesbania procumbens*	Sridevi et al., (2008) [136]
Carboxylates	*Pseudomonas fluorescens*	*Arachis hypogaea*	Dey et al., (2004) [137]
Salicylic acid	*Arthrobacter oxidans*	*Pinus* sp.	Barriuso et al., (2008) [138]
Stimulated plant growth, reduced Cd uptake	*Pseudomonas aeruginosa*	Indian mustard and pumpkin	Sinha and Mukherjee (2008) [139]
Hormone production			
Indole acetic acid	*Enterobacter chloacae*	*Oryza sativa*	Mehnaz et. al., (2001) [140]
Gibberellin	*Bacillus pumilus*	*Alnus glutinosa*	Gutierrez-Man˜ero et al., (2001) [141]
Influence on metal toxicity			
Increased Ni accumulation	*Bacillus subtilis*	*Brassica juncea*	Zaidi et al., (2006) [142]
Increased Cd accumulation	*Xanthomonas* sp.	*Brassica napus*	Sheng and Xia (2006) [143]
Reduction of Cr(VI) to Cr(III)	*Ochrobactrum intermedium*	*Helianthus annuus*	Faisal and Hasnain (2005) [144]
Stimulated plant growth, facilitated soil Pb mobilization, enhanced Pb accumulation	*Bacillus edaphicus*	Indian mustard (*Brassica juncea*)	Sheng et al., (2008) [145]

12.5.1.1 Bioremediation by microalgae

Algae are universally known to play a very important role in natural water purification process [24, 146]. Thus, the use of microalgae for removal of nutrients from different wastes has been described by a number of authors [21, 147-156].

Industrialization has led to increased emission of pollutants into ecosystems. Metal pollutants can easily enter the food chain if heavy metal-contaminated soils are used for production of food crops. Farm productivity has decreased in toxic metal polluted areas [157]. Accumulation of toxic metals, e.g., Hg, Cu, Cd, Cr and Zn in humans has several consequences such as growth and developmental abnormalities, carcinogenesis, neuromuscular control defects, mental retardation, renal malfunction and a wide range of other illnesses. Elevated levels of such metal ions are generally toxic and cause major damage to cell [158]. Conventional technologies, such as ion exchange or lime precipitation, are often ineffective and/or expensive, particularly for the removal of heavy metal ions at low concentrations (below 50 mg/L). Furthermore, most of these techniques are based on physical displacement or chemical replacement, generating yet another problem in the form of toxic sludge, the disposal of which adds a further burden on the techno-economic feasibility of the treatment process. In view of this, the development of new techniques is necessary to meet the environmental standards at affordable costs. They can provide solutions to the twin challenges of energy security and environmental pollution. They have a great potential for the removal of excess nitrogen and phosphorus from wastewater including the farm runoff. They can capture carbon dioxide in the flue gas from coal fired power plants, thereby reducing greenhouse gas, and also produce algal biomass, which can be converted into biofuel. *Chlorella, Scenedesmus* and *Spirulina* are the most widely used algae for nutrient removal.

The ability of algae to absorb metals has been recognized for many years [159]. In natural environments, algae play a major role in controlling metal concentration in lakes and oceans [160,161]. Algae possess the ability to take up toxic heavy metals from the environment, resulting in higher concentrations than those in the surrounding water [140, 162].

12.5.1.2 Bioremediation by N_2-fixing bacteria

N_2 fixation is an energy-consuming process and has been widely observed in symbiotic, associative and free living bacteria. This bacterial physiological feature plays a tremendous role in ecological sustenance, world agriculture and global nitrogen cycle. Nitrogen availability can limit microbial growth and affect ecosystem activity [163]. Although N_2 fixation is widely distributed among Bacteria and Archaea [164], this significant physiological feature has not been revealed in reductively dechlorinating bacteria.

Nitrogen is essential for the metabolism of microorganisms; besides it is also required for the biosynthesis of amino acids, proteins and nucleic acids (165-167). The soil rich in nitrogen has a good metabolic activity and good microbial biomass. Therefore, its presence in the soil for a good bioremediation process is necessary. However, some authors claim that biodegradation in negligible amounts of nitrogen is possible, but the efficiency is lower [168-170].The literature quotes different bacterial genera, which have the capacity to degrade hydrocarbons in poor concentration of nitrogen [169]. The genera *Pseudomonas* [168, 171], *Agrobacterium, Alcaligenes* [172], *Arthrobacter* [173], *Azotobacter* [174] have the capacity to fix nitrogen in soil with deficiency of nitrogen. The effects of bio-stimulation on petroleum hydrocarbon degradation have been investigated in different conditions [175, 176]. However, the effects of nitrogen on the biodegradation of hydrocarbon classes, i.e. aliphatic, aromatic and polar hydrocarbons, have not been completely studied in the Patagonian soil, which has a poor concentration of nitrogen and is an important oil production area.

Nitrogen and phosphorus are the nutrients that most frequently limit bioremediation [177]. The presence of petroleum products in the soil can widen the C:N ratio, therefore limiting available nitrogen for degradation processes [178]. In addition, microbes able to metabolize hydrocarbons will quickly immobilize the mineral nitrogen that is available, leaving unfavourable conditions for other microorganisms and growing plants [179]. The absence of sufficient nitrogen in the soil will, in turn, slow the degradation process resulting from microbial metabolism [177]. Therefore, adding nutrients in the form of either organic or inorganic fertilizers can stimulate contaminant degradation [180]. Free-living nitrogen-fixing bacteria can fix atmospheric nitrogen into a more usable form such as ammonia. Thus, the bioaugmentation of hydrocarbon polluted soil with free-living nitrogen-fixing bacteria may indirectly contribute to soil nitrogen by releasing nitrogenous biomass [181]. This study was undertaken to isolate hydrocarbon-degrading free-living, nitrogen-fixing bacteria from the soil and to determine their potential in bioremediation of hydrocarbon polluted soil.

12.5.2 Role of *Azolla* in Phytoremediation of Heavy Metals

Among various water pollutants, heavy metals are of major concern because of their persistent and bioaccumulative nature [182-185]. Water is an indispensable part for the sustenance of mankind and the increasing awareness about the environment, especially aquatic ecosystems, has attracted the attention of researchers worldwide. A definite need exists to develop a low cost and eco-friendly technology to remove pollutants, particularly heavy metals, thereby improving water quality. Phytoremediation offers an attractive alternative. Among these, *Azolla,* a free-floating, fast growing, and nitrogen-fixing pteridophyte seems to be an excellent candidate for removal, disposal, and recovery of heavy metals from the polluted aquatic ecosystems [186, 187].

Both living and dead biomass of *Azolla* have been exploited for the removal of heavy metals from industrial effluents and sewage water [188-191]. *Azolla* has great possibility of use in bioremediation of wastewaters and soils. There are three main aspects of use of *Azolla* in bioremediation. These areas are:

(a) For treatment of wastewaters rich in heavy metal pollutants.
(b) Treatment of domestic sewage effluents which are rich in N and P wastes.
(c) Bioremediation of saline soils.

A study in China stated tolerance of four *Azolla* species to Cu, Mn, Fe, Zn, Mo, Co, Cd, etc. under laboratory conditions and found that concentration capacity of *Azolla* for metals affected its growth only slightly without any detrimental effect or not at all [192].

12.6 CASE STUDIES ON BIOREMEDIATION WITH BIOFERTILIZER

12.6.1 Bioremediation of Sewage Wastewater Using Selective Algae for Manure Production

Phycoremediation is the process of employing algae for removing excess nutrient load from wastewater and subsequently diminishing the pollution load. It is an alternative technology of treating sewage wastewater compared to conventional treatment process in economical and sustainable way. The sewage wastewater of the IARI, New Delhi was treated with different microalgae, viz., *Chlorella minutissima, Scendesmus* spp. & BGA (Nostoc) and their consortium. Algae were very effective in reducing of BOD_5, COD, NO_3^-, NH_4, $PO4_3^-$ and TDS in sewage wastewater. Further, it has been observed that *Chlorella* has the best phycoremediation potential as well as manure production among all three microalgae and even better than consortium. Among the potential uses of algal biomass from such systems is its use as a slow release fertilizer. After 20 days microalgae were harvested using muslin cloth and fresh and dry weigh were determined. The maximum biomass was observed in *Scendesmus* spp. and *Chlorella minutissima,* while the percentage of nitrogen and phosphorus was highest in *Chlorella minutissima*. So *Chlorella minutissima* has the best manurial potential. The algae in nutrient-rich sewage wastewater offers a new option of applying algae to manage the nutrient load, and after phycoremediation the biomass itself can be utilized for manure application in agriculture, serving the dual roles of nutrient reduction and valuable manure feedstock production [193].

12.6.2 Bioremediation of Olive Oil Mill Wastewater: Chemical Alterations Induced by *Azotobacter vinelandii*

An environment-friendly bioremediation system of olive oil mill wastewater (OMWW) is studied with respect to its physicochemical characteristics and degradation efficiency on major characteristic constituents. The method exploits the biochemical versatility of the dinitrogen fixing bacterium *Azotobacter vinelandii* (strain A) to grow in OMWW at the expense of its constituents and to transform it into an organic liquid fertilizer. The system eliminates the phytotoxic principles from OMWW and concomitantly enriches it with an agriculturally beneficial microbial consortium along with useful metabolites of the latter. The end product, branded "biofertilizer", is used as a soil conditioner and liquid organic fertilizer. Growth of *A. vinelandii* in OMWW results in the decline of content of most of the compounds associated with phytotoxicity, and this is confirmed by the assessment of degradation yields. In parallel, during the process several other compounds noncommittally undergo degradation and biotransformation. More specifically, the biofertilization system is capable of achieving removal yields as high as 90 and 96% after 3 and 7 days of treatment, respectively [194].

12.6.3 Biofertilizers for Bioremediation of Pesticide Contaminated Soil

Biofertilizers such as *phosphoren, microbien cerealin,* and *azospirillum* were used by the General Organization of Agriculture Fund, Ministry of Agriculture, Egypt to degrade five selected pesticides representing different classes including organophosphate, carbamate and chlorinated organic compounds. There were differences in rates of biotransformation, suggesting the selective induction of certain metabolic enzymes. Inoculation of soil incorporated with malathion, fenamiphos, carbaryl, aldicarb and dieldrin resulted in 80-90% removal of malathion and fenamiphos within 8 days, and carbaryl and aldicarb within 11-15 days respectively. Dieldrin removal occurred slowly within 2 months. These data suggest that biofertilizers may act as potential candidates for soil inoculation to bioremediate pesticide contaminated soil. The production of CO_2 (soil respiration) was stimulated by some pesticides. In samples with microbien, about 2 times higher CO_2 production was measured [195].

12.6.4 Bioremediation of Petroleum Pollutants

Bioremediation tries to raise the rates of degradation, found naturally, to significantly higher rates. The two general approaches that have been tested for the bioremediation of marine oil spills are the application of fertilizer to enhance the abilities of the indigenous hydrocarbon-utilizing bacteria and the addition of naturally occurring adapted microbial hydrocarbon-degraders by seeding. Bioremediation, accomplished by the application of fertilizer to enhance the abilities of the indigenous hydrocarbon-utilizing bacteria, was successfully applied for the treatment of the Alaskan oil spill in Prince William Sound, Alaska [196].

12.6.5 Cyanobacterial Consortia for Bioremediation Purposes

Many studies have reported the ability of cyanobacteria to oxidize oil components and other complex organic compounds such as surfactants and herbicides [22, 197-199]. Among these cyanobacteria were the nonaxenic cultures of *Microcoleus chtonoplastes* and *Phormidium corium,* degrading n-alkane [200], *Oscillatoria* sp. and *Agmenellum quadruplicatum* oxidizing naphthalene to 1-naphthol [201, 202], *Oscillatoria* sp. strain JCM that oxidized biphenyl to 4-hydroxybiphenyl [203] and *Agmenellum quadruplicatum* that metabolized phenanthrene into trans-9,10-dihydroxy-9,10-dihydrophenanthrene and 1-methoxy-phenenthrene [204]. Cyanobacteria associated with aerobic organotrophic bacteria are more effective for the degradation of petroleum and other organic complexes [205-207]. Presence of cyanobacteria alongside with the aerobic organotrophs facilitated the degradation process, and both groups constituted ideal consortia for degradation of petroleum and other complex organic compounds [206].

12.6.6 Potential Use of Cyanobacteria Species in Phycoremediation of Municipal Wastewater

The treatment of municipal wastewater becomes effective by using microalgae which are superior, as a wide range of toxic and other wastes can be reduced with their help.

Locally available blue-green algae such as *Oscillatoria limosa* and *Nostoc commune* have ability to remove various nutrients to prevent further deterioration of water quality. The phycoremediation experiments were conducted at the Department of Environmental Sciences, University of Pune, India, using randomized complete block design with three replications of each treatment. The results of present investigation clearly indicated that both the algal species, viz., *Oscillatoria limosa* and *Nostoc commune* are highly efficient for removal of NO_3^{-2}, PO_4^{-2}, SO_4^{-2}, Cl^- and for reducing EC values. The average reduction was between 84 to 98%. The pollutant removal efficiency was increased with decreasing concentration of wastewater. Amongst the selected algae *Oscillatoria limosa* was the best candidate as compared to *Nostoc commune*. It was concluded that the cyanophycea members would be the best options for phycoremediation [208].

12.6.7 Bioremediation of Wastewater by Using Microalgae

Autotrophs play an important role in remediation of wastewater, particularly domestic waste, through their photosynthetic ability. Dominant and pollution tolerant algae such as *C. vulgaris* and *S. quadricauda* cultures in BBM were isolated from sewage wastewater treatment plant Bopodi from Pune city, and were used for treatment of the wastewater. To study the role of microalgae in wastewater, the following protocols were used, (i) wastewater treated with the culture of *C. vulgaris* and *S. quadricauda*; and (ii) wastewater treated without the culture of *C. vulgaris* and *S. quadricauda* (Control). Samples were periodically (every 5th day) analyzed for physico-chemical parameters such as pH, phosphate, nitrate, BOD and COD using standard methods. *C. vulgaris* showed the best removal capacity of nitrate and COD, while *S. quadricauda* shows BOD and phosphate reduction [209].

12.6.8 Microalgae for Bioremediation of Distillery Effluent

Distillery effluents, also referred to as spent wash/stillage/slop/vinasses, are one of the most environmentally aggressive industrial effluents. With the development of economies and resultant growth of distillery industries, large volume of spent wash is produced, which is likely to cause extensive soil and water pollution due to the presence of high amount of organic matter and dark brown coloured recalcitrant compounds. There have been many isolated studies for treatment of distillery effluents and related compounds using microalgae.

For treatment of wastewaters native microalgal strains are a favourable alternative to the traditional wastewaters treatment systems. A consortium of *Oscillatoria, Lyngbya* and *Synechocystis* decolorized melanoidin by 98% by absorption followed subsequently by degradation of the organic compounds. The microalgal strains like *Anabaena cylindrica, Phormidium foveolarum, P. valderianum, Synechococcus, Ankistrodesmus braunii and Scenedesmus quadricauda* have been reported instrumental in degradation of phenol and its derivatives, whereas the performances of *Phormidium ambiguum, Chroococcus minutus, Oscillatoria,* and *Anabaena azollae* were found satisfactory for degradation of lignin. *Phormidium ambiguum* and *Chroococcus minutus* were found to reduce lignin by over 73.0% from the pulp and paper mill wastes in 5 days, whereas *Phormidium, Oscillatoria,* and *Anabaena azollae* were able to degrade lignin by 89% and hemicellulose by 92% from coir waste [210].

The role of cyanobacteria in distilleries effluent was studied in Kanchipuram, Tamilnadu, India. Totally 12 species of cyanobacteria belonging to 6 genera were identified. *Nostoc muscorum* was found to be the most dominated genus in this effluent (heterocyst organism). The inoculation of cyanobacteria *Nostoc muscarum* resulted in removal of various chemicals as well as nutrients such as nitrogen, ammonia and phosphorus from the effluent and could be potentially employed for the treatment of distilleries effluent [211].

12.7 FUTURE SCOPE FOR BIOREMEDIATION BY BIOFERTILIZER PRODUCING ORGANISMS

Bioremediation using biofertilizer producing microorganisms, though a very new concept, is an advancing field of research. Different types of interactions like biotransformation, biosorption and binding of heavy metals with extracellular or intracellular organelles are effectively used and scientifically manipulated to perform with higher effectiveness. Designing of different types of bioreactors, within which microbes can accumulate or adsorb metals or radionuclides under controlled systems, is also being modernized. Several strains of microorganisms have the capacity of promoting the plant growths and reduce the environmental toxicants. However, more studies are needed to be done on application fields, so that we can better understand the effect of natural factors and co-toxicants present in concerned environment. Apart from choosing and developing such strains, there remains a considerable amount of work required to be done on effective and easy desorbtion of toxic contaminants from environment and regeneration of the biomass for further use. To attract more usage of biosorbent technology using plant growth promoting microorganism (PGPM), certain strategies have to be formulated and promoted in agricultural and environmental sectors. The education, awareness and biotechnological research in this area increased our crop production and decreased the environmental contaminants. With so much to be done in this rapidly growing field of biotechnology, hope that future promises us a better and healthier environment.

CONCLUSION

The population explosion and industrial development in the growing world has increased of pollution in soil and water. With advances in biotechnology, bioremediation has become one of the most rapidly developing fields of environmental restoration, utilizing microorganisms to reduce the concentration and toxicity of various chemical pollutants, such as petroleum hydrocarbons, polycyclic aromatic hydrocarbons, polychlorinated biphenyls, phthalate esters, nitro-aromatic compounds, industrial solvents, pesticides and metals. The term bioremediation relates to *in situ* treatment of contaminated sites using biological agents to remove toxic waste from the environment. A number of bioremediation strategies have been developed to treat contaminated wastes and sites. These bioremediation techniques are typically more economical and environment-friendly than traditional methods of physical, chemical and mechanical treatments. Since bioremediation is based on natural attenuation and thus it is more acceptable than other remedial technologies.

REFERENCES

1. Al-Masri, M.R. (2001). "Changes in biogas production due to different ratios of some animal and agricultural wastes", *Bioresource Technol.*, 77 (1), 97-100.
2. Masse, D.I., Croteau, F., Masse, L. and Danesh. S. (2004). "The effect of scale-up on the digestion of swine manure slurry in psychrophilic anaerobic sequencing batch reactor", *Trans ASAE*, 47, 1367-1373.
3. Alagwadi, A.R. and Gaur, A.C. (1988). "Associative effect of *Rhizobium* and phosphate solubilizing bacteria on the yield and nutrient uptake of chickpea", *Plant and Soil.*, 105(2), 241-246.
4. Gaur, A.C. (1990). "Phosphate Solubilizing Microorganisms as Biofertilizer", Omega Scientific Publishers, New Delhi, 29.
5. Gaind, S. and Gaur. A.C. (1991). "Thermotolerant phosphate solubilizing micro organisms and their interaction with mungbean". *Plant and Soil*, 133, 141-143.
6. Hedge, S.V., Munishamanna, K.B., Balakrishna, A.N. and Reddy, P.C.B. (1994). "Influence of phosphate solubilizing fungi and *Rhizobium* on growth and nitrogen fixation in cowpea and red gram". *J. Soil Biol. Ecol.*, 14, 113-118.
7. Rupela, O.P., Singleton, P.W., Wani, S.P. and P., Bhattacharyya (2004). "Production and quality control of microbial inoculants: some perspectives", *Proc. of Nat. Conf. Quality Control Biofert.*, pp. 8-21.
8. Dinesh Kumar, Y.K., Vishwanath, A.P., Vithalnavi, H.M., Atheek Rahaman, U.R. and Anand, M.R. (2008). "Influence of organic and chemical amendments on enzymatic activities during composting of organic residues", *J. Ecobiol.*, 22, 25-28.
9. Singal, R., Gupta, R., Kuhad, R.C. and Saxena. R.K. (1991). "Solubilization of inorganic phosphate by Basidiomycetous fungus *Cyathus*", *Indian J. Microbial.*, 31, 397-401.
10. VanderZee, F.P., Villaverde, S., Garcia, P.A. and Fdz-Polanco. F. (2006). "Sulfide removal by moderate oxygenation of anaerobic sludge environments", *Bioresour Technol.*, 98, 518-524.
11. Biswas, B.C., Yadav, D.S. and Satish, Maheshwari (1985). "Biofertilizers in Indian Agriculture", *Fertilizer News*, 30(10), 20-28.
12. Wani, S.P. and Lee, K.K. (1995). "Microorganisms as biological inputs for sustainable agriculture" In: *Organic Agriculture*, (Thampan, P.K. ed.) Peekay Tree Crops Development Foundation, Cochin, India, pp. 39-76.
13. Katyal, J.C., Venkatashwarlu, B. and Das, S.K. (1994). "Biofertilizer for nutrient supplementation in dryland agriculture" *Fertiliser News*, 39(4), 27-32.
14. Singh, B.K., Walker, A., Alun, J., Morgan, W. and Wright, D.J. (2004). "Biodegradation of Chlorpyrifos by Enterobacter Strain B-14 and its use in bioremediation of contaminated soils", *Appl. Environ. Microbial.*, 70(8), 4855-4863.
15. Singh, B.K., Walker, A. and Wright, D.J. (2006). 'Bioremedial potential of fenamiphos and chlorpyrifos degrading isolates: Influence of different environmental conditions", *Soil Biol. Biochem.*, 38, 2682-2693.

16. Zhu, J., Zhao, Y. and Qiu, J. (2010). "Isolation and application of a chlorpyrifos-degrading *Bacillus licheniformis* ZHU-1", *Afr. J. Microbiol. Res.,* 4, 2410–2413.
17. Kulshrestha, G. and Kumari, A. (2011). "Fungal degradation of chlorpyrifos by *acremonium* sp. strain (GFRC-1) Isolated from a laboratory-enriched red agricultural soil" *Biol. Fert. Soils,* 47, 219–225.
18. Liu, Z., Chen, X., Shi, Y. and Su, Z. (2012). "Bacterial degradation of chlorpyrifos by *Bacillus cereus*", *Adv. Mater. Res.*, 356, 676–680.
19. Zimmo, O.R., Al-Sa'ed, R.M., Van der Steen, N.P. and Gijzen, H.J. (2002). "Process performance assessment of algae-based and duckweed-based wastewater treatment systems", *Water. Sci. Technol.*, 45, 91–101.
20. Cerniglia, C.E. (1992). "Biodegradation of polycyclic aromatic hydrocarbons", *Biodegradation,* 3, 351–368.
21. De La Noue, J., Laliberte, G. and Proulx, D. (1992). "Algae and Waste Water", *Journal of Applied Phycology,* 4, 247–254.
22. Raghukumar C., Vipparty, V., David, J. and Chandramohan D. (2001). "Degradation of crude oil by marine cyanobacteria", *Appl. Microbiol. Biotechnol.*, 57, 433–436.
23. Muthukumaran, M., Raghavan, B.G., Subramanian, V.V. and Sivasubramanian, V. (2005). "Bioremediation of industrial effluents using microalgae", *India Hydrobiol.* 7, 105–122.
24. Olguin, E.J. (2003). "Phycoremediation: key issues for cost-effective nutrient removal processes", *Biotechnology Advance,* 22, 81–91.
25. Wilde, E.W. and Benemann, J.R. (1993). "Bioremoval of heavy metals by the use of microalgae", *Biotechnol. Adv.*, 11, 781–812.
26. Zayed, A.M. and Terry, N. (2003). "Chromium in the environment: Factors affecting biological remediation", *Plant Soil,* 249, 139–156.
27. Chu, K.H. and Hashim, M.A. (2004). "Quantitative analysis of copper biosorption by the microalgae *Chlorella vulgaris*", *Environ. Eng. Sci.*, 21, 139–147.
28. Travieso, L., Benitez, F., Weiland, P., Sánchez, E., Dupeyrón, R. and Domínguez, A.R. (1996). "Experiments on immobilization of microalgae for nutrient removal in wastewater treatments", *Bioresour. Technol.*, 55, 181–186.
29. Dominic, V.J., Murali, S. and Nisha, M.C. (2009). "Phycoremediation efficiency of three micro algae *Chlorella vulgaris, Synechocystis salina* and *Geloeocapsa gelatiosa*", *Academic Review,* XVI, 138–146.
30. Abdel-Raouf, N., Al-Homaidan, A.A. and Ibraheem, I.B.M. (2012). "Microalgae and wastewater treatment", *Saudi Journal of Biological Sciences,* 19, 257–275.
31. Witters, N., Mendelsohn, R.O., Van-Slycken, S., Weyens, N., Schreurs, E., Meers, F., Carleer, R. and Vangronsveld, J. (2012). "Phycoremediation a sustainable remediation technology? Conclusion from a case study. I: Energy production and carbon dioxide abatement", *Biomass and Bioenergy,* 39, 454–469.
32. Mallick, N. (2002). "Biotechnological potential of immobilized algae for wastewater N.P., and metal removal: a review", *Biometals,* 15, 377–390.

33. Christenson, L. and Sims, R. (2011). "Production and harvesting of microalgae for wastewater treatment, biofuels and bioproducts", *Biotechnology Advances*, 29, 686–702.
34. Kalin, M., Wheeler, W.N. and Meinrath, G. (2005). "The removal of uranium from mining waste water using algal/microbial biomass", *Journal of Environmental Radioactivity*, 78(2), 151–177.
35. Mu-oz, R. and Guieysse, B. (2006). "Algal-bacterial processes for the treatment of hazardous contaminants: a review", *Water Research.*, 40(15), 2799–2815.
36. Irisarri, P., Gonnet, S. and Monza, J. (2001). "Cyanobacteria in Uruguayan rice fields: diversity, nitrogen fixing ability and tolerance to herbicides and combined nitrogen", *Journal of Biotechnology*, 91(2-3), 95–103.
37. Das, D. and Vezirollu, T.N. (2001). "Hydrogen production by biological processes: a survey of literature", *International Journal of Hydrogen Energy*, 26(1), 13–28.
38. Tsygankov, A.A., Fedorov, A.S., Kosourov, S.N. and Rao, K.K. (2002). "Hydrogen production by cyanobacteria in an automated outdoor photobioreactor under aerobic conditions", *Biotechnology and Bioengineering*, 80(7), 777–783.
39. Lindblad, P., Christensson, K., Lindberg, P., Fedorov, A., Pinto F. and Tsygankov, A. (2002). "Photoproduction of H_2 by wild type *Anabaena* PCC 7120 and a hydrogen uptake deficient mutant: from laboratory experiments to outdoor culture", *International Journal of Hydrogen Energy*, 27(11-12), 1271–1281.
40. Cetinkaya, D.G., Aksu, Z., Ozturk, A. and Kutsal, T.A. (1999). "Comparative study on heavy metal biosorption characteristics of some algae", *Process Biochemistry*, 34(9) 885–892.
41. Doshi, H., Ray, A., Kothari, I.L. and Gami, B. (2006). "Spectroscopic and SEM studies on bioaccumulation of pollutants by algae", *Current Microbiology*, 53 148--157.
42. Barros, L.M., Jr, Macedo, G.R., Duarte, M. M. L., Silva, E. P. and Lobato, A.K. C.L. (2003). "Biosorption of cadmium using the fungus *Aspergillus niger*" *Brazilian Journal of Chemical Engineering*, 20(3), 229–39.
43. Volesky, B. (1986). "Biosorbent materials", In: *Biotechnology and Bioengineering Symposium*, Wiley, 16, 121–126.
44. Clemens, S. Kim, E.J., Neumann, D. and Schroeder, J.I. (1999). "Tolerance to toxic metals by a gene family of phytochelatin syntheses from plants and yeast", *The EMBO Journal*, 18(12) 3325–3333.
45. Rokhzadi, A., Asgharzadeh, A., Darvish, F., Nourmohammadi, G. and Majidi, E. (2008). "Influence of plant growth-promoting rhizobacteria on dry matter accumulation and yield of chickpea (*Cicer arietinum* L.) under field condition", Am-Euras. *J. Agric. Environ. Sci.*, 3(2), 253–257.
46. El-Yazeid, A.A., Abou-Aly, H.A., Mady, M.A. and Moussa, S.A., M. (2007). "Enhancing growth, productivity and quality of squash plants using phosphate dissolving microorganisms (bio phosphor) combined with boron foliar spray", *Res. J. Agric. Biol. Sci.*, 3(4), 274–286.

47. Lee, K.E. and Pankhurst, C.E. (1992). "Soil organisms and sustainable productivity", *Australian J. Soil Res.*, 30, 855–92.
48. Wani, S.P., Rego, T.G., Rajeshwari, S. and Lee, K.K. (1995). "Effect of legume-based cropping systems on nitrogen mineralization potential of Vertisol", *Plant Soil,* 175(2), 265–274.
49. Kannaiyan, S. (2002). "Biofertilizers for sustainable crop production", *Biotechnology of Biofertilizers,* Narosa Publishing House, New Delhi, India, p. 377.
50. Gothwal, R.K., Nigam, V.K., Mohan, M.K., Sasmal, D. and Ghosh, P. (2007). "Screening of nitrogen fixers from rhizospheric bacterial isolates associated with important desert plants", *Appl. Ecol. Environ. Res.*, 6(2), 101–109.
51. Bakulin M.K., Grudtsyna A.S. and Pletneva A. (2007). "Biological fixation of nitrogen and growth of bacteria of the genus *Azotobacter* in liquid media in the presence of Perfluoro carbons", *Appl. Biochem. Microbiol.*, 4, 399–402.
52. Gupta, A.K. (2004). "The complete technology book on biofertilizers and organic farming", National Institute of Industrial Research Press, India.
53. Wani S.P. (1990). "Inoculation with associative nitrogen-fixing bacteria: Role in cereal grain production improvement", *Indian J. Microbiol.*, 30, 363–393.
54. Triplett, E. (1996). "Diazotrophic endophytes: Progress and prospects for nitrogen fixation in monocots", *Plant and Soil,* 186, 29–38.
55. James, E.K., Olivares, F.L., Baldani, J.I. and Döbereiner, J. (1997). "*Herbaspirillum,* an endophytic diazotroph colonizing vascular tissue in leaves of *Sorghum bicolour* L. Moench", *Journal of Experimental Botany,* 48, 785–797.
56. Boddey, R.M., Da Silva, L.G., Reis, V., Alves, B.J.R. and Urquiaga, S. (2000). "Assessment of bacterial nitrogen fixation in grass species", In: E.W. Triplett (ed.), *Prokaryotic Nitrogen Fixation: A Model System for Analysis of a Biological Process,* Wymondham, UK: Horizon Scientific Press, pp. 705–726.
57. Malik, K.A., Bilal, R., Mehnaz, S., Rasul, G., Mirza, M.S. and Ali, S. (1997). "Association of nitrogen-fixing, plant growth-promoting rhizobacteria (PGPR) with kallar grass and rice", *Plant and Soil,* 194, 37–44.
58. James, E.K. (2000). "Nitrogen fixation in endophytic and associative symbiosis", *Field Crops Research,* 65, 197–209.
59. Bashan, Y. and Holguin, G. (1997). "*Azospirillum*-plant relationships: Environmental and physiological advances", *Canadian Journal of Microbiology.,* 43, 103–121.
60. Bashan, Y., Ream, Y., Levanony, H. and Sade, A. (1989). "Non-specific responses in plant growth, yield, and root colonization of noncereal crop plants to inoculation with *Azospirillum brasilense* Cd", *Canadian Journal of Botany,* 67, 1317–1324.
61. Benson, D.R. and Silvester, W.B. (1993). "Biology of *Frankia* strains, actinomycete symbionts of actinorhizal plants", *Microbiological Reviews,* 57, 293–319.
62. Smolander, A. and Sarsa, M.L. (1990). "*Frankia* strains of soil under *Betula pendula:* Behaviour in soil and in pure culture", *Plant and Soil,* 122, 129–136.

63. Roger, P.A., Zimmerman, W.J. and Lumpkin, T.A. (1993). "Microbiological management of wetland rice fields", In: Metting B (ed.) *Soil Microbial Ecology: Applications in Agricultural and Environmental Management,* Dekker, New York, pp. 417–455.
64. Watanabe, I. and Roger, P.A. (1984). "Nitrogen fixation in wetland rice fields", In: Subba Rao NS (ed.) *Current Developments in Biological Nitrogen Fixation.* Oxford, IBH, New Delhi, pp. 237–276.
65. Liu, C.C. (1979). "Use of *Azolla* in rice production in China", In: *Nitrogen and Rice*", IRRI, Manila, pp. 375–394.
66. De, P.K. (1936). "The problem of the nitrogen supply of rice. I. Fixation of nitrogen in the rice soil under waterlogged condition", *Indian J Agric Sci.,* 6, 1237–1245.
67. De, P.K. (1939). "The role of blue-green algae in nitrogen fixation in rice fields", *Proc R Soc London Ser B,* 127, 121–139.
68. Watanabe, I., Bai, K.Z., Berja, N.S., Espina, L.R., Ho, O. and Subidhi, R.P.R. (1981). "The *Azolla-Anabaena* complex and its use in rice culture". *IRRI Res Pap Ser,* 69.
69. Lumpkin, T.A. and Plucknett, D.L. (1982). "*Azolla* as a green manure: use and management in crop production", Westview tropical agriculture series no. 5. Westview, Boulder, Colo., pp. 230.
70. De, P.K. and Mandal, L.N. (1956). "Fixation of nitrogen by algae in rice soils", *Soil Sci.,* 81, 453–458.
71. Singh, R.N. (1961). "Role of blue-green algae in nitrogen economy of Indian agriculture", *Indian Council of Agricultural Research,* New Delhi.
72. Stewart., W.D.P., Fitzgerald, G.P. and Burns, R.H. (1968). "Acetylene reduction by nitrogen-fixing blue-green algae", *Arch Microbiol.,* 62, 336–348.
73. Shen, Z.H., Leu, S.E., Chen, K.Z. and Gi, S.A. (1963). "The initial experiment on *Azolla* nitrogen fixing ability", *Pedol Bull Peking,* 4, 45–48.
74. Peters, G.A., Toia, R.E., Jr., Lough, S.M. (1977). "The *Azolla-Anabaena azollae* relationship. V15 N_2 fixation, acetylene reduction and H_2 production". *Plant Physiol.,* 59, 101–102.
75. Singh, A.L. and Singh, P.K. (1987). "Influence of *Azolla* management on the growth, yield of rice and soil fertility. II. N and P contents of plants and soil", *Plant Soil,* 102, 49–54.
76. Fogg, G.E., Stewart, W.D.P., Fay, P. and Walsby, A.E. (1973). "The blue-green algae", Academic Press, London, pp. 459.
77. Roger, P.A. and Kulasooriya, S.A. (1980). "*Blue-green Algae and Rice*", IRRI, Manila, pp. 112.
78. Venkataraman, G.S. (1981). "Blue-green algae for rice production - a manual for its promotion". *Soils Bulletin No. 46.* FAO, Rome.
79. IRRI. (1987). "*Azolla* Utilisation". IRRI, Manila, pp. 296.
80. Wagner, G.M. (1997). "*Azolla*: a review of its biology and utilization", *Bot Rev.,* 63, 1–26.

81. Schachtman, D.P., Reid, R.J. and Ayling, S.M. (1998). "Phosphorus uptake by plants: From soil to cell", *Plant Physiology,* 116, 447–453.
82. Richardson, A.E. (2001). "Prospects for using soil microorganisms to improve the acquisition of phosphorus by plants", *Australian Journal of Plant Physiology,* 28, 897–906.
83. Alam S., Khalil S., Ayub N. and Rashid M. (2002). "In vitro solubilization of inorganic phosphate by phosphate solubilizing microorganism (PSM) from maize rhizosphere", *Intl. J. Agric. Biol.,* 4, 454–458.
84. Chen, Y.P., Rekha, P.D., Arunshen, A.B., Lai, W.A. and Young, C.C. (2006). "Phosphate solubilizing bacteria from subtropical soil and their tri-calcium phosphate solubilizing abilities", *Appl. Soil Ecol.,* 34, 33–41.
85. Fankem H., Nwaga D., Deubel A., Dieng L., Merbach W. and Etoa F.S. (2006). "Occurrence and functioning of phosphate solubilizing microorganisms from oil palm tree rhizosphere in Cameroon", *Afr J. Biotech.,* 5, 2450–2460.
86. Igual, J.M., Valverde, A., Cervantes, E. and Velázquez, E. (2001). "Phosphate-solubilizing bacteria as inoculants for agriculture: use of updated molecular techniques in their study" *Agronomie,* 21, 561–568.
87. Whitelaw, M.A. (2000). "Growth promotion of plants inoculated with phosphate solubilizing fungi", *Adv Agron.,* 69, 99–151.
88. Subbarao, N.S. (1988). "Phosphate solubilizing microorganism", In: *Biofertilizer in Agriculture and Forestry,* Regional Biofert. Dev. Centre, Hissar, India, pp. 133–142.
89. Kucey, R.M.N., Janzen, H.H. and Legett, M.E. (1989). "Microbially mediated increases in plant-available phosphorus", *Advances in Agronomy,* 42, 198–228.
90. Duponnois, R., Kisa, M. and Plenchette, C. (2006). "Phosphate solubilizing potential of the nemato fungus *Arthrobotrys oligospora*", *J. Plant Nutr. Soil Sci.,* 169, 280-282.
91. Bennett, P.C., Choi, W.J. and Rogera, J.R. (1998). "Microbial destruction of feldspars", *Mineral Management,* 8(62A), 149–150.
92. Barker, W.W., Welch, S.A., Chu, S. and Banfield, J.F. (1998). "Experimental observations of the effects of bacteria on aluminosilicate weathering", *Am Mineral.,* 83, 1551–1563.
93. Sheng, X.F., He, L.Y. and Huang W.Y. (2002). "The conditions of releasing potassium by a silicate-dissolving bacterial strain NBT", *Agric Sci China,* 1, 662–666.
94. Sheng, X.F., Xia, J.J. and Chen, J. (2003). "Mutagenesis of the *Bacillus edphicaus* Strain NBT and its effect on growth of chili and cotton", *Agric Sci China,* 2, 40–412.
95. Khandkar, M.R., Shinde, D.A., Kanekar, V.S., Jambley N.R. and Jain, N.K. (1985). "Growth, nodulation and yield of rainfed black gram as influenced by phosphate and sulphate fertilization in vertisol", *Indian J. Pl. Physiol.,* 28, 318–322.
96. Saraf, C.S. (1988). "Sulphur fertilization to pulses in yield and quality". *Proc. TSI-FAI Symp. Sulphur in Indian Agriculture,* New Delhi, pp. S II/2–1.

97. Mehta, U.R. and Singh. H.G. (1979). "Response of green gram to sulphur in calcareous soils", *Indian J. Agric. Sci.*, 49, 703–706.
98. Hanesklaus, S. and Schnug, E. (1992). "Baking quality and sulphur content of wheat", *Sulphur in Agriculture*, 16, 35–36.
99. Poorani, B. (1992). "Radiotracer studies on the interaction effects of application of P and S on available nutrients, yield and quality of green gram". M.Sc.(Ag.) Thesis, Tamil Nadu Agricultural University, Coimbatore.
100. Pasricha, N.S. and Fox, R.L. (1993). "Plant nutrient sulphur in the tropics and subtropics", *Adv. Agron.*, 50, 209–255.
101. Smith, S.E. and Read, D.J. (1997). "Mycorrhizal Symbiosis". London: Academic Press.
102. Rillig, M.C., Wright, S.F. and Eviner, V.T. (2002). "The role of arbuscular mycorrhizal fungi and glomalin in soil aggregation: Comparing effects of five plant species", *Plant and Soil*, 238, 325-333.
103. Carreiro, M.M., Sinsabaugh, R.L., Repert, D.A. and Parkhurst, D.F. (2000). "Microbial enzyme shifts explain litter decay responses to simulated nitrogen deposition", *Ecology*, 81, 2359–2365.
104. Sjöberg, G., Knicker, H., Nilsson, S.I. and Berggren, D. (2004). "Impact of long-term N fertilization on the structural composition of spruce litter and mor-humus", *Soil Biol Biochem.*, 36, 609–618.
105. Watkinson, S.C., Davison, E.M. and Bramah, J. (1981). "The effect of nitrogen availability on growth and cellulolysis by *Serpula lacrymans*", *New Phytol.*, 89, 295-305.
106. Andrén, O. and Lagerlöf, J. (1983). "Soil fauna (microarthropods, enchytraeids, nematodes) in Swedish agricultural cropping systems", *Acta Agriculturae Scandinavica*, 33, 33–52.
107. Weiss, B. and Larink, O. (1991). "Influence of sewage sludge and heavy metals on nematodes in an arable soil", *Biology and Fertility of Soils*, 12, 5–9.
108. Potter, D.A. (1993). "Pesticide and fertiliser effects on beneficial invertebrates and consequences for thatch degradation and pest outbreaks in turf grass". In: Racke KD and Leslie AR (eds.) *Pesticides in Urban Environments: Fate and Significance*, ACS Symposium Series No. 522, American Chemical Society, Washington, DC, pp. 331–343.
109. Ott, P., Hansen, S. and Vogtmann, H. (1983). "Nitrates in relation to composting and use of farmyard manures", In: Lockeretz W (ed) *Environmentally Sound Agriculture*, Praeger, New York, NY, pp. 145–154.
110. Mueler, J.G., Cerniglia, C.E. and Pritchard, P.H. (1996). "Bioremediation of Environments Contaminated by Polycyclic Aromatic Hydrocarbons", In: *Bioremediation: Principles and Applications*, Cambridge University Press, Cambridge pp. 125–194.
111. Vidali, M. (2001). "Bioremediation: An overview", *Pure Appl. Chem.*, 73(7), 1163–1172
112. McCutcheon, S.C. and Schnoor, J.L., (eds). (2003). "Phytoremediation - Transformation and Control of Contaminants", Wiley Interscience, pp. 985.

113. Glass, D.J. (2000). "Economical potential of phytoremediation". In: I Raskin, BD Ensley (Eds.), *Phytoremediation of Toxic Metals: Using Plants to Clean up the Environment.* John Wiley and Sons, NY., p. 304.
114. Koronelli, T.V., Komarova, T.I., Il'linskii, V.V., Kuz'min, Y.I., Kirsanov, N.B. and Yanenko, A.S. (1997). "Introduction of bacteria of the genus *Rhodococcus* into oil-contaminated tundra soil", *Appl. Biochem. Microbiol.*, 33, 198–201.
115. Borzenkov, I.A., Belyaev, S.S., Ibatullin, R.R., Pospelov, M.E. and Svitnev, A.I., (1998). "Clean-up method using bio-reparations for soil, natural aquatics and wastewaters contaminated with oil and petroleum products", *Russian Federation Patent* No. 2114071.
116. ICS-UNIDO (International Centre for Science and High Technology - United Nations Industrial Development organization)., 2001. "*Technologies for Remediation of Soils Contaminated with Crude Oil and Petroleum Products*", NIA-Priroda, Moscow, pp. 185.
117. Christofi, N., Ivshina, I.B., Kuyukina, M.S. and Philp, J.C. (1998). "Biological treatment of crude oil contaminated soil in Russia", In: *Contaminated Land and Groundwater: Future Directions* (Lerner, D.N., and Walton, N.R.G., Eds.). London, Geological Society, Engineering Geology Spec. Pub. 14, pp. 45–51.
118. Ivshina, I.B., Kuyukina, M.S., Philp, J.C. and Christofi, N. (1998). "Oil desorption from mineral and organic materials using biosurfactant complexes produced by *Rhodococcus* species" *World J. Microbiol. Biotechnol.*, 14, 307–312.
119. Ivshina, I.B., Kuyukina, M.S., Ritchkova, M.I., Philp, J.C., Cunningham, C.J. and Christofi, N. (2001). "Oleophilic biofertilizer based on a *Rhodococcus* surfactant complex for the bioremediation of crude oil-contaminated soil", *AEHS Contaminated Soil Sediment and Water: International Issue,* pp. 20–24.
120. Gunalan, P.H. and Fournier, C. (1993). "Effect of microbial competition on the survival and activity of 2,4-D-degrading *Alcaligenes xylosoxidans* sub sp. denitrificans added to soil" *Letters Appl. Microbiol*, 16, 178–181.
121. Van Veen, J.A., Van- Ovdrbeek, L.S. and Van Elsas, J.D. (1997). "Fate and activity of microorganisms introduced into soil", *Microbiol. Mol. Biol. Rev.*, 61, 121–135.
122. EL-Kabbany, S. (2002). "Evaluation of our biofertilizer for bioremediation of pesticide contaminated soil", The First Conf. of the Central Agric. Pesticide Lab., Sept. 3-5, Ministry of Agriculture and Land Reclamation, Agriculture Research Centre, Giza, Egypt.
123. Mulbry, W.W., Ahrens, E. and Karns, J.S. (1998). "Use of a field scale biofilter for the degradation of the organophosphate insecticide coumaphos in cattle dip wastes" *Pestic. Sci.*, 52, 268–274.
124. Karpouzas, D.G. and Walker, A. (2000). "Factors influencing the ability of *Pseudomonas putida* epI to degrade ethoprophos in soil" *Soil Biol. Biochem.*, 32, 1753–1762.

125. Khaled, A.O., Gamal, A.H., Ahmad, I.A. and Abdul Rahman, A.A. (2008). "Biodegradation kinetics of dicofol by selecting microorganisms", *Pestic. Biochem. Physiol.*, 91, 180–185.
126. Kanade, S.N., Ade, A.B. and Khilare, V.C. (2012). "Malathion degradation by *Azospirillum lipoferum* Beijerinck", *Sci. Res. Reporter*, 2(1), 94–103.
127. De-Bashan, L.E., Hernandez, J.P. and Bashan, Y. (2012). "The potential contribution of plant growth-promoting bacteria to reduce environmental degradation—A comprehensive evaluation", *Applied Soil Ecology*, 61, 171–189.
128. Abou-El-Naga, S.H. (1998). "Production of biofertilizers by General Organization for Agriculture Equalization Fund in Egypt (GOAEF)", The sixth International *Symposium on Nitrogen Fixation with Non-Legumes,* Ismailia-Egypt, 509.
129. Wu, S.C., Cheung, K.C., Luo, Y.M. and Wong, M.H. (2006). "Effects of inoculation of plant growth promoting rhizobacteria on metal uptake by *Brassica juncea*", *Environmental Pollution*, 140, 124–35.
130. Bianco, C. and Defez, R. (2009). "*Medicago truncatula* improves salt tolerance when nodulated by an indole-3-acetic acid-overproducing *Sinorhizobium meliloti* strain", *J Exp Bot.*, 60, 3097–3107.
131. Wani, P.A., Khan, M.S. and Zaidi, A. (2007). "Effect of metal tolerant plant growth promoting *Bradyrhizobium* sp. (vigna) on growth, symbiosis, seed yield and metal uptake by green gram plants", *Chemosphere*, 70, 36–45.
132. Ganesan, V. (2008). "Rhizoremediation of cadmium soil using a cadmium-resistant plant growth-promoting rhizopseudomonad", *Curr Microbiol,* 56, 403–407.
133. Idriss, E.E., Makarewicz, O., Farouk, A., Rosner, K., Greiner, R., Bochow, H., Richter, T. and Borris, R. (2002). "Extracellular phytase activity of *Bacillus amyloliquefaciens* FZB45 contributes to its plant growth-promoting effect", *Microbiology,* 148, 2097–2109.
134. Jahanian, A., Chaichi, M.R., Rezaei, K., Rezayazdi, K. and Khavazi, K. (2012). "The effect of plant growth promoting rhizobacteria (pgpr) on germination and primary growth of artichoke (*Cynara scolymus*)", *Int. J. Agric. Crop Sci.*, 4, 923–929.
135. Dimkpa, C., Svatos, A., Merten, D., B€uchel, G. and Kothe, E. (2008). "Hydroxamate siderophores produced by *Streptomyces acidiscabies* E13 bind nickel and promote growth in cowpea (*Vigna unguiculata L.*) under nickel stress", *Can J Microbiol.*, 54,163–172.
136. Sridevi, M., Kumar, M.G. and Mallaiah, K.V. (2008). "Production of catechol-type of siderophores by *Rhizobium* sp. isolated from stem nodules of *Sesbania procumbens* (Roxb.) W., and A.", *Res J Microbiol.*, 3, 282–287.
137. Dey, R., Pal, K.K., Bhatt, D.M. and Chauhan, S.M. (2004). "Growth promotion and yield enhancement of peanut (*Arachis hypogaea L.*) by application of plant growth-promoting rhizobacteria", *Microbiol Res.*, 159, 371–394.
138. Barriuso, J., Solano, B.R. and Gutiérrez Man˜ero, F.J. (2008). "Protection against pathogen and salt stress by four plant growth-promoting rhizobacteria isolated from Pinus sp. on *Arabidopsis thaliana*", *Phytopathology,* 98, 666–672.

139. Sinha, S. and Mukherjee, S.K. (2008). "Cadmium-induced siderophore production by a high Cd-resistant bacterial strain relieved Cd toxicity in plants through root colonization", *Curr. Microbiol.*, 56, 55–60.

140. Mehnaz, S., Mirza, M.S., Haurat, J., Bally, R., Normand, P., Bano, A. and Malik, K.A. (2001). "Isolation and 16S rRNA sequence analysis of the beneficial bacteria from the rhizosphere of rice", *Can J Microbiol.*, 47, 110–117.

141. Gutierrez-Man˜ero, F.J., Ramos-Solano, B., Probanza, A., Mehouachi, J., Tadeo, F.R. and Talon, M. (2001). "The plant growth-promoting rhizobacteria *Bacillus pumilus* and *Bacillus licheniformis* produce high amounts of physiologically active gibberellins", *Physiol Plant.*, 111,1–7.

142. Zaidi, S., Usmani, S., Singh, B.R. and Musarrat, J. (2006). "Significance of *Bacillus subtilis* strain S.J., 101 as a bioinoculant for concurrent plant growth promotion and nickel accumulation in *Brassica juncea*", *Chemosphere*, 64, 991–997.

143. Sheng, X.F. and Xia, J.J. (2006). "Improvement of rape (*Brassica napus*) plant growth and cadmium uptake by cadmium-resistant bacteria", *Chemosphere*, 64, 1036–1042.

144. Faisal, M. and Hasnain, S. (2005). "Bacterial Cr (VI) reduction concurrently improves sunflower (*Helianthus annuus* L.) growth", *Biotechnol Lett.*, 27, 943–947.

145. Sheng, X.F., Jiang, C.Y. and He, L.Y. (2008). "Characterization of plant growth-promoting *Bacillus edaphicus* NBT and its effect on lead uptake by Indian mustard in a lead-amended soil", *Can. J. Microbiol.*, 54, 417–422.

146. Han, S.Q., Zhang, Z.H. and Yan, S.H. (2000). "Present situation and developmental trend of wastewater treatment and eutrophication waters purification with alga technology", *Agro Environmental Development*, 63 (1), pp. 13–16.

147. Benemann, J.R., Weismann, J.C., Eisenberg, D.M., Koopman, B.L., Goebel, R.P., Caskey, P.S., Thomson, R.D. and Oswald, W.J. (1977). "A systems analysis of bioconversion with micro algae", Paper presented at symposium on clean fuels from biomass and wastes, Oriando, Florida.

148. Gupta, S.K. and Rao, P.V.S.S. (1980). "Treatment of urea by algae, activated sludge and flocculation algal bacterial system- A comparative study", *Indian Journal Environmental Health.*, 22, pp. 103–112.

149. Williams, W.D. (1981). "Ecological use of sewage. Some Approaches to Saprobiological Problems", *Environmental Pollution*, 58, 48–56.

150. Kunikane, S., Kaneko, M. and Maehara, R. (1984). "Growth and nutrient uptake of green alga, *Scenedesmus dimorphus*, under a wide range of nitrogen-phosphorus ratio(I). Experimental study", *Water Research*, 18(10), 1299–1311.

151. Sengar, R.M.S. and Sharma, K.D. (1987). "Tolerance of *Phormidium corium* and *Chlamydomonas* sp. against chemical elements present in polluted water of Yamuna river", *Proceeding National Academic Science (India)*, 57, pp. 56–64.

152. Tam, N.F.Y. and Wong, Y.S. (1989). "Wastewater nutrient removal by *Chlorella pyrenoidosa* and *Scenedesmus* sp.", *Environmental Pollution*, 58, 19–34.

153. Gantar, M., Obreht, Z. and Dalmacija, B. (1991). "Nutrient removal and algae succession during the growth of *Spirulina platensis* and *Scenedesmus quadricauda* on swine wastewater", *Bioresourse Technology*, 36, 167–171.
154. De-Bashan, L.E., Moreno, M., Hernandez, J.P. and Bashan, J. (2002). "Ammonium and phosphorus removal from continuous and semi-continuous cultures by the microalgae *Chlorella vulgaris* co-immobilized in alginate beads with *Azospirillum brasilense*", *Water Research*, 36, 2941–2948.
155. Queiroz, M.I., Lopes, E.J., Zepka, L.Q., Bastos, R.G. and Goldbeck, R. (2007), "The kinetics of the removal of nitrogen and organic matter from parboiled rice effluent by cyanobacteria in a stirred batch reactor", *Bioresource Technology*, 98, 2163–2169.
156. Rao, H.P., Kumar, R.R., Raghavan, B., G., Subramanian, V.V. and Sivasubramanian, V. (2011). "Application of phycoremediation technology in the treatment of wastewater from a leather-processing chemical manufacturing facility", *Water. Soil, Air*, 37(1), 7–14.
157. Gosavi, K., Sammut, J., Gifford, S. and Jankowski, J. (2004). "Macroalgal biomonitors of trace metal contamination in acid sulfate soil aquaculture ponds", *Sci. Total Environ.*, 324, 25–39.
158. Inouhe, M., Sumiyoshi, M., Tohoyanna, H. and Joho, M. (1996). "Resistance to cadmium ions and formation of a cadmium binding complex in various wild type yeasts", *Plant Cell. Physiol.*, 37: 341–346.
159. Megharaja, M., Ragusa, S.R. and Naidu, R. (2003). "Metal–algae interactions: implication of bioavailability. In: *Bioavailability, Toxicity and Risk Relationships in Ecosystems*", Naidu, R; Gupta, V.V.S.R; Rogers, S; Kookana, R.S; Bolan, N.S; Adriano, D.C., (eds.), Science Publishers, Enfield, New Hampshire, UK, pp. 109–144.
160. Sigg, L. (1985). "Metal transfer mechanisms in lakes; the role of settling particles", In: Stumm W (eds.) *Chemical processes in lakes*. Wiley, New York, pp. 283–310.
161. Sigg, L. (1987). "Surface chemical aspects of the distribution and fate of metal ions in lakes" In: W. Stumm (Editor), *Aquatic Surface Chemistry*. Wiley, New York, pp. 331–350.
162. Shamsuddoha, A.S.M., Bulbul, A. and Huq, S.M.I. (2006). "Accumulation of arsenic in green algae and its subsequent transfer to the soil–plant system", *Bangladesh Journal of Microbiology*, 2(2), 148–151.
163. Dixon, R. and Kahn, D. (2004). "Genetic regulation of biological nitrogen fixation", *Nat Rev Microbiol.*, 2, 621–631.
164. Raymond, J., Siefert, J.L., Staples, C.R. and Blankenship, R.E. (2004). "The natural history of nitrogen fixation", *Mol Biol. Evol.*, 21, 541–554.
165. Ferguson, S. (1998). "Nitrogen cycle enzymology", *Curr Opin Chem Biol.*, 2, 182-193.
166. Merrick, M. and Edwards, R. (1995). "Nitrogen control in bacteria". *Microbiol Rev*, 59, 604–622.
167. Richardson, D. and Watmough, N. (1999). "Inorganic nitrogen metabolism in bacteria". *Curr. Opin. Chem Biol.*, 3, 207–219.

168. Eckford, R., Cook, F., Saul, D., Aislabie, J. and Foght, J. (2002). "Free-living heterotrophic nitrogen-fixing bacteria isolated from fuel-contaminated Antarctic soil", *Appl Environ Microbiol.*, 68, 5181–5185.

169. Laguerre, G., Bossand, B. and Bardin, R. (1987). "Free-living dinitrogen-fixing bacteria isolated from petroleum refinery oily sludge", *Appl Environ Microbiol.*, 53, 1674–1678.

170. Piehler, M., Swistak, J., Pinckney, J. and Paerl, H. (1999). "Stimulation of diesel fuel biodegradation by indigenous nitrogen fixing bacterial consortia", *Microb Ecol.*, 38, 69–78.

171. Davis J., Coty, V. and Stanley, J. (1964). "Atmospheric nitrogen fixation by methane-oxidizing bacteria", *J Bacteriol.*, 88, 468-472

172. Prantera, M., Drozdowicz, A., Leite, S. and Rosado, A. (2002). "Degradation of gasoline aromatic hydrocarbons by two N_2-fixing soil bacteria", *Biotechnol Lett.*, 24, 85–89.

173. Efroymson, R. and Alexander, M. (1991). "Biodegradation by an *Arthrobacter* species of hydrocarbons partitioned into an organic solvent", *Appl Environ Microbiol.*, 57, 1441–1447.

174. Gradova, N., Gornova, I., Eddaudi, R. and Salina, R. (2003). "Use of bacteria of the genus *Azotobacter* for bioremediation of oil-contaminated soils" *Appl Biochem Microbiol.*, 39, 279–281.

175. Bento, F., Camargo, F., Okeke, B. and Frankenberger, W. (2005). "Comparative bioremediation of soils contaminated with diesel oil by natural attenuation, biostimulation and bioaugmentation", *Bioresour Technol.*, 96, 1049–1055.

176. Coulon, F., Pelletier, E. and Gourhant, L. (2005). "Effects of nutrient and temperature on degradation of petroleum hydrocarbons in contaminated sub-Antarctic soil", *Chemosphere*, 58, 1439–1448.

177. Riser-Roberts, E. (1998). "Remediation of petroleum contaminated soils: biological, physical, and chemical processes", Lewis Publishers, CRC Press, pp. 503.

178. Frick, C.M., Farrell, R.E. and Germida, J.J. (1999). "Assessment of phytoremediation as an in-situ technique for cleaning oil-contaminated sites", Report submitted to Petroleum Technology Alliance of Canada.

179. Newman, L.A. and Reynolds, C.M. (2004). "Phytodegradation of organic chemicals", *Curr. Opin. Biotechnol.*, 15, 1–6.

180. Huesemann, M.H. and Moore, K.O. (1993). "Compositional changes during landfarming of weathered Michigan crude oil-contaminated soil" *J. Soil Contam.*, 2, 245–246.

181. Perez-Vargas, J., Poggi-Varaldo, H.M., Calva-Calva, G., Rios-leal, E., Rodriguez-Varquez, R., Ferrera-Cerralto, R. and Esparza-Garcia, F. (2000). "Nitrogen-fixing bacteria capable of utilizing kerosene hydrocarbons as a sole carbon source", *Water Sci. Technol.*, 42, 407–410.

182. Rai, L.C., Gaur, J.P. and Kumar, H.D. (1981). "Phycology and heavy metal pollution", *Biological Reviews of the Cambridge Philosophical Society*, 56, 99–151.

183. Lokeshwari, H. and Chandrappa. G.T. (2007). "Effects of heavy metal contamination from anthropogenic sources on Dasarahalli tank, India", Lakes and Reservoirs: *Research and Management,* 12, 121–128.
184. Chang, J.S., Yoon, I.H. and Kim. K.W. (2009). "Heavy metal and arsenic accumulating fern species as potential ecological indicators in As-contaminated abandoned mines", *Ecological Indicators,* 9, 1275–1279.
185. Yadav, S.K., Juwarkar, A.A., Kumar, G.P., Thawale, P.R., Singh, S.K. and Chakrabarti, T. (2009). "Bioaccumulation and phyto-translocation of arsenic, chromium and zinc by *Jatropha curcas* L.: Impact of dairy sludge and biofertilizer", *Bioresource Technology,* 100, 4616–4622.
186. Arora, A., S., Saxena and D.K., Sharma (2006). "Tolerance and phytoaccumulation of chromium by three *Azolla* species", *World Journal of Microbiology & Biotechnology,* 22, 97–100.
187. Umali, L.J., Duncan, J.R. and Burgess. J.E. (2006). "Performance of dead *Azolla filiculoides* biomass in biosorption of Au from wastewater", *Biotechnology Letters,* 28, 45–49.
188. Bennicelli, R., Stezpniewska, Z., Banach, A., Szajnocha, K. and Ostrowski. J. (2004). "The ability of *Azolla caroliniana* to remove heavy metals (Hg(II), Cr(III), Cr(VI)) from municipal waste water", *Chemosphere,* 55, 141–146.
189. Upadhyay, A.R., Mishra, V.K., Pandey, S.K. and Tripathi. B.D. (2007). "Biofiltration of secondary treated municipal wastewater in a tropical city", *Ecological Engineering,* 30, 9–15.
190. Rai, P.K. (2008). "Phytoremediation of Hg and Cd from industrial effluent using an aquatic free floating macrophyte *Azolla pinnata*", *International Journal of Phytoremediation,* 10, 430–439.
191. Mashkani, S.G. and Ghazvini, P.T.M. (2009). "Biotechnological potential of *Azolla filiculoides* for biosorption of Cs and Sr: Application of micro-PIXE for measurement of Biosorption", *Bioresource Technology,* 100, 1915–1921.
192. Yong-huang, W. and Wei-Zhen, X. (1985). *Proc. C Workshop on Azolla use,* Fuzhou, Feejian, China, pp. 282.
193. Sharma, G.K. and Khan, S.A. (2013). "Bioremediation of sewage wastewater using selective algae for manure production", *International Journal of Environmental Engineering and Management,* 4 (6), 573–580.
194. Piperidou, C.I., Chaidou, C.I., Stalikas, C.D., Soulti, K., Pilidis, G.A. and Balis, C., (2000). "Bioremediation of olive oil mill wastewater: chemical alterations induced by *Azotobacter vinelandii*", *Journal of Agricultural and Food Chemistry,* 48(5), 1941-1948.
195. Atomic Energy Authority, Cairo (Egypt); 1555 p; 1999; p. 241-249; International conference on hazardous waste sources, effects and management; Cairo (Egypt); 12-16 Dec 1998; INIS-EG-165(V.1,2,3).
196. Ronald, M. and Atlas. (1995). "Bioremediation of petroleum pollutants", *International Biodeterioration & Biodegradation,* 311–321.

197. Yan, G.A., Jiang, J.W., Wu, G. and Yan, X. (1998). "Disappearance of linear alkylbenzene sulfonate from different cultures with *Anabaena* sp. HB 1017", *Bull Environ Contam Toxicol.,* 60, 329–334.
198. Radwan, S.S. and Al-Hasan, R.H. (2000). "Oil pollution and cyanobacteria", In: *The Ecology of Cyanobacteria* ed., Whitton, B.A., and Potts, M., The Netherlands: Kluwer Academic Publishers, pp. 307–319.
199. Mansy, A.E. and El-Bestway, E. (2002). "Toxicity and biodegradation of fluometuron by selected Cyanobacterial species", *World Journal of Microbiology and Biotechnology,* 18, 125–131.
200. Al-Hasan, R.H., Al-Bader, D.A., Sorkhoh, N.A. and Radwan, S.S. (1998). "Evidence for n-alkane consumption and oxidation by filamentous cyanobacteria from oil contaminated coasts of the Arabian Gulf", *Mar Biol.,* 130, 521–527.
201. Cerniglia, C.E., Gibson, D.T. and Baalen, C.V. (1979). "Algal oxidation of aromatic hydrocarbons: formation of 1-naphthol from naphthalene by *Agmenellum quadruplicatum,* strain PR-6", *Biochem Biophys Res Commun.,* 88, 50–58.
202. Cerniglia, C.E., Baalen, C.V.M and Gibson, D.T. (1980a). "Oxidation of biphenyl by the cyanobacterium, *Oscillatoria* sp., strain JCM" *Arch Microbiol.,* 125, 203–207.
203. Cerniglia, C.E., Gibson, D.T. and Baalen, C.V. (1980b). "Oxidation of naphthalene by cyanobacteria and microalgae", *J., Gen Microbiol.,* 116, 495–500.
204. Narro, M.L. (1985). "Oxidation of aromatic hydrocarbons by marine cyanobacteria" *Ph.D Thesis,* The University of Texas at Austin, Austin, TX.
205. Radwan, S.S., Al-Hasan, R.H., Salamah, S. and Al-Dabbous, S. (2002). "Bioremediation of oily sea water by bacteria immobilized in biofilms coating macroalgae", *Inter Biodeter Biodegrad,* 50, 55–59.
206. Abed, R.M.M. and Ko¨ster, J. (2005). "The direct role of aerobic heterotrophic bacteria associated with cyanobacteria in the degradation of oil compounds", *Int. Biodeterior Biodegrad,* 55, 29–37.
207. Sa´nchez, O., Diestra, E., Esteve, I. and Mas, J. (2005). "Molecular characterization of an oil-degrading cyanobacterial consortium". *Microb Ecol.,* 50, 580–588.
208. Hossein, A., Pejman, B., Kondiram, D. and Gorakh, P. (2014). "Potential use of cyanobacteria species in phycoremediation of municipal wastewater", *International Journal of Biosciences,* 4 (4), 105–111.
209. Kshirsagar, A.D. (2013). "Bioremediation of wastewater by using microalgae: an experimental study" *Int. J. Life Sc. Bt & Pharm. Res.,* 2 (3), 339–346.
210. Singh, N.K. and Patel, D.B. (2012). "Microalgae for bioremediation of distillery effluent", Farming for Food and Water Security Sustainable Agriculture Reviews, 10, 83–109.
211. Ganapathy Selvam, G., Baskaran, R. and Mohan, P.M. (2011). "Microbial diversity and bioremediation of distilleries effluent", *Journal of Research in Biology,* 3, 153–162.

13

Plant Growth Promoting Rhizobacteria and Their Potential Approach to Bioremediation of Toxic Metals

D.P. Samantaray, S. Mohapatra, A. Pradhan, A.K. Sethy and B.B. Mishra
Department of Microbiology, OUAT, Bhubaneswar, Odisha
E-mail: *dpsamantaray@yahoo.com*

ABSTRACT

Plant growth promoting rhizospheric bacteria are used for sustainable agriculture because they are potent for phosphate solubilization, production of HCN and catalase, production of indole acetic acid as well as toxic metal degradation, which enhances plant growth and productivity. Moreover, these bacterial species may prevent soil erosion on arid zones by improving growth of desert plants, which, in turn, leads to reduced dust pollution; on the other hand, PGPB promote plants that extract hazardous materials from soil. Rapid industrialization and modernization around the world have released toxic metals into the environment that adversely affects flora and fauna associated with them. Toxic metal contaminants accumulate in the agricultural soils to which they are applied, threaten food security and pose health risks to living organisms by their transfer within the food chain. Due to their persistent nature in soil, they may accumulate in both plants and animals over time. However, plant roots have limited ability to absorb toxic metals from soil, mainly because of high toxicity and low solubility. In order to overwhelm these problems, plant growth promoting rhizospheric bacteria are also employed for bioremediation of toxic metal from soil, sediment and water. PGPR may be accomplished by their direct effect on plant growth dynamics or indirectly by acidification, chelation, precipitation or immobilization of toxic metals in the rhizosphere. Thus, application of metal tolerant PGPR may be vital in enhancing the detoxification of toxic metal contaminated soils. These plant-microbe interactions can act as decontaminators by improving phytoremediation.

Keywords: Toxic Metal, Bioremediation and Phytoremediation.

13.1 INTRODUCTION

The world strives to move into population pressure, as well as royal food demand has been increased up to a great extent. The Indian land has gained the centre of attraction in the production of various foods during the centuries. Many researchers attempted and succeeded with their performance to protect the plants from the adverse effects of toxic metal with application of plant growth promoting bacteria as biofertilizer. Many of the

States had launched various industries for the production of biofertilizers especially for urban area, which are very much rich in metals. Soil pollution is a significant environmental problem and has been attracting considerable attention in recent years. The demand for food is growing worldwide, in both developed and rapidly developing countries like China and India. The plant growth stimulating agent biofertilizer and biocontrol agent biopesticides are termed PGPR (Fig. 13.1). Plant growth promoting bacteria (PGPB) are free-living, rhizospheric soil bacteria that are beneficial for plants [1]. PGPB promote plant growth in various ways, e.g., they directly affect the metabolism of the plants by providing substances that are usually in short supply. These bacteria are capable of fixing atmospheric nitrogen, solubilize phosphorus and iron and enhance production of plant growth hormones, improve the plant tolerance to stresses such as drought, high salinity, metal toxicity and pesticide load. One or more of these capabilities may contribute to an increased plant growth and development, which is greater than plants grown under standard cultivation conditions. Additionally, these bacteria are also used as biocontrol agents [2], which indirectly promote plant growth by preventing deleterious effects of phytopathogenic bacteria, fungi, nematodes and viruses.

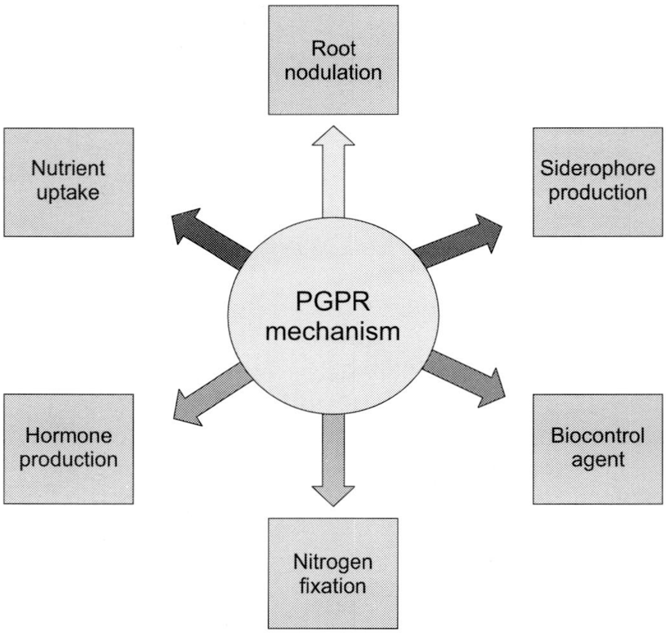

Fig. 13.1 Working principles of PGPR.

Many rhizospheric soil bacteria can stimulate the growth of crops, in the absence of a major pathogen, by directly affecting plant metabolism. These bacteria belong to diverse genera, including *Acetobacter, Achromobacter, Anabaena, Arthrobacter, Azoarcos, Azospirillum, Azotobacter, Bacillus, Burkholderia, Clostridium, Enterobacter, Flavobacterium, Frankia, Hydrogenophaga, Kluyvera, Microcoleus, Phyllobacterium, Pseudomonas, Serratia, Staphylococcus, Streptomyces Vibrio* and *Rhizobium*. In addition to their value as crop inoculants, the potential benefits of PGPB such as *Azospirillum* species have been used in bioremediation of wastewater treated with microalgae by increasing proliferation and metabolism of the microalgae [3, 4]. Although plants may suffer damage from excessive contact with toxic metals, they also require a range of toxic metals as essential micronutrients for normal growth and development. Among metals, some comprise elements that are essential for most redox reactions and are fundamental to normal cellular functions [5]. Thus metals stress condition results, a range of deficiency symptoms appear and growth is reduced. Notwithstanding, these toxic metals, when present in excessive amounts, interfere with cellular functions and alter normal metabolic processes to produce cellular injuries and potentially plant death [6] by displacing essential components in biomolecules, blocking essential biological functions of molecules and modifying enzyme and membrane transport function.

These toxic metals also generate oxidative stress that is mediated through generation of free radicals [7]. Some toxic metals directly affect biochemical and physiological processes through inhibition of photosynthesis and respiration leading to reduced growth [8]. Therefore, excessive toxic metal accumulation in plants may induce toxicity by modifying essential protein structure or replacing essential elements, which is manifested by chlorosis, growth impairment, browning of roots and inactivation of photosystems [9, 10]. Some microbial communities have the ability to sequester toxic metals and therefore may be useful for bioremediation of contaminated areas [11, 12]. Moreover, plant-associated bacteria can be potentially used to improve phytoextraction activities by altering the solubility, availability and transport of toxic metals and nutrients as well, by reducing soil pH and releasing chelators [13]. Among the metabolites produced by PGPR, siderophores play a significant role in metal mobilization and accumulation [14].

13.2 PGPR AND BIOREMEDIATION

PGPR are divided into two groups, such as plant growth promotors as biofertilizer [15] and biocontrol agent as biopesticides [16]. PGPR can increase the availability of nutrients by enzymatic nutrient mobilization from organic matter and production of siderophores [17]. Bacteria producing extracellular degrading enzymes are major decomposers of organic matter. They contribute to the soil aggregation and nutrient availability [18]. PGPR can also promote the root growth. This can be caused by the ability of most rhizobacteria to produce phytohormones like indole-3-acetic acid, cytokinins, gibberellins, and ethylene which promote cell division and cell enlargement, extension of plant tissue and other morphological changes of roots [19]. Plant growth is also dependent upon soil fertility; however the fertility percentage is very less in toxic metal contaminated sites due to rapid industrialization or anthropogenic activities. Thus, these contaminated sites need detoxification or remediation by PGPR prior to crop plantation. Bioremediation is a biological process to detoxify toxic metals by means of biological sources present in the environment. It has obvious advantages over physicochemical remediation methods due to several merits: cost-effective, convenient, complete degradation of organic pollutants and no collateral destruction of the site material or its indigenous flora and fauna [20].

Extensive use of PGPR for remediation of toxic metals such as cadmium, chromium, copper, lead, mercury, nickel and zinc is the most promising approach of bioremediation.

13.2.1 Bioremediation of Organic Contaminants by PGPR

In contrast with inorganic compounds, microorganisms can degrade and even mineralize organic compounds in association with plants [21]. Hence discovery of effective pathways for degradation and mineralization of organic compounds may play an important role in the future. Application of PGPR increased the removal of organic pollutant [22, 23] such as polycyclic aromatic hydrocarbon and creosote, probably by enhancing plants germination and survival in soils that were heavily contaminated and by stimulating the plants to grow faster and accumulate more root biomass. Ethylene is an important plant growth factor, while excessive ethylene promoted by stresses can depress growth [24]. PGPR also have a positive impact on plant growth by consuming amino cyclopropanecarboxylic acid (ACC), the immediate precursor of ethylene, through synthesis of 1-aminocyclopropane-1-carboxylate deaminase (ACC deaminase) to decrease the ethylene production in stressed plants [25]. The rhizosphere seems to be a promising environment for the bioremediation of contaminated soils, but many bacteria capable of degrading certain kinds of organic pollutants cannot survive and achieve bioremediation in the soil environment, because they are not competitive enough compared to other indigenous organisms. Mean while, many bacteria that are robust in the rhizosphere do not show or show only limited ability in degrading organic pollutants. With the development of molecular biology, the genetically modified rhizobacteria having the contaminant-degrading gene are constructed to conduct the bioremediation in rhizosphere.

The molecular mechanisms of some pollutants such as trichloroethylene [26] and polychlorinated biphenyls [27], have been studied clearly. Another crucial problem to be solved is to select a suitable strain for genre combination and inoculation into the rhizosphere. The following criteria should be considered; (a) the strain should be stable after cloning and the target gene should have a high expression; (b) the strain should be tolerant to the toxic metal and can survive in rhizospheric regions [27, 28].

Table 13.1 Bioremediation of organic contaminants by PGPR

Bacteria	*Plant*	*Compound*	*Role of PGPR*	*References*
Pseudomonas fluorescens 279	Wheat	Trichloroethylene (TCE)	Degradation of TCE	[28]
Azospirillum lipoferum strains 15	Wheat	Crude oil	Promote development of root system	[29]
Pseudomonas fluorescens F113	Alfalfa	Polychlorinated biphenyls	Metabolization of PCBs	[30]
Pseudomonas putida Flav 1-1	*Arabidopsis*	Polychlorinated biphenyls	Utilization of secondary metabolites	[31]

Contd...

Contd...

Enterobacter cloacae CAL2	Tall fescue	Petroleum hydrocarbons	Promoted plant growth in presence of PHs (Petroleum hydrocarbons)	[32]
Azospirillum brasilense Cd	Tall fescue	Polycyclic aromatic hydrocarbons	Increased plant tolerance to PAHs (Polycyclic aromatic hydrocarbons)	[22]
Enterobacter cloacae CAL 2	Tall fescue	Polycyclic aromatic hydrocarbons	Promoted plant growth under stress condition	[23]

13.2.2 Bioremediation of Toxic Metals by PGPR

The toxic metal contamination has great effects on the microbial communities in soils in several ways such as reduction of total microbial biomass [33], decreased numbers of specific populations [34], and shifts in the microbial community structure [35]. The rhizosphere, with high concentration of nutrients exuded from the roots, attracts more bacteria than the bulk soils. These bacteria in turn facilitate the growth of the plant. Recently, this phytobacteria system has proved to be more effective in removing toxic metals (Table 13.2) than its ingredients. Moreover, PGPR can be categorized into two groups based on their residing sites: iPGPR, the symbiotic bacteria, which live inside the plant cells produce nodules, and ePGPR, the free-living rhizobacteria, which live outside the plant cells and do not produce nodules, but still prompt the plant growth. The best known iPGPR are rhizobia, which are restricted to nodules in leguminous plants [36]. However, a wild-type strain of *Azospirillum brasilense* Sp245 has been proven to be capable of colonizing both interior and exterior of the plant root, while other strains can colonize the root surface only [37]. They compared the responses of Sp245 and other strains to several toxic metals like Co^{2+}, $Cu^{2+}$and Zn^{2+} and observed that the toxic metal uptake by endophytic strain Sp245 is higher than the nonendophytic. They supposed

Table 13.2 Bioremediation of toxic metals by PGPR

Bacteria	*Plant*	*Toxic metal*	*Role of PGPR*	*References*
Azotobacter chroococcum HKN-5	Brassica	Lead, zinc	Stimulated plant growth	[39]
Bacillus megaterium HKP-1	Brassica	Lead, zinc	Protected plant from toxicity	[39]
Bacillus subtilis SJ-101	Brassica	Nickel	Facilitated Ni accumulation	[40]
Kluyvera ascorbata SUD165	Indian mustard	Lead, zinc	Decreased plant growth inhibition by toxic metals	[41]
Kluyvera ascorbata SUD165/26	Canola	Nickel, lead, zinc	No increase of metal uptake	[41]
Mesorhizobium huakuii sub sp. rengei B3	Astragalus sinicus	Cadmium	Increased ability of cells to bind Cd^{2+}	[43]

the dissimilarities in their behaviour may be due to different adaptation abilities of these strains to the stress conditions owing to their different ecological status.

Toxic metal pollutants should be either completely removed from the contaminated soil or transformed or immobilized in ways that render them safe. For survival under metal-stressed environment, PGPR have developed a range of mechanisms by which they can immobilize, mobilize or transform toxic metals, thereby rendering them inactive [38]. The various mechanisms are exclusion metal ions that are kept away from target sites, extrusion metals that are pushed out of the cell through chromosomal or plasmid mediated events, accommodation metals that form complexes with metal binding proteins (metallothioneins), biotransformation, in which the toxic metal is reduced to less toxic forms and methylation and demethylation.

13.2.3 PGPR for Enhancing Phytoremediation

Several industrial operations such as smelting, mining, metal forging, manufacturing of alkaline storage batteries and combustion of fossil fuels release toxic metals into the soil [42]. In soil, these metals exist in both bioavailable and non-bioavailable forms [44] and their mobility depends on whether they precipitate in soil as positively charged ions or are associated with negatively charged salt components of soil. The toxicity of the noxious metals within soils of high CEC is generally low, even when metal concentrations are high [45]. However, under oxidizing and aerobic conditions, metals are usually present as soluble cationic forms, whereas under reducing or anaerobic conditions they exist as sulphide or carbonate precipitates. At low soil pH, bioavailability increases due to the presence of metals as free ionic species, and just the opposite occurs when soil pH is high. The soil mobility and bioavailability of the following metals usually proceed in this order: Zn > Cu > Cd > Ni [46]. When metals coexist with contaminants, the rate of absorption and accumulation into food webs and ultimately into animal and human diets may be accentuated. Phytoremediation is an in situ bioremediation process that uses green plants and microorganisms associated with them to extract, sequester or detoxify pollutants. Plants have the capacity to take up, accumulate, degrade or eliminate metals, pesticides, solvents, crude oil and many industrial contaminants. Phytoremediation is a clean, cost-effective, environment-friendly technology, especially for treating large and diffused areas that are contaminated. There are many successful examples where phytoremediation has been employed and where it has been documented to work well for remediating contaminated industrial environments [47]. Depending on the method used and nature of the contaminant involved, phytoremediating areas where toxic metals and other inorganic compounds exist, utilize various methods such as phytoextraction, phytostabilization, phytostimulation, phytovolatilization, rhizovolatilization, phytodegradation and rhizofiltration respectively [48].

13.2.4 PGPR Activity in Phytoextraction

Phytoextraction also referred to as phytoaccumulation, phytoabsorption, and phytosequestration, is an in situ technology that uses plants to remove contaminants from the environment by concentrating them within the biomass of the plant. The process has been studied mostly for extraction of toxic metals from soils and water containing moderate concentrations of contaminants. PGPB improve rates of phytoextraction either by enhancing the growth of plant or increasing metal mobilization by microbial metabolites [49]. The PGPB *Kluyvera ascorbata* can provide protection to plants against metal toxicity due to the presence of enzyme ACC deaminase. Which protects Indian mustard *B. juncea* and rapeseed *B. campestris* against nickel, lead and zinc toxicity probably by reducing the stress caused by high ethylene content. Root elongation of rapeseed *Brassica napus* is also stimulated by the IAA that is synthesized by PGPB in soil contaminated with cadmium [50] by unidentified rhizobacterium on Indian mustard roots [51]. In cadmium and lead contaminated soil, *Arthrobacter mysorens* can improve plant growth in the rhizosphere of barley and prevent accumulation of toxic metal in the plant tissues. Several *Bacillus* strains isolated from chromium contaminated soil reduced the highly toxic, mutagenic, and carcinogenic Cr^{6+} to the less toxic Cr^{3+}[52]. Similarly, *Rhodococcus erythropolis,* isolated from a metal contaminated site in the Himalayas, had plant growth promoting traits and an ability to reduce chromate at low temperatures of approximately 10°C [53].

CONCLUSION

In this chapter, we briefly described the potentiality of PGPR with respect to bioremediation of toxic metal. Plant roots have limited ability to absorb these metals from soil because of high toxicity and low solubility in the soil solution. The phytoavailability of metal is closely tied to the soil properties and the metabolites that are released by PGPR. The role played by PGPR may be accomplished by their direct effect on plant growth dynamics or indirectly by acidification, chelation, precipitation or immobilization of toxic metals in the rhizosphere. Phytoremediation of contaminated lands assisted by PGPB is a promising avenue to overcome the limitation imposed on using plants alone that until now did not yield commercial application despite the large volume of knowledge on that topic. The recent research of PGPR on the remediation of contaminated soils shows a brilliant prospect for the successive studies. However, combined use of PGPR and specific contaminant-degrading bacteria can successfully remove toxic metals.

REFERENCES

1. Bashan, Y. and de-Bashan., L.E. (2005). Bacteria/plant growth-promotion. In *Encyclopedia of Soils in the Environment*. Vol. 1, D., Hillel, pp. 103–115. Elsevier, Oxford, UK.
2. Bashan, Y. and Holguin, G. (1998). Proposal for the division of Plant Growth-Promoting Rhizobacteria into two classifications: biocontrol-PGPB (Plant Growth-Promoting Bacteria) and PGPB. *Soil Biol. Biochem.*, 30, 1225–1228.
3. de-Bashan, L.E., Hernandez, J.P., Morey, T. and Bashan, Y. (2004). Microalgae growth-promoting bacteria as "helpers" for microalgae: a novel approach for removing ammonium and phosphorus from municipal wastewater. *Water Res.*, 38, 466–474.
4. Hernandez, J.P., de-Bashan, L.E. and Bashan, Y. (2006). Starvation enhances phosphorus removal from wastewater by the microalga *Chlorella* spp. co-immobilized with *Azospirillum brasilense. Enzyme Microb. Technol.*, 38, 190–198.

5. Sirko, A. and Brodzik, R. (2000). Plant ureases: roles and regulation. *Acta Biochim Pol*, 47, 1189–1195.
6. Ochiai, E.I. (1977). *Bioinorganic Chemistry: An Introduction.* Allyn and Bacon, Boston, pp. 218–262.
7. Seth, C.S., Chaturvedi, P.K. and Misra, V. (2008). The role of phytochelatins and antioxidants in tolerance to Cd accumulation in *Brassica juncea* L. *Ecotoxicol Environ Saf.*, 71, 76–85.
8. Vangronsveld, J. and Clijsters, H. (1994). Toxic effects of metals. In: Farago MG (ed.) *Plants and the Chemical Elements.* VHC-Verbgsgesellschaft, Weinheim, Germany, pp. 149.
9. Shaw, B.P., Sahu, S.K. and Mishra, R.K. (2004). Heavy metal induced oxidative damage in terrestrial plants. In: Prasad MNV (ed.) *Heavy Metal Stress in Plants: from Biomolecules to Ecosystems.* Narosa Publishing House, New Delhi, India, pp. 84–126.
10. Gorhe, V. and Paszkowski, U. (2006). Contribution of arbuscular mycorrhizal symbiosis to heavy metal phytoremediation. *Planta*, 223, 1115–1122.
11. Hallberg, K.B. and Johnson, D.B. (2005). Microbiology of a wetland ecosystem constructed to remediate mine drainage from a heavy metal mine. *Sci Total Environ*, 338, 53–66.
12. Umrania, V.V. (2006). Bioremediation of toxic heavy metals using acidothermophilic autotrophs. *Bioresour Technol*, 97, 1237–1242.
13. Ma, Y., Prasad, M.N.V., Rajkumar, M. and Freitas, H. (2011). Plant growth promoting rhizobacteria and endophytes accelerate phytoremediation of metalliferous soils. *Biotechnol Adv*, 29, 248–258.
14. Rajkumar, M., Ae, N., Prasad, M.N.V. and Freitas, H. (2010). Potential of siderophore-producing bacteria for improving heavy metal phytoextraction. *Trends Biotechnol*, 28, 142–149.
15. Vessey, J.K. (2003). Plant growth promoting rhizobacteria as biofertilizers. *Plant Soil*, 255, 571–586.
16. Whipps, J.M. (2001). Microbial interactions and biocontrol in the rhizosphere. *J Exp Bot*, 52, 487–511.
17. Anderson, T.A., Guthrie, E.A. and Walton, B.T. (1993). Bioremediation in the rhizosphere: plant roots and associated microbes clean contaminated soil. *Environ Sci Technol*, 27, 2630–2636.
18. Johansen, J.E. and Binnerup, S.J. (2002). Contribution of *Cytophaga*-like bacteria to the potential of turnover of carbon, nitrogen, and phosphorus by bacteria in the rhizosphere of barley (*Hordeumvulgare* L.). *Microb Ecol*, 43, 298–306.
19. Salisbury, F.B. (1994). The role of plant hormones. In: Wilkinson RE (ed.) *Plant Environment Interactions.* Marcel Dekker, New York, USA, pp. 39–81.
20. Timmis, K.N. (1999). Pieper DH. Bacteria designed for bioremediation. *Trends Biotechnol*, 1999:17:200–4.

21. Saleh, S., Huang, X.D., Greenberg, B.M. and Glick, B.R. (2004). Phytoremediation of persistent organic contaminants in the environment. In: Singh A., Ward O., editors. Soil Biology: Vol. 1. *Applied Bioremediation and Phytoremediation*. Berlin: Springer-Verlag; pp. 115–34.

22. Huang, X.D., El-Alawi, Y., Penrose, D.M., Glick, B.R. and Greenberg, B.M. (2004a). Responses of three grass species to creosote during phytoremediation. *Environ Pollut*, 130, 453–63.

23. Huang, X.D., El-Alawi, Y., Penrose, D.M., Glick, B.R. and Greenberg, B.M. (2004b). A multiprocess phytoremediation system for removal of polycyclic aromatic hydrocarbons from contaminated soils. *Environ Pollut*, 130, 465–76.

24. Morgan, P.W. and Drew, C.D. (1997). Ethylene and plant responses to stress. *Physiol Plant*, 100, 620–30.

25. Reed, M.L.E., Warner, B. and Glick, B.R. (2005). Plant growth-promoting bacteria facilitate the growth of the common reed *Phragmites australis* in the presence of copper or polycyclic aromatic hydrocarbons. *Curr Microbiol*, 51, 425–9.

26. Bradford, M.M. (1976). A rapid and sensitive method for the quantitation of microgram quantities of protein utilizing the principle of protein-dye binding. *Anal Biochem*, 72, 248–54.

27. Brazil, G.M., Kenefick, L., Callanan, M., Haro, A., de-Lorenzo, V. and Dowling, D.N. (1946). Construction of a rhizosphere pseudomonad with potential to degrade polychlorinated biphenyls and detection of bph gene expression in the rhizosphere. *Appl Environ Microbiol*, 61, 1946–52.

28. Yee, D.C., Maynard, J.A. and Wood, T.K. (1998). Rhizoremediation of trichloroethylene by a recombinant, root-colonizing *Pseudomonas fluorescens* strain expressing toluene ortho-monooxygenase constitutively. *Appl Environ Microbiol*, 64, 112–8.

29. Muratova, A.Y., Turkovskaya, O.V., Antonyuk, L.P., Makarov, O.E., Pozdnyakova, L.I. and Ignatov, V.V. (2005). Oil-oxidizing potential of associative rhizobacteria of the genus *Azospirillum*. *Microbiology*, 74, 210–5.

30. Villacieros, M., Whelan, C., Mackova, M., Molgaard, J., Sanchez-Contreras, M. and Lloret, J. (2005). Polychlorinated biphenyl rhizoremediation by *Pseudomonas fluorescens* F113 derivatives, using a *Sinorhizobium meliloti* nod system to drive bph gene expression. *Appl Environ Microbiol*, 71, 2687–94.

31. Narasimhan, K., Basheer, C., Bajic, V.B. and Swarup, S. (2003). Enhancement of plant-microbe interactions using a rhizosphere metabolomics-driven approach and its application in the removal of polychlorinated biphenyls. *Plant Physiol*, 132, 146–53.

32. Huang, X.D., El-Alawi, Y., Gurska, J., Glick, B.R. and Greenberg, B.M.A. (2005). Multi-process phytoremediation system for decontamination of persistent total petroleum hydrocarbons (TPHs) from soils. *Microchem J.*, 81, 139–47.

33. Fliessbach, A., Martens, R. and Reber, H.H. (1994). Soil microbial biomass and microbial activity in soils treated with heavy metal contaminated sewage sludge. *Soil Biol Biochem*, 26, 1201–5

34. Chaudri, A.M., McGrath, S.P., Giller, K.E., Rietz, E. and Sauerbeck, D.R., Enumeration of indigenous *Rhizobium leguminosarum biovar trifolii* in soils previously treated with metal-contaminated sewage sludge. *Soil Biol Biochem,* 25, 301–9

35. Gray, E.J. and Smith, D.L. (2005). Intracellular and extracellular PGPR: commonalities and distinctions in the plant-bacterium signaling processes. *Soil Biol Biochem,* 37, 395–412

36. Sriprang, R., Hayashi, M., Ono, H., Takagi, M., Hirata, K. and Murooka, Y. (2003). Enhanced accumulation of Cd^{2+} by a *Mesorhizobium* sp. transformed with a gene from *Arabidopsis thaliana* coding for phytochelatin synthase. *Appl Environ Microbiol,* 69, 1791–6

37. Kamnev, A.A., Tugarova, A.V., Antonyuk, L.P., Tarantilis, P.A., Polissiou, M.G. and Gardiner, P.H.E. (2005). Effects of heavy metals on plant-associated rhizobacteria: comparison of endophytic and non-endophytic strains of *Azospirillum brasilense. J Trace Elem Med Biol,* 19, 91–5

38. Nies, D.H. (1999). Microbial heavy metal resistance. *Appl Microbiol Biotechnol,* 51, 730–750

39. Wu, S.C., Cheung, K.C., Luo, Y.M. and Wong, M.H. (2006b). Effects of inoculation of plant growth-promoting rhizobacteria on metal uptake by *Brassica juncea. Environ Pollut,* 140, 124–35

40. Zaidi, S., Usmani, S., Singh, B.R. and Musarrat, J. (2006). Significance of *Bacillus subtilis* strain SJ-101 as a bio inoculant for concurrent plant growth promotion and nickel accumulation in *Brassica juncea. Chemosphere,* 64, 991–7

41. Burd, G.I., Dixon, D.G. and Glick, B.R. (2004). Plant growth-promoting bacteria that decrease heavy metal toxicity in plants. *Can J Microbiol,* 46, 237–45

42. Giller, K.E., McGrath, S.P. and Hirsch, P.R. (1989). Absence of nitrogen fixation in clover grown on soil subject to long-term contamination with heavy metals is due to survival of only ineffective *Rhizobium. Soil BiolBiochem,* 21, 841–848

43. McGrath, S.P., Chaudri, A.M. and Giller, K.E. (1995). Long-term effects of metals in sewage sludge on soils, microorganisms and plants. *J Ind Microbiol,* 14, 94–104

44. Sposito, F.G. (2000). The chemistry of soils. In: Maier R.M., Pepper IL, Gerba CB (eds.) *Environmental Microbiology.* Academic, Press, London, pp. 406

45. Roane, T.M. and Pepper, I.L. (2000). Microorganisms and metal pollution. In: Maier RM, Pepper IL, Gerba CB (eds.) *Environmental Microbiology.* Academic, Press, London, pp. 55

46. Lena, Q.M. and Rao, G.N. (1997). Heavy metals in the environment. *J Environ Qual.* 26, 264

47. Suresh, B. and Ravishankar, G.A. (2004). Phytoremediation— a novel and promising approach for environmental clean-up. *Crit Rev Biotechnol,* 24, 97–124

48. Glick, B.R., Cheng, Z., Czarny, J. and Duan, J. (2007). Promotion of plant growth by ACC deaminase containing soil bacteria. *Eur J Plant Pathol,* 119, 329–339.

49. Lebeau, T., Braud, A. and Jezequel, K. (2008). Performance of bioaugmentation-assisted phytoextraction applied to metal contaminated soils: a review. *Environ. Pollut.*, 153, 497–522.
50. Burd, G.I., Dixon, D.G. and Glick, B.R. (1998). A plant growth-promoting bacterium that decreases nickel toxicity in seedlings. *Appl. Environ. Microbiol*, 64, 3663–3668.
51. Belimov, A.A., Kunakova, A.M. and Gruzdeva, E.V. (1998). Influence of soil pH on the interaction of associative bacteria with barley. *Mikrobiologiya*, 67, 561–568.
52. Khan, M.S., Zaidi, A., Wani, P.A. and Oves, M. (2009). Role of plant growth promoting rhizobacteria in the remediation of metal contaminated soils. Environ. *Chem. Lett.*, 7, 1–19.
53. Trivedi, P., Pandey, A. and Sa, T. (2007). Chromate reducing and plant growth promoting activities of psychrotrophic *Rhodococcus erythropolis* MTCC 7905. *J. Basic Microbiol.*, 47, 513–517.

14

Potential Characteristic Features of Biofertilizers

Sandip Das and Rabindranath Bhattacharyya*

Department of Biological Sciences, Presidency University, Kolkata, West Bengal, India

E-mail: *rabindranath.dbs@presiuniv.ac.in*

ABSTRACT

With the progression of technology, man has become more and more conscious of the scientific endeavours in relation to human welfare. It has now become clear that synthetic chemicals used in the agriculture have a harmful impact on the ecosystem, but still the chemicals rule the agro industry. Although a century has passed since the development of the first bioinoculant, but as compared to the general maturity to mankind, bioinoculants still lag far behind. Biofertilizers are one of the best tools of our modern agricultural science. Biofertilizers are applied in the agricultural field as a replacement to our conventional fertilizers. In the present chapter, we will discuss the history and the future of bioinoculants with special reference to rhizobia inoculants and its constraints.

Keywords: Microorganisms, Bioinoculants, Biofertilizer, Rhizobia, Constraints.

14.1 INTRODUCTION

Organic farming has emerged as a significant priority area globally in sight of the growing demand for safe and healthy food and long-term sustainability and concerns on environmental pollution associated with unsystematic use of agrochemicals. Indiscriminate use of synthetic fertilizers has led to the pollution and contamination of the soil, has polluted water basins, destroyed microorganisms and friendly insects, making the crop more prone to diseases and reduced soil fertility. Though the use of chemical inputs in agriculture is inevitable to meet the growing demand for food in world, there are opportunities in selected crops and niche areas where organic production can be encouraged to tape the domestic export market. Biofertilizers play a very significant role in improving soil fertility by fixing atmospheric nitrogen, both in association with plant roots and without it, solubilizing insoluble soil phosphates and producing plant growth substances in the soil. They are in fact being promoted to harvest the naturally available, biological system of nutrient mobilization [15]. In context of both the cost and environmental impact of chemical fertilizers, excessive dependence on the chemical fertilizers is not a viable strategy in long run because of the cost, both in domestic resources and foreign exchange, involved in setting up of fertilizer plants and sustaining the production. In this context, organic manures (biofertilizers) would be the viable option for farmers to increase productivity per unit area. Due to various constraints the progress in the field of biofertilizer production technology always remained below satisfaction in Asia [8].

What is Biofertilizer?

The term biofertilizer or 'microbial inoculants' can be generally defined as a preparation containing live or latent cells of efficient strains of nitrogen fixing, phosphate solubilizing or cellulolytic microorganisms used for application in seed, soil or composting areas with the objective of increasing the numbers of such microorganisms and accelerating certain microbial processes to augment the extent of the availability of nutrients in a form which can be assimilated by plant [9]. The term may be used to include all organic resources for plant growth which are rendered in an available form for plant absorption through microorganisms or plant associations or interactions.

Knowledge of applied microbial inoculums as biofertilizer is a long history which passes from generation to generation of farmers. It started with culture of small-scale compost production that has evidently proved the ability of biofertilizer. This is recognized when the cultures accelerate the decomposition of organics residues and agricultural by-products through various processes and gives healthy harvest of crops [1]. In Malaysia, the use of industrial scale microbial inoculants started in the late 1940s. *Bradyrhizobium* inoculant as biofertilizer peaked up in 1970s. Government research institute, the Malaysian Rubber Board (MRB), has been conducting research on *Rhizobium* inoculums for leguminous cover crops in the inter rows of young rubber trees in the large plantations. Apart from that, University Putra Malaysia (UPM) also has conducted many experiments since 1980s on mycorrhiza and initiated the research to evaluate the contribution of nitrogen from *Azospirillum* to oil palm seedlings. Mycorrhiza inoculums are the biofertilizers that are increasingly being utilized and accepted in agriculture industry of Malaysia. Biofertilizers are produced on a large scale mainly for supplying nutrient, amelioration of toxic effect in soils, root pest and disease control, improved water usage and soil fertility [1]. Since the substrates for inoculate are abundant such as the mine sands and agricultural wastes, the production cost is cheaper and it is environmentally safe. Biofertilizer perceived to be more expensive than the chemical fertilizers due to the lack of skills and technology to produce biofertilizer products from abundant wastes. Besides, the effect on the crops is slow, compared to chemical fertilizers. As biofertilizers contain living organisms, their performance therefore depends on surrounding. So, special care, such as storage or mixing with powders, is also needed to handle microbial inocula to make them remain effective for extended use. Different constraints of biofertilizers like short shelf life, lack of suitable carrier materials, susceptibility to high temperature, problems in transportation and storage still need to be solved in order to obtain effective inoculation.

Why to Explore Biofertilizers

1. **Demand is much higher than the availability:** It is estimated that by 2020, to achieve the targeted production of 321 million tonnes of foodgrain, the requirement of nutrients will be 28.8 million tonnes, while their availability will be only 21.6 million tonnes being a deficit of about 7.2 million tonnes [12].
2. **Growing concern about environmental hazards:** Indiscriminate use of chemical fertilizers, however, has deteriorated the soil fertility, destroyed soil microbial activity, disturbed environmental balance and ecological soundness.
3. **Depleting feedstock/fossil fuels (energy crisis) and increasing cost of fertilizers:** This is becoming unaffordable by small and marginal farmers.
4. **Depleting soil fertility:** Due to widening gap between nutrient removal and supplies.
5. **Increasing threat to sustainable agriculture.**

Besides above facts, the long term use of biofertilizers is economical, eco-friendly, more efficient, productive and accessible to marginal and small farmers over chemical fertilizers [14].

14.2 IMPORTANT FEATURES OF SOME BIOFERTILIZERS

(i) Nitrogen fixers, *Rhizobium*: *Rhizobium* belongs to the family Rhizobiaceae, is symbiotic in nature, and fixes nitrogen 50-100 kg/ha in association with legumes only. It is useful for pulse legumes like chickpea, red gram, pea, lentil, black gram, etc., oilseed legumes like soybean and groundnut and forage legumes like berseem and lucerne. Successful nodulation of leguminous crops by *Rhizobium* basically depends on the availability of compatible strain for a particular legume. It colonizes the roots of specific legumes to form tumour-like growths called root nodules, which act as factories of ammonia production. *Rhizobium* has an ability to fix atmospheric nitrogen in symbiotic association with legumes and certain non-legumes like *Parasponia*. Population of *Rhizobium* in the soil depends on the presence of legume crops in the field. In absence of legumes, the population decreases. Artificial seed inoculation is often needed to restore the population of effective strains of the *Rhizobium* near the rhizosphere to hasten nitrogen fixation. For effective nodulation each legume requires a specific species of *Rhizobium* [15].

(ii) *Azospirillum*: *Azospirillum* is heterotrophic and associative in nature. Its nitrogen-fixing ability is about 20-40 kg/ha, and it also produces growth regulating substances. Although there are many species under this genus like *A. amazonense, A. halopraeferens, A. brasilense*, but global distribution and benefits of inoculation have been proved mainly with *A. lipoferum* and *A. brasilense*. *Azospirillum* form associative symbiosis with many plants, particularly those having the C4-dicarboxylic pathway of photosynthesis, because they grow and fix nitrogen on salts of organic acids such as malic, aspartic acid [2]. Thus, it is mainly recommended for maize, sugar cane, sorghum, pearl millet, etc.

The *Azospirillum* colonizing the roots not only remains on the root surface, but a sizeable proportion of them also penetrate into the root tissues and lives in harmony with the plants. They do not, however, produce any visible nodules or out growth on root tissue.

(iii) *Azotobacter:* This bacterium belongs to the family Azotobacteriaceae, and is aerobic, free living, and heterotrophic in nature. Azotobacters are present in neutral or alkaline soils and *A. chroococcum* is the most commonly occurring species in arable soils. *A. vinelandii, A. beijerinckii, A. insignis* and *A. macrocytogenes* are other reported species. The number of *Azotobacter* rarely exceeds 104 to 105 per gram of soil due to lack of organic matter and presence of antagonistic microorganisms in soil. The bacterium produces anti-fungal antibiotics, which inhibit the growth of several pathogenic fungi in the root region, thereby preventing seedling mortality to a certain extent. The population of *Azotobacter* is generally low in the rhizosphere of the crop plants and in uncultivated soils. The occurrence of this organism has been reported from the rhizosphere of a number of crop plants such as rice, maize, sugar cane, bajra, vegetables and plantation crops [12].

(iv) Blue-Green Algae (Cyanobacteria) and *Azolla:* These belong to eight different families. These are phototrophic in nature and produce auxin, indole acetic acid and Gibberellic acid, and have nitrogen-fixing ability of 20-30 kg/ha in submerged rice fields. They are referred as "paddy organisms" as they are abundant in paddy. Nitrogen is the key input required in large quantities for low land rice production. Soil nitrogen and biological nitrogen fixation by associated organisms are major sources of nitrogen (N_2) for low land rice. The 50-60% N_2 requirement is met through the combination of mineralization of soil organic N_2 and BNF by free living and rice plant associated bacteria [11]. To achieve food security through sustainable agriculture, the requirement for fixed nitrogen must be increasingly met by biological nitrogen fixation rather than by industrial nitrogen fixation. Blue-green algae forms symbiotic association capable of fixing nitrogen with fungi, liverworts, ferns and flowering plants, but the most common symbiotic association has been found between a free floating aquatic fern, the *Azolla* and *Anabaena azollae* (BGA) (Figs. 14.1 and 14.2). *Azolla* contains 4-5% N_2 on dry basis and 0.2-0.4% on wet basis and can be the potential source of organic manure and nitrogen in rice production. The important factor in using *Azolla* as biofertilizer for rice crop is its quick decomposition in the soil and efficient availability of its nitrogen to rice plants. Besides nitrogen fixation, these biofertilizers also contribute significant amounts of different macronutrients, viz., P, K, S, Zn, Fe, Mb and other micronutrients. The fern forms a green mat over water with a branched stem, deeply bilobed leaves and roots. The dorsal fleshy lobe of the leaf contains the algal symbiont within the central cavity. *Azolla* can be applied as green manure by incorporating in the fields prior to rice planting. The most common species *A. pinnata* can be propagated on commercial scale by vegetative means in India. It may yield on an average about 1.5 kg/meter^2 in a week. India has recently introduced some species of *Azolla* for their large biomass production; these include *A. caroliniana, A. microphylla, A. filiculoides* and *A. mexicana.*

(v) Phosphate solubilizers: The soil bacteria belonging to the genera *Pseudomonas* and *Bacillus* and Fungi are more common in this respect. Several reports have examined the ability of different bacterial species to solubilize insoluble inorganic phosphate

Fig. 14.1 Cultivation of BGA.

Fig. 14.2 Cultivation of *Azolla.*

compounds, such as tricalcium phosphate, dicalcium phosphate, hydroxyapatite, and rock phosphate. Among the bacterial genera with this capacity are *Pseudomonas, Bacillus, Rhizobium, Burkholderia, Achromobacter, Agrobacterium, Microccocus, Aereobacter, Flavobacterium* and *Erwinia*. There are considerable populations of phosphate solubilizing bacteria in soil and in plant rhizospheres. These include both aerobic and anaerobic strains, with a prevalence of aerobic strains in submerged soils. A considerably higher concentration of phosphate solubilizing bacteria is commonly found in the rhizosphere in comparison with non-rhizosphere soil.

(vi) Phosphate absorbers (Mycorrhiza): The term "Mycorrhiza" denotes "fungus roots". It is a symbiotic association between host plants and certain group of fungi at the root system, in which the fungal partner is benefited by obtaining its carbon requirements from the photosynthates of the host, and the host in turn is benefited by obtaining the much needed nutrients, especially phosphorus, calcium, copper, zinc, etc., which are otherwise inaccessible to it, with the help of the fine absorbing hyphae of the fungus. These fungi are associated with majority of agricultural crops, except with those crops belonging to families of Chenopodiaceae, Amaranthaceae, Caryophyllaceae, Polygonaceae, Brassicaceae, Commelinaceae, Juncaceae and Cyperaceae [12].

(vii) Zinc solubilizers: The nitrogen fixers like *Rhizobium, Azospirillum, Azotobacter*, BGA and phosphate solubilizing bacteria like *Bacillus magaterium, Pseudomonas striata*, and phosphate mobilizing mycorrhiza have been widely accepted as biofertilizers. However, these supply only major nutrients, but a host of microorganisms that can transform micronutrients are present in soil that can be used as biofertilizers to supply micronutrients like zinc, iron, copper, etc. The zinc can be solubilized by microorganisms such as *Bacillus subtilis, Thiobacillus thioxidans* and *Saccharomyces* sp. These microorganisms can be used as biofertilizers for solubilization of fixed micronutrients like zinc. Results show that a *Bacillus* sp. (Zn solubilizing bacteria) can be used as a biofertilizer for zinc, or in soils where native zinc is higher or in conjunction with insoluble cheaper zinc compounds like zinc oxide, zinc carbonate and zinc sulphide instead of costly zinc sulphate [8].

14.3 POTENTIAL ROLE OF BIOFERTILIZERS IN AGRICULTURE

Biofertilizers play a major role in improving soil fertility, and thereby enhance final yield. It improves soil biota and minimizes the sole use of chemical fertilizers. Under temperate conditions, inoculation of *Rhizobium* improved number of pods per plant, number of seed per pod and 1000-seed weight (g), and thereby yield over the control. It is an established fact that the efficiency of phosphate fertilizers is very low (15-20%) due to their fixation in different soil pH, and unfortunately both soil types are predominating in India, accounting for more than 34% acidity affected and more than seven million hectares of productive land salinity/alkaline affected [16]. Therefore, inoculation with phosphate solubilizing bacteria and other useful microbial inoculants in these soils becomes mandatory to restore and maintain the effective microbial populations for solubilization of chemically fixed phosphorus and availability of other macro- and micronutrients to harvest good sustainable yield of various crops [8].

14.4 ESTIMATED DEMAND AND SUPPLY OF SOME IMPORTANT BIOFERTILIZERS IN INDIA

The twelve-monthly requirement and production of different biofertilizers have clearly shown remarkable gap in this area. Thus a strategy for judicious combination of chemical fertilizers and biofertilizers will be economically viable and ecological useful. It should be recommended that biofertilizers are not a substitute, but a supplement to chemical fertilizers for maximizing not only the yield but also agro system stability.

14.5 CONSTRAINTS IN BIOFERTILIZER TECHNOLOGY

Despite having various potential activities, biofertilizers did not get farmers' acceptance adequately. There are various factors influencing the adoption of biofertilizers. By knowing the different constraints faced by the farmers in use of biofertilizers, the extent of adoption of biofertilizers can be increased by overcoming these constraints. The constraints can be broadly classified into four major groups, which are as follows:

1. Production Constraints.
2. Market level Constraints.
3. Field Level Constraints.
4. Resource Constraints.

1. **Production Constraints:** Despite significant refinement in this technology over the years, the progress in the field of biofertilizer production technology is below satisfaction due to the following:
    - **Unavailability of appropriate and efficient strains:** As biofertilizers are not only crop-specific but soil-specific too, lack of region-specific strains is one of the major constraints. Moreover, the selected strains should have competitive ability over other strains, nitrogen-fixing ability over a range of environmental conditions, and ability to survive in broth and in inoculant carrier.
    - **Mutation during fermentation:** Biofertilizers tend to mutate during fermentation and thereby raise the production and quality control cost. Extensive investigation on this aspect is urgently needed to eliminate such undesirable changes.
    - **Unavailability of a suitable carrier:** Unavailability of a suitable carrier due to which the shelf life of biofertilizers is short is a major constraint. Peat of a good quality (> 75% carbon) is a rare commodity in India; Nilgiri peat is of poor quality (<50% carbon). According to the availability and cost at production site, choice is only with lignite and charcoal in India.

    As per the suitability, the order is peat >lignite > charcoal > FYM > soil > rice husk. Good quality carrier must have good moisture holding capacity, they must be free from toxic substances, sterilisable and have a readily adjustable pH (6.5-7.0). Under Indian conditions where extremes of soil and weather conditions prevail, there is yet no suitable carrier material identified capable of supporting the growth of biofertilizers. Better growth of bacteria is obtained in sterile carrier and the best method is gamma irradiation for sterilization (while using autoclave, lime mixed lignite is filled up to two-thirds capacity of steel trays for one to two hours for three days and sterilized at 121°C) for carrier material.
2. **Market Level Constraints:** It includes following constraints:
    - **Lack of awareness of farmers:** Despite considerable efforts in recent years, majority of farmers in India are not aware of biofertilizers, and their usefulness in increasing crop yields sustainably.

- **Seasonal and unassured demand:** The biofertilizer use is seasonal and both production and distribution is done only in few months of year, and as such production units, particularly private sectors, are not sure of their demand.
- **Inadequate and Inexperienced staff:** Because of inadequate staff and that too not technically trained who can attend to technical troubles, farmers are not given proper instructions about the application aspects.
- **Lack of quality assurance:** The sale of poor quality biofertilizers through corrupt marketing practices results in loss of faith among farmers; to regain the faith is very difficult and challenging.

3. **Field Level Constraints:** Different field level constraints are as follows:
    - **Edaphic factors:** Among soil and climatic conditions, high soil fertility status, unfavorable acidity and alkalinity, high and low nitrate level, high and low temperature, drought, deficiency of P, Cu, Co, Mo or presence of toxic elements affect the microbial growth and crop response.
    - **Native microbial population:** Antagonistic microorganism already present in the soil competes with microbial inoculants and many times does not allow their effective establishment by out-competing the inoculated population.
    - **Faulty inoculation techniques:** Majority of the marketing sales personals do not know proper inoculation techniques. Biofertilizers, being living organisms, require proper handling, transport and storage facilities.
4. **Resource Constraints:**
    - **Limited resource generation for biofertilizer production:** The investment in biofertilizer production unit is very low. But keeping in view of the risk involved largely because of short shelf life and no guarantee of off take of biofertilizers, the resource generation is very limited.

14.6 FUTURE PROSPECTS

Bioinoculants have now long been in the market, but their present status in comparison to the chemical application in the agriculture is very scanty. Continuous use of chemicals is disturbing the environmental balance and ecological soundness. Therefore, there is an urgent need to develop a definite correlation between agriculturists, microbiologists and farmers. To develop this correlation and prevent our environment from the ill effects of the chemicals, biological products should be developed that are reliable and act as suitable alternatives to chemicals, leading to sustainable development. Recent research on Plant Growth Promoting Rhizobacteria (PGPR) and bacterial antagonists to phytopathogens has opened new and exciting horizons for inoculation industry [6]. The encapsulation of microorganisms into a polymer matrix is also an upcoming method in the bacterial inoculation technology. The objective of this technique is to trap beneficial microorganisms in a matrix. The immobilized microbial cells are already in use in fermentation industry [7]. One of the major challenges for the inoculant industry is to find or develop an improved carrier that provides high shelf life, high number of viable cells, protection against soil environment, convenience to use and cost effectiveness [13]. The encapsulated bacterial formulations in bioinoculants temporarily protect the encapsulated microorganisms from the soil environment and microbial competition and felicitate the gradual release of microorganisms for the colonization of plant roots. The main objective of encapsulating bacteria is to increase their survival time during storage. The lowered water content in the final product is responsible for long-term survival during storage. In this way, the bacteria in the formulation remain inactive, resistant to environmental stresses, insensitive to contamination and more compatible with the fertilizer application [3, 10]. Usually, inoculants are being used for legume crops and to a certain degree for cereal crops. Fresh alternatives should be exploited for the use of bioinoculants for other high value crops such as vegetables, fruits and flowers. This will not only increase the field of the inoculants but also create confidence among the farmers for their use. Both PGPR and antagonistic organisms can be used with great effect for vegetables, fruits and flower crops. Their biological formulations will also preserve the natural flavour of vegetables and fruits which is destroyed by the chemicals. Another portion which can revolutionalize the inoculant industry is the development of bioinoculants which carry out more than one function. Inoculant development not only increases the fertility of the soil but also acts as a PGPR or antagonist against phytopathogens that is a bioinoculant with triple action. This requires development of special carriers which can support and carry more than one microorganism simultaneously in the fields. In case of rhizobia, bioformulations can be developed which contain more than one type of strains or species of the root nodulating bacteria. In a wide varity of legume crops this will increase the chances of nodulation by a single formulation. Such type of inoculants will be much more cost effective and competent in comparison to the chemicals. Synthetic costly carriers can be very effective and cheaper if they can support more than one microorganism simultaneoulsy. Finally marketing is key to the success of any product [10]. The farmers must have a comprehensive knowledge about the properties, application, and the advantages of the bioinoculants. Workshops should be established for the farmers in the villages by the government where trained personnel will help them in gaining the knowledge. With increasing awareness and research work, eco-friendly economically feasible biological inoculants should be developed and promoted for sustainable development.

- **Use of Liquid Biofertilizers (Breakthrough in Biofertilizer Technology):** Liquid biofertilizers are special liquid formulations containing not only the desired microorganisms and their nutrients but also special cell protectants or chemicals that promote formation of resting spores or cysts for longer shelf life and tolerance to adverse conditions [5].

Biofertilizers manufactured in India are mostly carrier-based, and in the carrier-based biofertilizers, the microorganisms have a shelf life of only six months. They are not tolerant to UV rays and temperatures more than 30°C. The population density of these microbes is only 10^8 c.f.u/mL at the time of production. This count reduces day by day. In the fourth month it reduces to 10^6 c.f.u/mL, and at the end of 6 months the count is almost nil. Owing to this, the carrier-based biofertilizers were not effective and did not become popular among the farmers. These defects are rectified and fulfilled by the introduction of liquid biofertilizers. The shelf life of the microbes in these liquid biofertilizers is two years. They are tolerant to high temperatures (55°C) and ultraviolet radiations. The count is as high as

109 c.f.u/mL, which is maintained constant up to two years. So, the application of 1 mL of liquid biofertilizers is equivalent to the application of 1 kg of 5-months-old carrier based biofertilizers (1000 times). Since these are liquid formulations, the application in the field is also very simple and easy. They are applied using hand sprayers, power sprayers, fertigation tanks and as basal manure [4].

CONCLUSION

In India several factors, i.e., accumulation of toxic elements, waterlogging, loss of nutrition and soil erosion affect soil fertility remarkably. Proper application of biofertilizer plays a significant role towards holistic well-being of the soil ecosystem. But different constraints like production level (unavailability of most effective strain and its carrier), market level (lack of alertness of farmers related to organic agriculture), field-level (different agro-climatic zone related problems) and resource level (low investment in production unit) hinder the growth of biofertilizers. By overcoming these constraints there are optimistic signs for green agriculture and it will motivate larger section of farming community to opt for organic agriculture.

REFERENCES

1. Abdul Halim N.B. (2009). "Effects of using enhanced biofertilizer containing N-fixer bacteria on patchouli growth". Thesis. Faculty of Chemical and Natural Resources Engineering University Malaysia Pahang., 145.
2. Arun K.S. (2007). "Biofertilizers for sustainable agriculture. *Mechanism of P-solubilization*". 6th edn., Agribios Publishers, Jodhpur, India, 196–197.
3. Basha, Y. and Carillo, A. (1996). "In: *New Horizons in Agriculture: Agroecology and Sustainable Development*". (eds. J. Perez-Moreno and R. Ferrera-Cerrato), Colegio De postgraduados en ciencian agricolas, Montecillo, Mexico., 125–155.
4. Bhatttacharyya, P. and Kumar, R. (2000). "*Liquid Biofertilizer-current Knowledge and Future Prospect*". National seminar on development and use of biofertilizers, biopesticides and organic manures. Bidhan Krishi Viswavidyalaya, Kalyani, West Bengal, November 10–12.
5. Hegde, S.V. (2008). "Liquid biofertilizers in Indian agriculture". *Biofertilizer News Letter,* 17–22.
6. Kloepper. J.W., Lifshitz, R. and Zablotowicz, R.M. (1989). "Free-living bacteria inocula for enhancing crop productivity, *Trends Biotechnol. 7,* 39–44.
7. Lin, J.E. and Wang, H.Y. (1991). "Degradation of pentachlorophenol by Non-immobilized, immobilized and co-immobilized *Arthobacter* Cells", J. Ferment. Bioengin. 72, 311–314.
8. Mishra, D.J., Singh Rajvir, U.K., Mishra and Shahi Sudhir Kumar (2013). "Role of biofertilizer in organic agriculture: a review." *Research Journal of Recent Sciences* ISSN 2277, 2502.
9. Mohammadi, K. and Yousef S. (2012). "Bacterial biofertilizers for sustainable crop production: A review." *J Agric Biol Sci,* 7, 307–316.

10. Arora, N.K., Kumar, V. and Maheshwari, D.K. (2001). "Constraints, development and future of the bioinoculants with special reference to rhizobial inoculants." *Innovative Approaches* in *Microbiology*, 241–254.
11. Roger, P.A. and Ladha J.K. (1992). "Biological N_2 fixation in wetland rice fields: estimation and contribution to nitrogen balance". *Plant Soil*, 141, 41–5.
12. Samoon, H.A., Showkat A., Dar, B., Zehra, S., Sheraz Mahdi and S.A., GI Hassan (2010). "Biofertilizers in organic agriculture." *Journal of Phytology* 2, 10.
13. Smith, R.S. (1995). "In: *Nitrogen Fixation: Fundamentals and Applications*", Kluwer Academic Publishers, 653–657.
14. Venkataraman, G.S. and Shanmugasundaram, S. (1992). *Algal Biofertilizers Technology for Rice*. DBT Centre for BGA Biofertilizer, Madurai Kamraj University, Madurai, 1–24.
15. Venkatashwarlu B. (2008). "*Role of biofertilizers in Organic Farming: Organic Farming in rainfed Agriculture*": Central Institute for Dryland Agriculture, *Hyderabad*, pp. 85–95.
16. Wani, S.P. and Lee K.K. (2002). "Population dynamics of nitrogen-fixing bacteria associated with pearl millet (*P. americanum* L.)," In: *Biotechnology of Nitrogen Fixation in the Tropics*, University of Pertanian, Malaysia, 21–30.

15

Integration of Eco-friendly Biotechnological Tools for Sustainable Environmental vis-a-vis Bioresource Management

Susanta Kumar Chakraborty
Department of Zoology, Vidyasagar University, Midnapore (West) – 721102, West Bengal, India
E-mail: *susantachakraborty@yahoo.com*

ABSTRACT

Sustainable bioresource management ensures a sound balance between optimum bioproductivity for human use and renewal of bioresources. This should take into account the merits and demerits of different environmental activities related to biodiversity management and utilization, of the past and present, so that the needs of the moment are satisfied as well as the carrying capacity of natural resources are kept in reserve for fulfilling the future requirement. This can be achieved if the pragmatic principles of *ecology* are properly adhered to, in devising of biotechnological tools giving due care to the mode of availability of bioresources, agroclimatic conditions and socio-economic needs. In this context, the present paper has attempted to highlight the possibilities of integration of several ecofriendly biotechnologies like vermitechnology, sericulture, aquaculture, biodiversity documentation, joint forest management, integrated pest management, mushroom culture, poultry, duckery etc., developed based on local bioresource availability and peoples participation citing the research outcomes of the case studies undertaken in the Midnapore district (West Bengal). Integration implies the application of waste materials generated in one kind of biotechnological activity as resource base for others. The development and standardization of several people-friendly and eco-friendly technologies in tune with the agro-climatic conditions of the region and their dissemination for implementation corresponding with the aspirations of grassroot people based on their traditional knowledge have resulted in self-employment generation, eco-restoration and biodiversity documentation. The prime objective of such action oriented research study involving local peoples, especially the rural women by way of contributing social, environmental and economic benefits has been for the development of a more sustainable society, which promoted in the Brundtland Report in 1987, in Agenda 21 of the Earth Summit in Rio de Janeiro in1992, the Report of the World Summit on Sustainable Development, held in Johannesburg in 2002 and which has been widely accepted in the environmental policies of many countries of the world.

Keywords: Sustainable Bioresource Management, Ecorestoration, Rural Development Through Biotechnology, Biodiversity Conservation.

15.1 INTRODUCTION

Biotechnology has its roots in the distant past, and with the passage of the time, because of the revolutionary development in the different fields of biotechnology, controversies exist with regard to its definition. The most simple and recent definition for biotechnology is the "The application of scientific and engineering principles to the processing of materials by biological agents to provide goods and services. Agents include a wide range of biological substances such as enzymes, whole cell of the microbes or multi-cellular organisms. The words "goods" and "services" cover processes such as waste and water treatments along with providing natural resources like foods, fodders, materials for shelter, medicines etc. [10, 17]. Biotechnology is the "integration of natural sciences and engineering in order to achieve the application of organisms, cells, parts there off and molecular analogues for products and services [46]. Biotechnology, broadly defined, includes any technique that uses living organisms (or parts of organisms) to make or modify products. The definition is applicable to both traditional and modern concept of biotechnology. Traditional biotechnology embraces conventional age-old processes for the production of cheese, wines, honey, silk, fish, curd, organic compost etc, while modern biotechnology includeecombinant DNA technology, tissue culture, cell fusion techniques etc. The Department of Biotechnology, India, has identified 13 different branches of biotechnology [5, 23].

The emphasis on biotechnology and its contribution to food security, poverty alleviation, environment management and economic upliftment has been increasing during last couple of decades, even in the developing countries, like India. Innovations and advancement of knowledge in the sustainable management, production and use of biological resources (microorganism, plants, animals), will provide the basis for new, sustainable, safer, eco-efficient and competitive biotechnologies and products for agriculture, fisheries, food, health, forest-based and related industries. In order to address challenges in the fields of agriculture, forestry, food safety and security, community health and disease control, fisheries, animal husbandry, soil fertility and biofertilisers, several research priorities have been identified for sustainable production and management of biological resources from land, forest and aquatic environments utilizing the knowledge bases pertaining to the subjects such as genomics, proteomics, metabolomics, systems biology, bioinformatics and converging technologies for microorganisms, plants and animals, including exploitation and sustainable uses of their biodiversity.

The responsible use of biotechnology to get economic, social and environmental benefits is inherently attractive and determines a spectacular evolution of research from traditional fermentation technologies(cheese, bread, beer making, animal and plant breeding) to modern techniques(recombinant DNA technology, biochemistry, molecular biology) to provide efficient synthesis of low toxicity products, renewable bioenergy, and yielding of new methods for environmental monitoring. The earlier decades of 21st century witnessed the emergence of biotechnology as a key driver of sustainable environmental management [10]. The distinct role of environmental and rural biotechnologies in present time is being emphasized considering their contribution towards new solutions and directions in the remediation of contaminated environment, minimizing of waste release, ecorestoration of ecodegraded habitats and rejuvenation of bioresource bases. To ensure such ecological changes, innovative new biotechnological tools have to be developed with the sharing and integration of biological, molecular, biochemical and ecological knowledge. This is expected to improve the understanding of existing biological processes in order to increase their efficiency, productivity and flexibility.

The prime objective of the present paper is to review the possibility of standardization of several ecofriendly and rural-based biotechnological tools in tune with the agroclimatic set ups of the region and also to integrate these technologies in order to derive maximum benefits towards sustainable bioresource management.

15.2 PHYSIOGRAPHY OF STUDY AREAS

Erstwhile, undivided Midnapore District not only represented the largest District of India but also possesses a mosaic of habitat diversities in the form of coastal belt in the east, lateritic hilly forest tracts in the west, freshwater riverine networks in the north and depressed flood affected areas having a number of wetlands in the south. All those unique eco-regions are endowed with varieties of biodiversity components, based on which a number of biotechnological tools have been developed in order to address the problems of eco-degradation, deforestation, biodiversity depletion, energy crisis, poverty and pollution. Some selected study areas in the West Midnapore District, especially in Binpur-II and Jamboni Blocks of Jhargram subdivision have been identified to take up long term studies on organic waste recycling through vermitechnology and sericulture development with organic inputs through the empowerment of rural women. The entire area characterized by undulating hilly lateritic tracts and endowed with forest vegetation is drought-prone where the forest cover has been depleting steadily during the last four decades. Soil erosion coupled with land degradation has converted many areas to wasteland, characterized by deep ground water level and low agriculture productivity. Therefore, objectives of the present study have been oriented to ensure sustainable natural resource management utilizing eco-friendly technologies involving rural organized women with an idea to provide them a scope for self-employment alongside helping them to ensure ecorestoration of ecodegraded wasteland [15, 37].

15.3 BACKGROUND OF DIFFERENT ACTORS

Under the auspices of the Centre for Women's Development Studies (C. W. D. S), New Delhi and with the support from the Zilla Parishad of Midnapore, rural poor women of the study sites formed several *Samities* (Gramin Mahila Shramik Unnayan Samity), each with 60 members in the year, 1996 in order to undertake a cooperative venture for earning their livelihood. Initially, each *samity* was able to procure a chunk of 15 to 20 acres of wasteland through self-donation and from the Midnapore Zilla Parisad. They planted the land with arjun and asan trees, the host plants of tasar silk worms taking subsidies from the respective Government Department and the Panchayat.Women then volunteered their labour to protect and maintain the plants organizing themselves into village-based groups. Samities afterwords diversified other income generating agricultural and environmental activities with the adoption of a number of sustainable, ecofriendly, user-friendly technologies like vermitechnology, mulberry sericultural,aquaculture etc. in accordance with local demands and utilizing the locally available raw materials,on undergoing collaborative research venture with the researchers of Vidyasagar University, Midnapore (West), West Bengal, India[15, 37]

15.4 BIOTECHNOLOGY FOR BIODIVERSITY CONSERVATION — A NEW VISTA FOR HUMAN WELFARE

Biodiversity, the totality of genes, species and ecosystem encounter interactions within, between and among various levels to maintain the self-sustaining substances and functional attributes of it within the scale of time and space. The survival of human being is entirely dependent on goods and services from biodiversity—the food, medicine, clothing and shelter—besides contributing towards culture and aesthetic sense of human being. Conservation and sustainable use of biodiversity is prerequisite for sustainable environmental management and development.

Based on the interdependence of Biotechnology and Biodiversity, the entire world can be divided into three categories—(i) Biodiversity rich and Biotechnology poor e.g., India, Brazil, Malaysia, etc; (ii) Biodiversity poor and Biotechnology poor e.g. countries in Gulf and Arab; (iii) Biodiversity poor but Biotechnology rich e.g., U.K. Canada, Japan, Germany etc. There is a possibility for the emergence of a fourth category i.e., Biodiversity rich and Biotechnology rich. (India, China, Brazil etc.)

From the Indian point of view, serious emphasis is being made for the conservation of biodiversity in tune with global biodiversity conservation strategy as the country's ecological, economical, social, cultural and political future are dependent on it. Conserving biodiversity means taking steps to protect genes, species, habitats and ecosystems [17].

15.5 VERMITECHNOLOGY FOR ORGANIC FARMING: AN ECO-FRIENDLY APPROACH

Organic farming, usage of eco-friendly manures such as green manures, urban wastes, rural wastes, etc., can bring sustainability to agriculture. Vermicomposting, being an eco-friendly, economical and efficient novel technology, can be applied to utilize agricultural and other organic wastes. Vermitechnology is a common biotechnological tool where organic solid wastes are recycled to organic manure using suitable species of earthworms as bioreactors. Presently, the importance of earthworms in tropical agriculture has been realized in a newer dimension because of evil effects of large scale application of chemical fertilizers and pesticides leading to extensive loss of soil biodiversity [1, 2, 3, 26, 47].

The increase in human population and urbanization has led to food scarcity. To increase food production, agriculturalists are tempted to use more amount of chemical fertilizers and pesticides frequently, which cause several hazards to soil microflora, which also in turn affect fertility. The accumulation of chemicals by biomagnification in plants and their products causes several diseases in human beings. Infant disease such as methanoglobinaemia is caused due to nitrogen containing fertilizers. Potential environmental benefits of vermitechnology include reduction of noxious qualities of organic wastes, elimination/reduction of harmful microorganisms, conversion of agro-wastes into highly valued fertilizer, and production of food and feed from food discards [26, 47]. Vermicompost envisages the soil fertility for years together without affecting the food quality [12].

Advantages of vermicompost

1. The N.P.K. contents of vermicompost are almost 8 times more than farm yard manure (FYM) (Table 15.1).
2. All micronutrients are present.
3. The C:N ratio is reduced to a favourable condition (20:1 or less).
4. The duration of composting is around two months.
5. It maintains soil texture and water conservation.
6. The locally available organic wastes are recycled.
7. Vermicompost can be used as feed in aquaculture[9].

Table 15.1 Nutrient contents of vermicompost along with other organic manures

Nutrients	*Vermicompost*	*Farm yard wastes (FYW)*
Nitrogen (N)	0.5%	0.18%
Phosphorus (P)	0.57%	0.2%
Potassium (K)	3.14%	0.5%

Vermicompost, technology, by providing employment to millions of youth, can eliminate dependence on chemicals; can convert wastes into fertilizers; can bring waste land under cultivation, can feed hungry citizen; and can make a country green and prosperous in a span of just a few years [26]. This technique also helps conserve the biodiversity, which is the need of the hour. Apart from providing self-employment opportunities for the weaker section and profitable agricultural waste utilization, it will also help maintaining the environmental/ecological balance [1]. Application of vermicompost alone or in combination with other chemical fertilizers is expected to trigger the growth of phytoplankton, zooplankton and other aquatic organisms and thereby will generate higher aquatic biological productivity [2, 3, 9, 12].

15.6 STEPS IN VERMICULTURE TECHNOLOGY

15.6.1 Earthworm for Composting

Earthworms constitute more than 80 percent of soil invertebrate biomass. They feed on a variety of organic waste materials and produce "Vermicastings". It contains more microorganisms, organic matters and inorganic minerals in the form that can be used by plants. Nearly 10-15 percent primary production is channelized through earthworms. In absolutely ideal conditions of comfort in ground up and moist food, the herd will recycle their own weight in wastes every 24 hours [13]. Vermicomposting is an ideal and appropriate method for disposal of nontoxic solid and liquid organic wastes [12, 13, 26, 42, 43, 47].

15.6.2 Selection of Suitable Species of Earthworms

3200 species of earthworms belonging to 240 genera and more than one dozen families under the Class Oligochaeta of Phylum Annelida are present all over globe. Of these 509 species under 67 genera and 13 families have been found to occur in the Indian subcontinent. Majority of these genera (47) are endemic [28]. From West Bengal, India, 63 species under 26 genera and 8 families have been recorded [27], whereas 17 species of earthworm belonging to 13 genera and 6 families have been found to occur in Midnapore district [11].

Epigeic earthworm species efficiently enhance the rate of organic manure production through biodegradation/mineralization compared to other species. Two commonly occurring exotic epigeic category of earthworm used for vermicomposting are *Eudrillus eugeniae, Eisenia foetidae,*while indigenous category includes *Perionyx excavatus, Lampito mauritii* and *Dravida willsil.*

15.6.3 Culture Maintenance

Epigeic earthworms remain active throughout the year under favourable conditions. Conducive moisture levels, temperature, food, space and oxygen are essential for their survival and biomass production. Temperature below 32°C and moisture level around 40% is ideal condition for earthworm activity.

15.6.4 Raw Materials for Vermicomposting

Though earthworms can digest a diverse range of organic residues and yield rich vermicompost, it is better to use pre-digested organic wastes for worms to act faster and produce high quality compost. The pre-digested waste is an ideal medium for the worms to act on. The pre-digested materials will be converted into quality vermicompost within 30 days. The composite organic wastes should be degraded using diluted fresh cow dung slurry, which is to be sprinkled over several layers of heap. The heap has to be kept moist by regular sprinkling of water in order to maintain the moisture content around 40%, and the entire composting materials have to be turned two to three times at an interval of ten days with an object to ensure steady supply of oxygen [1].

15.6.5 Methods Adopted in Vermicomposting

(a) **Waste Materials Required for Vermicomposting:** Animal's (cattle, sheep and goat) dung, poultry wastes, sericulture wastes, agricultural wastes (stem, leaves, pulse and vegetable waste); various types of forest leaf litter, terrestrial and aquatic weeds, municipal garbage, biogas slurry and waste materials of paper industry were mainly used in this study (1).

(b) **Preliminary treatment of composting materials:** Initially after collection, nonbiodegradable components like plastics, stones, glasses, ceramics and metals were discarded. Heavily contaminated wastes were also eliminated. Before composting, selected organic matters were heaped and large lumps were broken into pieces. These separated matters were then spread in a layer upto 30 cm, and was exposed to sun for a day for cleaning several unwanted organisms and also for removing foul smell. [1]

(c) **Selection of suitable earthworm species:** Indigenous local verms mainly *Lampito mauritii* and *Perionyx excavatus* have been found to be effective because of their active habit surface dwelling behaviour and power of prolific reproduction. These are used as inocula in vermicompost preparation. These two species of earthworms have been found to be distributed ubiquitously throughout the Midnapore district [11].

(d) **Vermiculture:** Vermiculture is prepared in earthen pots containing mixtures of cow dung, pond soil, grind dried leaves, poultry droppings and goat dung in the ratio of 6:1:1:1:1. Previously isolated indigenous earthworms like *Lampito mauritiii* and *Perionyx excavatus* were grown in this mixture maintaining proper humidity and temperature for 30-45 days. [13, 42, 43].

15.7 SERICULTURE—A RURAL BASED COTTAGE INDUSTRY

Mulberry silkworm is not the traditionally reared silkworm in the district. However, experimental cultivation of the mulberry silkworms have been successfully undertaken by the researchers associated with Vidyasagar University [2]. Mulberry is a perennial crop which continues to grow and produce leaves throughout the year in tropics. The continuous production of mulberry for a long time results in gradual reduction of leaf yield and quality. This lacuna can be improved through application of organic manures, fertilizers and biofertilizers with ideal water management [2, 20, 24]. Attempt has been made to study the potentials of different organic residues along with biofertilizers, 50% nitrogenous fertilizers and 33% cut in phosphorus fertilizers under lateritic soil conditions of Bhalukkhulia in Jhargram subdivision of West Midnapore district in West Bengal on morphological parameters, leaf yield and quality of S1635 mulberry [18, 19, 20].

Thus from the present study, it is inferred that application of organic manures along with biofertilizers and application of reduced dose of chemical fertilizers is highly beneficial for mulberry cultivation, which minimized the cost of mulberry cultivation (9.26%) and is eco-friendly. Thus, this improved package is recommended for new mulberry plantation under semi-irrigated lateritic soil conditions of West Midnapur and adjoining areas of similar agroclimatic situation [18, 19, 20].

15.8 AQUACULTURE AND FISHERY

Aquaculture and fishery represent two most challenging sectors involving lot of innovative biotechnological tools in order to meet the demand of cheap sources of protein on one hand, and generation of selfemployment scopes on the other [16, 29, 31]. In the process of standardizing new technologies, attention should be made to ensure sustainability by virtue of selecting the inputs from organic world [3, 9, 16, 40].

The rivers and an array of wetlands of South West Bengal are already under great ecological stresses, which adversely affects feeding, spawning and nursery habitats of fishes. Due to the discharge of agricultural, industrial and municipal waste water, habitat destruction, wanton destruction, usage of non-restricted nylon thread fishing gears, fishing during breeding seasons, siltation as a result of deforestation, over exploitation, multifarious water use, human population pressure, and introduction of exotic fish species are being considered as major problems for the declining of fishery resources [34, 36, 38, 41]. Some of the recent studies indicate that different human activities harm the river productivity enormously [38,41]. For ensuring good quality seeds of cultivable fish, it is expedient to introduce an enduring system of producing seeds of the fishes through induced breeding, as an alternative to the present unreliable dependence on wild seed collections for supply to fish farmers engaged in fish culture. For conservation and rehabilitation of declining fish species, induced spawning appeared to be very much effective. Artificial breeding of rare and endangered fish species is being practiced to make available these species in their natural habitats and thereby restoring the gene bank of threatened species. It has been possible to achieve success through hypophysation technique of some threatened fish species to produce pure seeds [35].

As it is difficult to meet such a high demand of seed supply through natural breeding, techniques for reliable spawning and fingerling production of carps and catfishes have been evolved [21, 22]. The seed production of cultivable carps and catfish is mainly based on the induced breeding, and its success depends largely upon the availability of well-maintained brood stocks [35] and inducing agents such as carp pituitary extract (CPE) [13-14] or natural as well as synthetic hormonal preparations (Ovaprim and Ovatide) of the gonadotropin releasing hormone (GnRH) [21, 22, 25, 29, 39, 40].

15.9 HYBRIDIZATION TECHNIQUE

The intergeneric hybridization between Indian and Exotic major carps was found to be sterile due to aneuploid nature of their genome [29, 29, 40, 48]. The sterile nature of triploid individuals of common carp (Cyprinus carpio) was also reported to be due to the aneuploid nature of their genome [44]. However, it was noticed that crosses between triploid carps also yielded fertile offspring [44].

15.9.1 Induced Breeding in Fishes

The major breakthrough was achieved in the year 1957, when the carps could be bred successfully for the first time in the world, in India, by injection of fish pituitary hormones, and a very useful technique was developed to produce millions of quality seed for culture, [21, 22]. The success revolutionized prevailing aquaculture practices and the technique was effectively introduced all over India and also in many other countries in the world.

Owing to some limitations in pituitary gland extracts, several other synthetic hormones analogues (Ovaprim and Ovatids) are now being used in the enhancement of aquaculture production [35, 39]. Aquaculture faces many challenges over the next decade, notably, combating diseases and epizootics, broodstock improvement and domestication, development of appropriate feeds and feeding mechanisms, hatchery and grow-out technology, as well as water-quality management. These all present considerable scope for biotechnological and other technology interventions. Aquaculture biotechnology can be described as the scientific application of biological concepts that enhance the productivity and economic viability of its various industrial sectors. Endocrine regulation of reproduction has been effectively applied across a broad range of cultured fish species, however, advancements have been slow with shrimp and molluscs. Recent research has shown that there is potential for chemical treatment of shrimp Gonad Inhibiting Neurohormone (GIH) that could promote reproduction without the negative side effects of eye stalk ablation [34]. Research on shrimp GIH isolation is still ongoing but elucidation of the structure and function of shrimp GIH, using peptide biotechnology-based approaches, shows promise for countering the reproductive inhibitory effects of GIH.

15.9.2 Feed Technologies

Currently, one of the most heated debates concerning aquaculture development is the use of fishmeal and other animal proteins in aquafeeds [32]. Although fishmeal is used for its high quality protein content, it has several disadvantages, including high cost and instability of supply. Wild fish catches are on the decline and there are increasing environmental concerns (eutrophication, pollution associated with excess nutrient waste), ethical concerns over feeding fish to non-piscivorous fish, and social concerns over using aquatic protein to feed fish that could be used for human nutrition (especially in nutritionally deficient areas of the world). Although, the major user of fishmeal is terrestrial agriculture, and the salmon, bream, bass and shrimp farming sectors are using species that would not normally be used for human consumption, the concerns of consumers provide a strong impetus to find ways to replace fish meal with vegetable protein from more sustainable sources. Biotechnology offers opportunities for development of alternatives to fishmeal, especially plant-based protein sources, by enhancing production and processing techniques. Other technologies also offer potential for enhancing the efficacy of feed delivery.

Plant protein has significant potential for addressing the problem of phosphorus pollution, since plants do not contain the high levels of phosphorus found in animal protein. The use of plant protein in aquafeeds also helps reduce pressure on wild fish stocks. Research in this area is focusing on the investigation of various plant species and plant-animal protein mixes as new sources for protein for aquafeeds of major fishery resources [25, 29, 32].

15.10 BIODIVERSITY DOCUMENTATION: FISHERY RESOURCES

People's participation in all four open water fisheries management systems viz. research system, extension system, client system and support system is more important than the products or processes put to use. Local participation is not the only new criterion by which the management of riverine fisheries needs to be judged. It is equally important that the problems be approached holistically taking into account the full range of human and community potentials. Through a series of interactions with the local fishermen community, it was noted that these people maintain a tradition of knowledge brought down to them through generations which has turned them into a repository of traditional expertise on fishes and fishes environment. Fisheries management of open waters necessarily involves both individual and group action where the need for participatory approach is probably maximum. In fact, participatory approach is indispensable for successful management of fishery resources of threatened aquatic landscape.

Selected Participatory Rural Appraisal (PRA) techniques were carefully designed and applied to extract primary information relating to the population status of fishes from different categories of water bodies. The techniques like Matrix Ranking, Rank-Based Quotient (RBQ) and Trend Analysis (Table 15.2) were simultaneously used for the recording trend analysis of the mode of depletion of fishery resources [38, 41]. Through several interactions in the form of group meetings, it was possible to involve local communities to document fishery resources to understand indigenous fishing technologies and also to assess the ecological characteristics of six major rivers viz. Subarnarekha, Kansai, Shilabati, Rupnarayan, Dwarakeswar and Keleghai and their associated wetlands of South West Bengal during two consecutive years. An important objective of the present research programme was to document the perceptions and knowledge of the local communities regarding different categories of water mass from where they used to gather resources. So far, 106 fish species and 11 prawn species from different habitats (river, canals, beels, small ditches and wetlands) of South West Bengal have been reported [36].

Table 15.2 Different techniques for quantification of biodiversity

Sl. No	*Technique*	*Purpose*
1.	PRA	To extract information relating to fish population and their present status
2.	Matrix Ranking	To isolate the perceived causes and their relative values in causing the fish population decline
3.	RBQ	It was made to re-rank the identified causes in terms of their vertical and horizontal dimensions
4.	Trend Analysis	To assess the trend of depletion or enhancing of fish abundance

15.11 BIOLOGICAL CONTROL

Large scale application of chemical pesticides and fertilizers has lead different types of environmental degradation. An alternate approach to control pest and disease is the biological control with the use of microorganisms and insect's predators to control pests and diseases. The most successful biocontrol agent is *Bacillus thuringiensis*, a spore forming bacterium containing crystalline protein inclusions. The proteins are highly toxic to insect pests but specific in their activity. They have been widely used for over thirty years against Lepidopteran (caterpillars) pests. The commercial application of these biopesticides is usually with formulation of spores and crystalline inclusions of disrupted *B. thuringiensis* and are applied at 4-6 gm/ha.

In recent years, the toxin genes have been isolated and sequenced and recombinant DNA based products produced. These genes have also been incorporated into various plant species and expressed in the plant tissues. This now offers the farmers an option to either direct apply microbial biopesticides or protect the plant directly by toxic gene expression in the plant tissues [4].

15.12 APICULTURE

Bees are crucial in all flowering ecosystems and provide essential ecosystem services in the form of pollination. Both native/wild bees and the honeybees are important when pollination is concerned. Wild trees and shrubs rely solely on the poorly studied native/wild bees for pollination. Besides, apiculture using only honeybees is a major source of revenue generation at the village level. Honey and beeswax find wide spread application in industries as diverse as food to pharmaceuticals. Apiculture is especially a suitable means of employment generation for the women of rural societies and promises income generation at a massive scale if it is done scientifically with regular technological up-gradation. Apiculture is a traditional industry in the district with people mainly rearing *Apis cerana indica* and *Apis mellifera*. The industry is however yet to grow into a large one with people rearing bees in purely domestic setups with little scientific organizational approach. It has been found that apicultural practices are being impeded due to unscientific rearing practices and agriculture. Rampant pesticide application on flowering crops during the effective foraging hours of the honeybees is greatly adding to their mortality and discouraging the migratory beekeepers from frequenting the farmlands of the district. Besides, lack of scientific knowhow among the farmers and non-co-operation from the village level authorities like the Panchayats and Divisional Offices is also adding to the burden of the migratory beekeepers [7]. Although thus news sounds discouraging, the fact that apiculture is gaining popularity as a means of income generation among the rural woman and unemployed youth is cause for elation. It is the interest of the rural communities that the government at the local levels introduces scientific bee-keeping practices and imparts training to encourage the growth of this rural industry which is a certain means of rural social upliftment. Integration of bee keeping with horticulture/orchard can yield more honey for the beekeepers and also derive benefit from the pollination activities of the bees. Several fruit plants are known to get benefit from insect pollination like cashew, tea, papaya, coconut, squash, pumpkin, cucumber, litchi, apples, watermelons, sunflower etc [8, 45].

15.13 MEDICINAL AND AROMATIC PLANTS

Forests along with scrub lands, being the repository of diverse medicinal plants, are to be protected and conserved through peoples participation for alleviating poverty and also boosting of rural economy [6, 14, 30]. Out of 15,000 different medicinal plant species in India, 2000 plants are used in Ayurveda, 700 in Unani, 600 in Siddha, 450 in Homoepathy and 30 in modern medicine [6].

Most of the medicinal and aromatic plant species used are not cultivated. Those are collected from the natural habitat over the centuries, and due to indiscriminate collection many species have become threatened, endangered or vulnerable. A network of four GeneBanks has been setup for ex-situ conservation of rare, endangered and commercially important medicinal and aromatic plants.

15.14 FORESTRY

Research activities for Sustainable Agriculture and Forestry will foster systems-wide approaches with an emphasis on agro-ecological research, integrated concepts and production systems with a higher degree of diversity (e.g. agro-forestry systems, intercropping). Food security is assessed against global drivers like climate change. These broader research topics will be complemented by targeted activities addressing strategically important crops, genetic resources, and specific plant/animal pests and diseases and animal welfare issues. Peoples participation in the management system should be ensured in tune with the model of Joint Forest Management, Arabari Forest, West Midnapore, West Bengal, India [14].

15.15 MUSHROOM CULTIVATION

Mushroom cultivation is a cost effective and user friendly avenue of rural income generation. Mushroom has a good market demand. It is a good source of proteins. Besides, the wastes generated can be recycled through vermicomposting and organic manuring. Button mushroom(*Agaricus bisporus*) is a popular variety of mushroom which contributes about 90% of total country's production as against its global share of about 40%. It requires cereal straw for substrate preparation and thereby recycling and reuse waste biomass for generating nutritional and medicinal bioproducts. [33, 49]

15.16 ANIMAL HUSBANDRY

Animal husbandry plays an important role in the rural economy of India, especially in terms of GDP from livestock sector during 2003-2004 which was Rs. 772 billion with 27% share in the agriculture and allied sector. Some priorities of Government of India are genetic upgradation with the creation of germplasm bank, extension of dairy development, provision of nutritious feed and fodders, improvement of small ruminants and management of pasture lands.

Animal husbandry provides ample potential in the district due to importance in commercialization of agriculture, willingness of the farmer to explore new avenues for investments due to pressure on land, and limited capacity of the primary sector to absorb the ever increasing rural population.

The crossbred cattle population is hardly 1% of the total population and therefore, enormous scope exists for genetic improvement of cattle. The improved layers are less than 5% of the population. Mostly traditional poultry birds are reared in the backyard. The per capita availability of the milk is only 125 grams per day. Considering the huge market inside and outside the district, the dairy sector needs to be planned in a scientific manner to have the desired growth [50].

CONCLUSION

It appears from the above details that research on different aspects of biotechnology had gained tremendous momentum during last three decades all over the world.

Harmonizing of sustainable development with conservation of biodiversity is also now being considered as an important concern for human beings in order to meet the demand for food, health, environment and other needs of growing world population. But the world has been facing one serious problem with regard to making a balance between application of modern biotechnological tools and ensuring sustainability because several evil effects of biotechnology on environment, in general, and human society in particular.

Moreover, the benefits generated from the application of biotechnology have still been restricted to a small portion of human society and a fractional part of the world environment because of the non-involvement of greater section of human population vis-à-vis rural community endowed with poverty, illiteracy and different social constraints. Here lies the significance of the involvement of large scale people's participation in order to spread the benefits of biotechnological development to different strata of human society through environmental restoration, higher production of food, fiber, fuel and biodiversity conservation.

Initiative was taken to develop joint partnership venture with rural community, especially the rural women of Midnapore district, to standardize the existing biotechnology in tune with the agro-climatic condition of a particular region and to disseminate the same in the field for environmental management and self-employment. From such a joint research efforts, during last 15 years, it has been possible to ecorestore a considerable area of wasteland by their conversion into a green covered area with large scale plantation with Arjun and Assan trees for the cultivation of Tasar silkworm. Solid waste recycling through vermitechnology has been standardized in tune with the prevailing environmental conditions (high temperature, low rainfall, non-fertile wasteland etc) of the area, and mass production of vermicompost is being done by the organized women community, which has not only enabled the women groups to earn money, but has also enhanced their confidence and status in the rural society (1). Besides, it has been possible to document icthyofaunal resources of South West Bengal [36] along with the determination for their conservation status through standardizing different Participatory Rural Appraisal (PRA) methods with direct participation of rural fishermen community [38,41]. From the experience and knowledge gained from the above-mentioned research ventures, it is believed that biotechnology can only produce fruitful results for the all round upliftment of human society if a blend of traditional and modern biotechnological tools is made taking into consideration the joint participation of grass-root people with the researchers, because this rural community maintained a tradition of knowledge brought down by them through generations which has turned them into a repository of traditional expertise in many fields of biotechnology for environmental restoration and biodiversity conservation.

REFERENCES

1. Anon (2001). Mass production of vermicompost with the participation of rural women of Eastern region of Midnapore district, WB, through the recycling of locally available solid waste. UNDP-UNIFEM sponsored Research Project. PI. Dr. Susanta K. Chakraborty, VU, Midnapore, 721102, WB.

2. Anon (2005). Studies on the effect of vermicompost developed from indigenous vermitechnology for promoting sericulture through the empowerment of rural women of South West Bengal, India. DBT, GOI sponsored Research Project. PI. Dr. Susanta K., Chakraborty, VU, Midnapore, 721102, WB.
3. Anon (2009). Impact of vermicompost produced by indigenous vermitechnology on aquaculture. UGC, GOI sponsored Research Project. PI. Dr. Susanta K. Chakraborty, VU, Midnapore, 721102, WB.
4. Anantha Krishnan, T.N. (1998). Some technological inputs for increased efficiency in biological control. (In: Technology in biological control, Publ. Oxford & IBH New Delhi): 7–14.
5. Arai, K. (2006). Towards biotechnology: The mission of IUBMB in the 21st century. IUBMB Life. 58: 267–268.
6. Baishya, R.A., Sarma, J. and Begum. A. (2013). Forest based medicinal plants rendering their services to the rural community of Assam,India.*International Journal of Applied Biology and Pharmaceutical Technology*. **4(4):** 10–20.
7. Bhattacharyya, M. (2008). Toxicity of pesticides on honeybees: Focus on Midnapore districts: A study. M.Sc. Dissertation. Dept. of Zoology, Vidyasagar University: 70–71.
8. Bhattacharya, M. and Chakraborty, S.K. (2014). Luffa cylindrical as a host plant for pollinator bees—a study based in West Midnapore, West Bengal, India. *Journal of Entomology and Zoology*. 2(3): 21–26.
9. Bhunia, G., Pradhan, P., Chanda, G.K. and Chakraborty, S.K. (2008). Comparative study of water quality parameters and plankton production after the application of different fertilizers in cemented aquaculture tanks and natural water bodies in Midnapore, West Bengal. *Proceedings of Zoological Research in Human Welfare.* 45: 429–439.
10. Canter, C.R. (2000). Biotechnology in the 21st century. Trends in Biotechnology. 18: 6–7.
11. Chanda, G.K., Dey, B., Pati, B.R. and Chakraborty, S.K. (2003). Earthworms biodiversity of Midnapore District, W.B., India. *Journal of Nature Conservators* 15(1): 17–30.
12. Chanda G.K., Bhunia G. and Chakraborty S.K. (2011). The effect of vermicompost and other fertilizers on cultivation of tomato plants. *Journal of Horticulture and Forestry*. 3(2): 42–45.
13. Chanda G.K., Bhunia G. and Chakraborty S.K. (2013). Assessment of decomposition rate and recovery number of earthworm Perionyx excavates Perrier during vermicomposting by combined index(CBI). *Global Advanced Research Journal of Environmental Science and Toxicology* 2(8): 183–187.
14. Chakraborty, S.K. and Majumder, S. (1995). Management of forest ecosystem of Arabari Forest, Midnapore, West Bengal: A review of problems and prospects. *Journal of Save the Environment* 111(11): 52–62.

15. Chakraborty, S.K. and Pati, B.R. (1998). Transfer of ecofriendly technology for sustainable environmental management in agriculture through the empowerment of rural women in western part of Midnapore District,West-Bengal. *Proceedings of International Symposium on Decentralisation: Devolution and Participation*: Pd2/1-Pd2/7.
16. Chakraborty, C., Chatterjee, T.K. and Chakraborty, S.K. (1999). Development of indigenous larval rearing technology for the production of freshwater prawn, Macrobrachium rosenbergii(de Man). In. Waste Recycling and Resource Management in the developing world. (Edited by B.B. Jana, Banerjee. R.D. Guterstam. B., and Heeb. J. and Published by the Kalyani University): 149–155.
17. Chakraborty, S.K. (2003). Biodiversity and Conservation (IN: Perceptions and Issues in Environmental Studies. Edited by A.D. Mukhopadhyay and Published by Vidyasagar University): 267–285.
18. Chakraborty, B., Chanda A. and Chakraborty S.K. (2008). Effect of integrated nutrient supply on growth, leaf yield and field performance of mulberry (*Morus alba*) under semiirrigated laterite soil condition of West Midnapore, District. *Journal of Environment and Society* 52: 221–228.
19. Chakraborty, B., Chanda Asit Kumar and Chakraborty, S.K. (2015). Effect of bio-organic amendments on the infestation of major pests and foliar disease, leaf productivity in mulberry(*Morus alba L*). *American Journal of Experimental Agriculture* 7(1): 10–16
20. Chakraborty, B. and Chakraborty, S.K. (2015). Seasonal variation of leaf yield in mulberry raised under bio and organic treatments in drought prone lateritic zone of West Midnapur,West Bengal. *Journal of Biological Science* 1(1): 1–8.
21. Chaudhuri, H. (1960). Experiments on induced spawning of Indian carps with pituitary injections. *Indian Journal of Fisheries* 7: 20–48.
22. Chaudhuri, H. and Singh, S.B. (1984). Induced breeding of carps. Indian council of agricultural research, New Delhi 1-82.
23. DBT, Task Force Recommendation. (1993). Department of Biotechnology, Government of India, New Delhi.
24. Earanna, N. and Govindan, R. (2002). Role of biofertilizers in Mulberry production-A Review. *Indian Journal of Sericulture* 41(2): 92–99.
25. Forster, J. and R., Hardy (2001). Measuring efficiency in intensive aquaculture. World *Aquaculture* 32(2): 41–45.
26. Gupta, P.K. (2003). Why vermicomposting? In: Vermicomposting for sustainable agriculture, Agrobios (India), Agro House, Jodhpur, 14–25.
27. Halder, K.R. (1998). Annelida: Oligochaeta: Earthworms; State Fauna Series 3: Fauna of West Bengal, Zoological Survey of India: 22–39
28. Julka, J.M. (1993). Earthworm resources of India and their utilization in vermiculture. Earthworm Resource and Vermiculture: 51–56.
29. Liao, I.C. and N.H., Chao. (1997). Developments in aquaculture biotechnology in Taiwan. *Journal of Marine Biotechnology* 5: 16–23.

30. Mazid, Mohd. Khan, Tagi Ahmed and Mohammad, F. (2012). Medicinal plants of rural India; A review of use by Indian Folks. *Indo Global Journal of Pharmaceutical Sciences* 2(3): 286–304.

31. Mondal, B. Datta, C. and Laha, U.K. (2012). Freshwater ornamental fish farming as a tool for socioeconomic development of backward women self-help groups (SHG's) to restore livelihood security. *International Journal of Advanced Biological Research* 2(1): 95–98.

32. Mendoza, R., A. De Dios, C., Vasquez, E., Cruz, D., Ricque, C., Aguilera and J. Montemayor (2001). Fishmeal replacement with feather-enzymatic hydrolyzates co-extruded with soya-bean meal in practical diets for the Pacific white shrimp (Litopenaeus vannamei). *Aquaculture Nutrition* 7(3): 143–151.

33. Mehta, B.K., Jain, S.K., Sharma, G.P., Doshi, A. and Jain, H.K. (2011). Cultivation of button mushroom and its processing: An techno-economic feasibility. *International Journal of Advanced biotechnology and Research* 2(1): 201–207.

34. Mishra, S.S., Pradhan, P. and Chakraborty, S.K. (2001). Status of fishing technologies for inland water bodies of Southern West Bengal. *Vidyasagar University Journal of Biological Sciences* 7: 45–59.

35. Mishra, S.S., Pradhan, P., Dutta, N.C. and Chakraborty, S.K. (2001). Studies on the performance of ovatide-on breeding of Indian Major Carps. *Journal of Inland Fishery Association* 28: 125–129.

36. Mishra, S.S., Pradhan, P., Kar, S. and Chakraborty, S.K. (2003). Ichthyofaunal diversity of Midnapore, Bankura and Hoogly Districts, South West Bengal, Records of the Zoological Survey of India. Occasional Paper No-220: 1–65.

37. Mishra, S.S., Pradhan, P. and Chakraborty, S.K. (2006). Rural development through participatory bioresource management-case study from Midnapore District. *In Perspectives of Rural Development Problems nad Prospects: Edited by Rajatkanti Das and Published by Indian Anthropological Society and Department of Anthropology, Vidyasagar University*: 50–60.

38. Mishra, S.S., Acharya, S. and Chakraborty, S.K. (2009). Development of tools for assessing conservation categories of siluroid fishes of fresh water and brackish water wetlands of South West Bengal, India. *Journal of Environmental Biology of Fishes*: 84: 395–407.

39. Nandeesha, M.C., Das, S.K., Nathianiel, D.E. and Varghes, T.J. (1990). Project report on breeding of carps with Ovaprim in India. Spl. Publ. No. 4, AFSIB., Mangalore, pp. 41.

40. Naylor, R.L., R.J., Goldburg, J.H., Primavera, N., Kautsky, M.C.M., Beveridge, J., Clay, C., Folke, J., Lubchenko, H., Mooney and C., Troell (2000). Effect of aquaculture on world fish supplies. *Nature*, 405(6790): 1017–1024.

41. Patra, M.K., Acharjee, S.K. and Chakraborty, S.K. (2005). Conservation categories of siluroid fishes in North-East Sundarbans, India. *Biodiversity and Conservation* 14: 1863–1876.

42. Pattanayak, S., Dasgupta, R., Chakravorty, P.P. and Chakraborty, S.K. (2014). Changes in growth ratr and activities of α-amylase and cellulose of *P. excavatus,* an epigeic earthworm in decomposing leaf litter of five tree species. *The international Journal of Science and Technoledge* 2(11): 68–71.
43. Pattanayak, S., Dasgupta, R., Chakravorty, P.P. and Chakraborty, S.K. (2014). Degradation of leaf litter of five tree species by *Perionyx excavatus* with relation to their nutrient and anti-nutrient content. *Agriculture, Forestry and Fisheries* 3(2): 128–132.
44. Reddy, P.V.G.K., G.V., Kowtal and M.S., Tantia (1990). Preliminary observations on induced polyploidy in Indian major carps, *Labeo rohita* (Ham.) and *Catla catla* (Ham.). *Aquaculture,* 87: 279–287.
45. Savoor, P.R. (1998). Pollinator management: An ecofriendly green revolution eludes India. *Current Science* 74(2): 121–125.
46. Trehan, K. (1990). Cell Biotechnology. Wiley Eastern Ltd. New Delhi.
47. Tripathi, Y.C., Hazaria, P., Kaushik, P.K. and Kumar, A. (2005). Vermitechnology and waste management. In: Verms and vermitechnology, (edi: A. Kumar) A.P.H. Publishing corporation, New Delhi 9-21.
48. Wu, C., Y. Ye and R., Chen. (1986). Genome manipulation in carp (*Cyprinus carpio L.*) *Aquaculture,* 54: 57–61.
49. Yaoqi Zhang, Wei Geng. Yueqin Shen,Yanlingwang and Yu Cheng Dai. 2014. Edible mushroom cultivation for food security and rural development in China: Bio-innovation, technological dissimination and marketing. *Sustainability* 6: 2961–2973.
50. www.baif.org.in—report of planning commission(2007-2012).

Index